MICROCOMPUTER APPLICATIONS IN MANUFACTURING

MICROCOMPUTER APPLICATIONS IN MANUFACTURING

A. GALIP ULSOY
University of Michigan, Ann Arbor

WARREN R. DeVRIES
Rensselaer Polytechnic Institute

JOHN WILEY & SONS
New York   Chichester   Brisbane   Toronto   Singapore

Copyright © 1989, by John Wiley & Sons, Inc.

All rights reserved. Published simultaneously in Canada.

Reproduction or translation of any part of
this work beyond that permitted by Sections
107 and 108 of the 1976 United States Copyright
Act without the permission of the copyright
owner is unlawful. Requests for permission
or further information should be addressed to
the Permissions Department, John Wiley & Sons.

Library of Congress Cataloging in Publication Data

Ulsoy, Ali Galip.
Microcomputer applications in manufacturing / A. Galip Ulsoy,
Warren R. DeVries.
p. cm.

Bibliography: p.
Includes indexes.

1. Manufacturing processes—Data processing. 2. Microcomputers.
I. DeVries, W. R. (Warren Richard) II. Title.
TS183.U45 1989
670'.28'5—dc19   88-32123
                                                           CIP

Printed in Singapore

10 9 8 7 6 5 4 3

Preface

Microcomputer Applications in Manufacturing is the outgrowth of notes written for a course of the same name taught in the Department of Mechanical Engineering and Applied Mechanics at the University of Michigan since 1980 [DeRU83]. The course is designed for students in mechanical engineering at the senior or first-year graduate level with an interest in manufacturing. It is intended to provide a foundation in the real-time application of microcomputer systems to problems in manufacturing. The book should also prove to be a useful reference for practicing engineers interested in manufacturing automation.

The assumed background is one that is typical of senior-level mechanical engineering students: a scientific programming language (e.g., FORTRAN), an introductory automatic control course, and familiarity with the manufacturing processes treated in the case studies, problems, and examples (e.g., machining, metal forming, and welding). In most curricula this background material is covered in several separate courses. Here the coverage of this background material is only with the depth required for the text; it is intended to supplement, not replace, courses in manufacturing processes, electrical engineering, programming, system modeling, design, and control. Instead, the coverage is aimed at the integration of that basic material through design of both hardware and software for manufacturing automation. Our assumption is that the best design solutions will come from engineers who know both the process and computer technology and who can intelligently interact with experts in various basic disciplines.

This textbook was written because we were unable to find an existing text that provided the breadth of coverage required for such a course, while still having adequate depth. We have taken a case-study approach, where four realistic manufacturing engineering problems are studied in detail. We have also provided background material on a variety of basic topics needed to undertake the problems posed in the case studies. The case studies are designed to be taught in conjunction with the background material. The background material is also appropriate for numerous other case studies of engineering interest and can be covered in 30 to 40 hours of lectures. We have considered hardware and software issues, as well as the underlying theoretical material. The material is best taught in conjunction with a laboratory, which could include some of or all the setups described in the case studies or other similar laboratory exercises. A separate book describing the hardware and assembly language programming of the particular laboratory microcomputer system is strongly recommended. The material in the book should be supplemented by additional reading, and many suggested books and papers are included in the Bibliography for a variety of topics from assembly language programming to estimation theory.

Chapter 1 provides motivation and background, including a brief historical review of manufacturing and computer technologies. Chapter 2 provides an overview of computers in engineering, and in particular microcomputer systems for data acquisition, analysis, and control. These two chapters constitute Part I Introduction, whereas Part II Microcomputer Operation and Programming contains three chapters. Chapter 3 reviews digital logic, Boolean algebra, and number systems. Chapter 4 describes microcomputer architecture and programming, including an introduction to assembly language programming. Chapter 5 considers interfacing of microcomputers to physical systems, with topics such as sensors and actuators, digital-to-analog and analog-to-digital conversions, digital input and output, and interrupts. Part III Methods for Data Analysis and Control also contains three chapters. Chapter 6 introduces techniques for data acquisition and analysis, including statistical techniques and spectral analysis. Chapter 7 introduces mathematical modeling and system identification, including differential and difference equations and their solution, and least squares parameter estimation. Chapter 8 is an introduction to the topic of digital control using microcomputers for both single-input single-output and multi-input multi-output systems. Part IV Case Studies contains four chapters, each a separate manufacturing case study. Chapter 9 considers the use of a microcomputer as an aid in setting up a rolling mill. Chapter 10 addresses the problem of acquiring and analyzing force and spindle-speed data in a turning operation. Chapter 11 considers the control of a stepping-motor-driven x-y table typical of many machine tools. The final case study in Chapter 12 is the speed control of a dc motor using a microcomputer. This is representative of many servo-control problems associated with machine tools and robots. The four case studies illustrate how the material in previous chapters can be applied to realistic manufacturing engineering problems. Hardware and software design is emphasized in the case studies, and each includes specific programming objectives. Problems are provided at the end of each chapter, as well as an extensive Bibliography at the back of the book. The two appendixes provide complete program listings for the four case studies; Appendix A includes a description of an 8088-based laboratory system, and Appendix B provides a description of a Z80-based laboratory system.

We would like to thank the many excellent students who have taken the course and several colleagues who have taught the course (Elijah Kannatey-Asibu, Bob Keller, Yoram Koren, and Jeff Stein); all have provided us with useful feedback for improvements to it. We also appreciate the able assistance of our graduate teaching assistants, particularly Jerry Raski, Tsu Ren Ko, and Paul Koenig. Our thanks also go to Steve Culp who helped to set up the experiments and keep them running and to Susan Glowski and Carol Richmond who assisted in the preparation of the manuscript.

A. GALIP ULSOY
WARREN R. DEVRIES

January 1988

Contents

PART I INTRODUCTION   1

1 Motivation and Background   3
    1.1 Motivation   3
    1.2 Historical Review of Manufacturing Technology   7
    1.3 Historial Review of Computer Technology   8
    1.4 Summary   9
    1.5 Problems   10

2 Microcomputer Systems for Data Acquisition and Control   13
    2.1 Computers in Engineering   13
    2.2 Structure and Operation of Microcomputer Systems   14
    2.3 Programming Microcomputer Systems   17
    2.4 Summary   21
    2.5 Problems   21

PART II MICROCOMPUTER OPERATION AND PROGRAMMING   23

3 Digital Logic, Boolean Algebra, and Number Systems   25
    3.1 Logic Gates and Boolean Algebra   25
    3.2 Simple Logic Circuits   29
    3.3 Number Systems and Arithmetic   33
    3.4 Summary   39
    3.5 Problems   39

4 Microcomputer Architecture and Operation 45
4.1 Microprocessor Architecture 45
4.2 Instructions and Operations 51
4.3 Summary 61
4.4 Problems 61

5 Interfacing 65
5.1 Sensors and Actuators 65
5.2 Digital Input and Output 70
5.3 Conversions Between Digital and Analog Domains 80
5.4 Scheduling Service to I/O Devices 85
5.5 Summary 96
5.6 Problems 96

PART III METHODS FOR DATA ANALYSIS AND CONTROL 107

6 Data Acquisition and Analysis 109
6.1 Basic Concepts 109
6.2 Some Statistical Techniques 113
6.3 Spectral Analysis 117
6.4 Summary 122
6.5 Problems 122

7 System Modeling and Identification 131
7.1 Modeling Dynamic Systems 131
7.2 Solution of Linear Differential Equations 137
7.3 Transformation of Differential to Difference Equations 140
7.4 Model Structure and Order 143
7.5 Parameter Estimation 145
7.6 Summary 150
7.7 Problems 150

8 Digital Control 155
8.1 Computer Control Systems 155
8.2 Digital Control of SISO Systems 157

8.3	Digital Control of MIMO Systems	163
8.4	Summary	167
8.5	Problems	168

PART IV CASE STUDIES   171

9 Setting Clearances on a Rolling Mill   173
9.1	Description of the Engineering Problem	173
9.2	Programming Objectives	177
9.3	Summary	178
9.4	Problems	179

10 Monitoring a Turning Process   181
10.1	Description of the Engineering Problem	181
10.2	Programming Objectives	184
10.3	Summary	190
10.4	Problems	190

11 Control of a Stepping-Motor-Driven x-y Table   193
11.1	Description of the Engineering Problem	195
11.2	Programming Objectives	208
11.3	Problems	213

12 Speed Control of a DC Motor   215
12.1	Description of the Engineering Problem	215
12.2	Programming Objectives	218
12.3	Problems	221

Bibliography   223

Appendix A Description of an 8088-Based System   229

Appendix B Description of a Z80-Based System   331

Index   367

Part I
INTRODUCTION

Chapter 1
MOTIVATION AND BACKGROUND

This book focuses on the design and implementation of computer-based systems for data acquisition, analysis, and control in manufacturing. The contemporary manufacturing engineer must have an understanding of the manufacturing processes and the computer systems interacting with them. This book is intended to provide the manufacturing engineer with the basic knowledge and skills that are required to work with microcomputer systems in a manufacturing environment. The approach adopted is the use of selected case studies (see Part IV) and actual hands-on laboratory experience (see Appendixes A and B). Basic facts and methods related to the operation of microcomputer systems, programming, data analysis, and control are introduced (see Parts II and III) as needed to support the applications discussed in the case studies. References for further in-depth reading are provided in the Bibliography.

In this chapter the motivation for such applications is first discussed. Subsequently, a brief historical review is presented for the areas of both manufacturing technology and microcomputer technology.

1.1 MOTIVATION

The motivation for the use of microcomputers in various manufacturing operations is an economic one [LuCA87]. For many products, manufacturing costs are a significant fraction of the total cost and have an important effect on profitability. During the past decade, computer systems—in particular microcomputer systems—have evolved into a powerful, reliable, and inexpensive tool. The ways in which computers can affect engineering design and manufacturing are numerous and complex [Huds82]. However, we can consider the following simple classification:

1. Business and management functions (e.g., reporting, inventory, accounting)
2. Engineering analysis and design (e.g., simulation, optimization, engineering drawings)
3. Manufacturing operations (e.g., production planning, process planning and scheduling)
4. Data acquisition and analysis (e.g., process monitoring and diagnostics, statistical quality control, product inspection)
5. Process control [e.g., computer numerically controlled (CNC) machine tools, adaptive control systems, robots]

Computers are used extensively in all the activities listed. The business and management functions, such as accounting and payroll, have long been available on large computer systems. Microcomputers are also useful for such functions owing to improved software for word processing, spreadsheets, business graphics, and data base management functions.

Computer-aided design (CAD) systems are rapidly becoming accepted by industry for generating engineering drawings. The first generation of commercial CAD systems were limited to computer-aided drafting, but this is rapidly changing with the incorporation of powerful engineering analysis software for finite-element analysis, kinematic and dynamic simulation of mechanical systems, design optimization, and various specialized mechanical analysis and design programs. Potential benefits of CAD systems include improved designs and reduced time to complete the designs. These CAD systems are often implemented on large mainframe computers, large minicomputers, or, more recently, engineering workstations that are part of a larger network. These new engineering workstations are microcomputers, often with specialized hardware for graphics, floating-point arithmetic, and network communications.

These developments, together with significant advances in networking and communications, lead naturally to the goal of computer-integrated manufacturing (CIM). The idea here is that the design information available in the CAD system data base be directly transferred to the computers controlling production on the factory floor. Conversely, data collected by computers on the factory floor can be made available to aid in the design and management functions. This link between computer-aided manufacturing (CAM) systems and CAD systems not only provides for quick and error-free transfer of engineering information, but also opens up new possibilities for integrating the design and manufacturing functions. A large distributed and/or hierarchical computer system is required to implement CIM; computers of various sizes and capabilities must be integrated through a communications network and a shared data base (see Fig. 1.1).

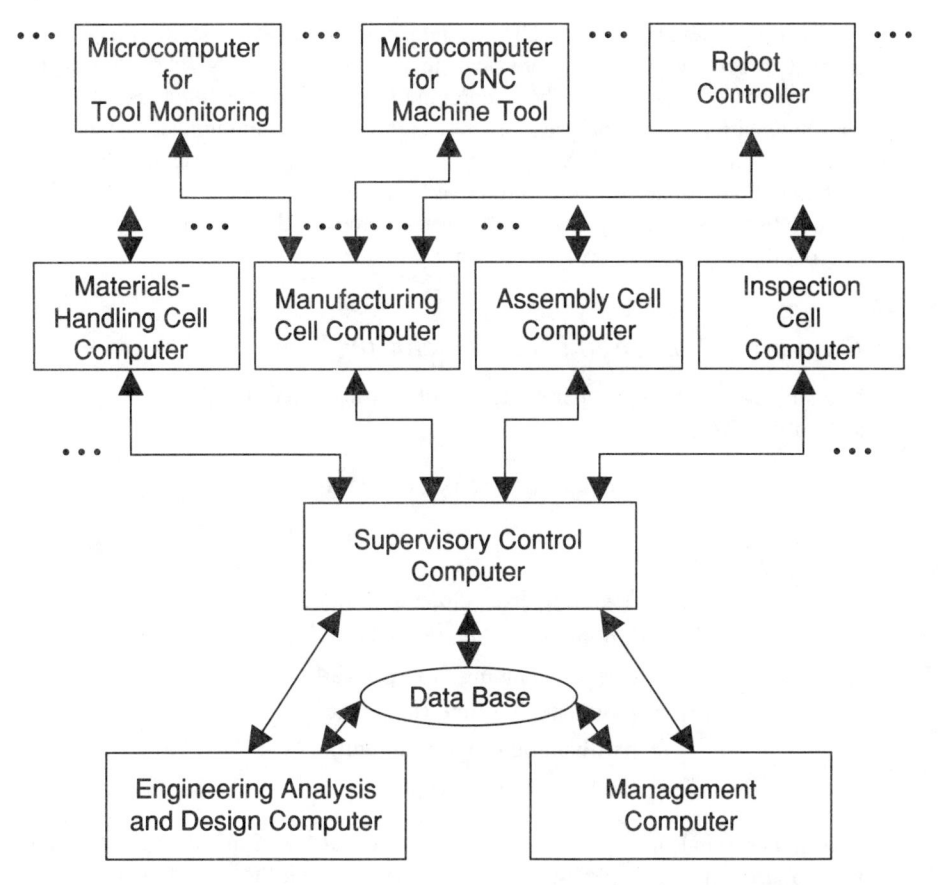

Figure 1.1 A distributed and hierarchical computer-integrated manufacturing (CIM) system

A CIM system can be expected to provide a variety of important benefits. The most significant of these is the efficiencies that can be obtained through integration and coordination of a diverse number of functions such as scheduling, inventory, marketing, design, and manufacturing. Consistent and reliable operation of the manufacturing plant and all its components is another important benefit arising from the continuous monitoring and reporting features. This can mean less downtime and less expensive repairs. The CIM system can also be expected to provide considerable improvements in product quality through inspection and process control. The flexibility afforded by CIM systems to handle a variety of changing product designs, or product mixes, is another very significant economic benefit. Increased production rates, and reduced labor costs, are also expected benefits, but perhaps not the most significant ones. In fact, the shift from unskilled to skilled labor and the related retraining costs may be more significant than the actual reduction of labor costs. Most importantly, however, a highly automated manufacturing plant helps to keep the products more competitive in the marketplace and prevents layoffs and plant shutdowns.

The lowest level of such a CIM system must necessarily be computers that interact directly with various physical devices on the factory floor; this level is the focus of this book. These computers must perform functions such as data acquisition, data analysis, process control, and communication with the other computers in the CIM system. Microcomputers, because of their low cost and high reliability, are attractive candidates at this level of the CIM system [KoBu83]. Example 1.1 discusses some of the potential benefits of microcomputer application to a machining process.

Example 1.1 Economics of Machining

In a single-point turning operation, the total cost per part (C) can be expressed as [Gilb50; Kore83],

$$C = C_1 t + C_2(t/T) + C_1 t_1 (t/T) + C_1 t_2 \qquad (1.1.1)$$

where

C_1 = operating cost (direct and indirect) (\$/min)
C_2 = tooling cost (\$/cutting edge)
t = direct machining time per part (min/unit)
t_1 = tool changing time (min/edge)
t_2 = part handling time (min/unit)
T = the tool life (min/edge)

The first term in Eq. 1.1.1 corresponds to the actual machining cost per part, the second term represents the tooling cost per part, the third term is the tool changing cost, and the fourth term gives the part loading and unloading cost. Thus, in Eq. 1.1.1 the term (t/T) is the number of tool edges required per workpiece. For this example, consider the various ways in which a microcomputer-based data acquisition, analysis, and control system might be used to reduce the total cost per part:

1. The direct labor component of C_1 can be reduced through automation and reduction of labor content (e.g., one operator may attend to several CNC machine tools rather than one conventional machine tool); however, the indirect component of C_1 will probably increase.
2. t can be reduced through process control (generally termed "adaptive control") by on-line manipulation of machine feeds and speeds based on in-process measurements (e.g., force, torque, motor current).
3. t_1 and t_2 can be reduced by use of microcomputer-controlled robots or automatic tool changers.
4. T can be accurately estimated on line for each cutting edge from process measurements (e.g., force, acoustic emissions). Thus, the statistically "worst case" value of T need not be used as the basis for tool replacement.

Of course, it is only fair to mention the disadvantages of microcomputer automation too: the increased capital expenditures; the need for skilled technicians, programmers and engineers; and the introduction of new components with their attendant maintenance and repair problems.

<<>>

1.2 HISTORICAL REVIEW OF MANUFACTURING TECHNOLOGY

Manufacturing processes date back to the earliest civilizations with the production of various articles made from wood, ceramics, stone, and metal [Kalp84]. The mechanization of manufacturing began with the industrial revolution during the eighteenth century. The origin of this revolution can be traced to the invention of machines for spinning and weaving in the textile industry in England and to the cotton gin developed by Eli Whitney in the United States. Whitney's introduction of the concept of jigs and fixtures to produce interchangeable parts was a significant development in that manufacture moved from the realm of artisans to factory workers. The milling machine, engine lathe, and numerous other machine tools were introduced in the early 1800s; other important inventions (e.g., the Watt steam engine) of this same era were in fact built by these machine tools, jigs and fixtures.

The early 1900s saw the work of Frederick W. Taylor and others provide a scientific basis for both manufacturing processes and management. Technology was the driver for a number of other changes: the electric motor began to replace steam power as the prime mover; there were developments in materials such as metals (e.g., steels), nonmetals (e.g., plastics), and tool materials (e.g., high-speed steel and carbides); and an understanding of fly-ball governors, cams, and electrical devices. These developments led to the introduction of mechanisms for fixed automation, such as the automatic screw machine and turret lathe, and the transfer line for mass production. This manufacturing technology was pretty much the state of the art through the 1940s.

Starting in the late 1940s, machine tools with simple automatic controls, such as plug-board controllers and the copying machine that used servos to follow and magnify the tracing motions on a model, gradually evolved into numerically controlled (NC) machine tools. The first NC machine tool was a three-axis milling machine developed initially by the Parsons Company (Traverse City, Michigan) and completed in 1952 at the MIT Servomechanism Laboratory under a U.S. Air Force contract. Although this machine used vacuum tubes and hard-wired logic, it defined numerical control to mean machining based on coded numerical information. With advances in computer technology (see Section 1.3), an on-board computer replaced hard-wired logic, the computer lending its name to NC to give the computer numerically controlled (CNC) machine tools of the late 1960s. Concurrent with the development and commercial availability of CNC machine tools was the development and sale of industrial robots;

today these two technologies are "partners" in many manufacturing applications [Kore85].

During the 1970s and 1980s, the emphasis has been on: (1) developing intelligent manufacturing systems by making full use of the computing available at the process; (2) horizontal CIM through improved coordination of computer-controlled devices on the factory floor; and (3) vertical CIM by coordination between manufacturing, design, and management functions. However, another important feature of this period has been the role of new materials such as plastics, composites, ceramics, and their fabrication. Some of the processing technologies used for this are lasers for heat treatment, cutting, and welding; near-net shape forming; and powder metallurgy. These developments can be expected to continue. This period has also seen the maturation of advanced cutting tool technology, which has opened up the way for high-speed machining. High-speed machining, at speeds greater than 600 meters per minute, leads to significant improvements in metal removal rate and surface finish and to reduced cutting forces [FlKL84, Koma85], but requires improved machine-tool technology that can be obtained in part by a combination of process knowledge and the application of microcomputer technology.

1.3 HISTORICAL REVIEW OF COMPUTER TECHNOLOGY

Computing devices have appeared throughout history in various forms [Anon74; Osbo78]. However, the origins of the modern digital computer can be traced to ENIAC (Electronic Numerical Integrator and Calculator), developed in 1946 at the University of Pennsylvania. This was a vacuum-tube device that was programmed using wired plug boards. At the same time, John Van Neumann suggested the idea of storing in the computer, as coded digits, the instructions (i.e., the program) in the same way that the data itself is stored. The earliest stored-program computer was the EDSAC, which was completed in 1949 at Cambridge University. The first commercial machine of this type was the Univac I, developed in 1951. Although these early computers were mainly computing devices, they were immediately considered as potential components in aircraft and missile systems. However, these early computers were very unreliable, consumed too much power, were large and expensive, and were fairly limited in storage capacity and computational speed; little to recommend them for control applications.

Serious attempts at using digital computers for real-time data acquisition and control functions did not emerge until the mid- to late 1950s [AsWi84]. These computers typically had an addition time of 1 millisecond and a mean time between failure (MTBF) of the central processing unit of about 50 to 100 hours. These components were also expensive and typically had to serve several processes or tasks. They typically operated in a supervisory mode, performing tasks such as reporting, scheduling, production planning, and calculating set points for several analog regulators. The hardware interrupt feature also emerged during this early period.

In the early to mid-1960s, digital computers had an addition time of about 100 microseconds and the MTBF was around 1000 hours. During this period digital computers were used not only in a supervisory mode, but also to directly control several processes. This was termed direct digital control (DDC). There were still problems, however, with cost and reliability.

During the decade from 1965 to 1975, smaller, faster, more reliable and lower-cost computers, termed *minicomputers*, emerged. These minicomputers could typically perform additions in 1 microsecond, and their central processing units had a MTBF greater than 20,000 hours. A typical minicomputer of this period had a 16-bit word length and on-board memory of 8- to 256-thousand bytes (8 binary digits or bits). During this period, minicomputers found widespread application for data acquisition, analysis, and control in various industrial plants (e.g., chemical plants, oil refineries, mineral processing plants). It was also during this period that CNC machine tools were introduced and the first industrial robots began to appear, both using this fast and more reliable computer technology.

It was with the development of the *microcomputer* in the early to mid-1970s, however, that the digital computer became an attractive alternative for nearly all data acquisition, analysis, and control tasks. These devices were just as fast, powerful, and reliable as existing minicomputers at from one-tenth to one-hundredth the cost. During the 1980s, microcomputers have become even more attractive for both simple and demanding automation tasks. Because of developments in microelectronics technology, such as very large scale integration (VLSI), the current trend toward more powerful microcomputer hardware can be expected to continue. Equally important, however, will be developments in communications technology and standards [e.g., Ethernet, local area networks, manufacturing automation protocol (MAP)], microcomputer software (e.g., expert systems and artificial intelligence), and perhaps, most importantly, the availability of a new generation of engineers familiar with this powerful new tool.

1.4 SUMMARY

Microcomputer technology is rapidly evolving. These advances have led to the development of an engineering tool that can have a great economic impact on all phases of engineering, including manufacturing. The economic benefits are a result of the efficiencies achieved through integration of the design, manufacturing, and management functions; through the flexibility provided by the use of reprogrammable devices; through the improved product quality obtained from monitoring, inspection and control; through increased production rates; through reduction of downtime; and so forth.

Manufacturing processes evolved rapidly during the early part of this century and currently represent a relatively mature technology. Emerging processes include powder metallurgy, laser processing, and high-speed machining. However, the most significant advances expected in manufacturing processes during the next decade will be a result of automation. This will require knowledge of the processes in the form of mathematical models and formal production rules. Intelligent manufacturing processes, which incorporate functions for monitoring, control, and decision making, will be developed to take full advantage of computer technology and advances in control theory, artificial intelligence, and signal processing.

Although computing devices have been used for about a century, it was the development of the digital computer in the 1950s that initiated the current information revolution. The maturation of the microcomputer, a powerful, reliable, and low-cost device, during the past decade has opened up a host of potential engineering applications, many of them in manufacturing.

1.5 PROBLEMS

1. Look at advertisements and product announcements in trade journals such as *Design Engineering*, *Manufacturing Engineering*, *Mechanical Engineering*, and *Control Engineering* to prepare a list of CAD/CAM products available in each of the following categories: (a) business and management, (b) engineering analysis and design, (c) data acquisition and analysis, (d) process control, (e) manufacturing technology.

2. Use the cost expression given by Eq. 1.1.1 with the following values:

$$C_1 = 1.00 \ (\$/min)$$

$$C_2 = 2.00 \ (\$/cutting\ edge)$$

$$t = 10.0 \ (min/unit)$$

$$t_1 = 0.5 \ (min/edge)$$

$$t_2 = 2.0 \ (min/unit)$$

$$T = 30.0 \ (min/edge)$$

Compare and discuss the economic impact of each of the following hypothetical changes due to automation:

a. An automatic tool changer is added to the system and reduces the tool changing time t_1 by 80%, it also increases C_1 by 5%.

b. A robot is introduced for part handling and reduces the handling time t_2 by 50% while increasing C_1 by 10%.

c. A process control system is introduced for on-line manipulation of feeds and speeds. This results in a reduction of 50% in t, an increase of 5% in C_1, and a decrease of 10% in T because of the higher feeds and speeds.

d. The three systems in a, b, and c are all used together.

3. Read trade journals in manufacturing to summarize the most important developments in manufacturing technology over the last 2 to 3 years.
4. Read trade journals in computers to summarize the most important developments in computer technology over the last 2 to 3 years.
5. Prepare a short essay (1–2 pages) on how computer technology (or a particular development in computer technology) will affect a particular aspect of manufacturing, or how a development in manufacturing technology (e.g., processing ceramic superconductors) will change computer technology. Cite references to support your claims.

Chapter 2
MICROCOMPUTER SYSTEMS FOR DATA ACQUISITION AND CONTROL

This chapter discusses the use of microcomputer systems in engineering, particularly for data acquisition, analysis, and control. The basic structure and operation of such systems is also described, along with an introduction to their programming.

2.1 COMPUTERS IN ENGINEERING

A variety of computer systems are used widely in virtually every phase of engineering. Traditionally, computer systems have been classified as:

- Supercomputers
- Mainframes
- Minicomputers
- Microcomputers

As we go through this list from top to bottom, costs can decrease by a factor of a million. There are also corresponding decreases in speed and capacity, but these are not as dramatic as the decrease in price. Only the most computationally intensive engineering problems require the use of supercomputers (e.g., large deformation analyses such as an auto crash, some climatic studies). Mainframes are used for certain large engineering analysis programs (e.g., simulation of large-scale dynamical systems, finite element stress and flow analyses, optimization studies) as well as for many management and business functions. Until recently, minicomputers were required to handle most engineering analysis and design tasks as well as data acquisition, analysis, and control functions. However, with the rapid development of microcomputers, these less-expensive devices are now able to handle many of the functions that previously required a minicomputer or mainframe. This has enabled more engineers to have direct access to computers and to engineering software. Advances in networking and communications have also encouraged the use of microcomputers by allowing engineers to access more powerful machines as they are needed. Thus, one can expect microcomputer systems to play an increasingly important role in all phases of engineering.

Although most engineers have some familiarity with the use of computers in engineering analysis and design, many are not familiar with their use for data acquisition, analysis, and control. There are numerous examples of computer based systems for data acquisition, analysis, and control. Many have direct application to manufacturing:

- Complex industrial plants use computer data acquisition and analysis systems to monitor production and provide reports at regular intervals (e.g., mineral processing plants, oil refineries).
- Microprocessor-based systems are used for product inspection and quality control (e.g., light bulbs, printed circuit boards).
- Many control systems are microcomputer based, particularly as part of a distributed and hierarchical control system (e.g., steel mills, chemical plants).
- Computer numerically controlled (CNC) machine tools and robots are reprogrammable and provide the flexibility needed for making and handling different parts.

For these, and many other applications, microcomputer systems offer a powerful, cost effective, and reliable engineering tool. The primary reasons for selecting computer-based systems in these applications are *flexibility* (they are reprogrammable) and *intelligence* (they are capable of complex logical and arithmetic operations as well as communication).

2.2 STRUCTURE AND OPERATION OF MICROCOMPUTER SYSTEMS

A microcomputer-based system for data acquisition, analysis, and control will generally consist of the following interrelated parts:

1. The physical system being analyzed, monitored, or controlled (e.g., a machine tool, a part-handling system).
2. The microprocessor (e.g., 8088, Z80, 6502, 68000) and required peripherals (e.g., disk drives, terminal).
3. The interface (e.g., digital input/output ports, analog-to-digital converters).
4. The development software (e.g., editor, compiler) and the application software (e.g., control algorithms, data acquisition routines).

Microcomputer applications require a good understanding all these interrelated parts. For applications in manufacturing, the physical system must be understood from basic studies in mechanics, materials, and manufacturing processes. This book will introduce the additional topics required to understand microprocessor operation, interfacing, and programming. In this section an overview of the hardware and operation of microcomputer-based systems is presented. The next section presents an overview of the software and programming.

The microprocessor and components of the interface, including any sensing and actuating elements, are typically electronic devices. Electrical signals are characterized by the pair of variables *voltage* and *current*, whose product is power. The important characteristics of the signal are its power and its information content. The information is contained in one of these two variables, usually the voltage. Electrical signals are characterized as being *analog* or *digital*. Analog signals are nonquantized whereas digital signals are quantized to one of two levels, values, or states (e.g., HI or LO, 0 or 1, TRUE or FALSE, ON or OFF).

A *pulse* is a signal that goes from one level to another and back again (see Fig. 2.1). In digital signals only two levels are defined, and the value of the signal may be undefined as it switches from one level to the other. This is an important consideration in digital circuit design. One of two solutions is commonly used to handle this undefined signal problem. In *asynchronous* circuits there is a control circuit that determines the order in which signals change. Another solution is to specify the times during which the signal output is said to be valid. This latter approach leads to *synchronous* circuit design and requires an additional signal called the *clock*. The clock signal is a continuous train of pulses at a fixed or adjustable clock frequency.

Information in digital systems can be coded in either *serial* or *parallel* form, as illustrated in Figure 2.2. Serial information is transmitted as a group of pulses

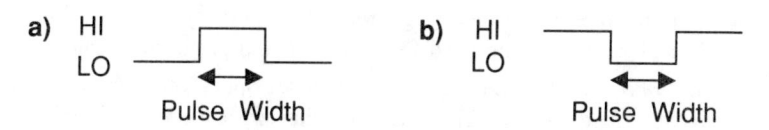

Figure 2.1 Electric pulses: (a) high pulse; (b) low pulse

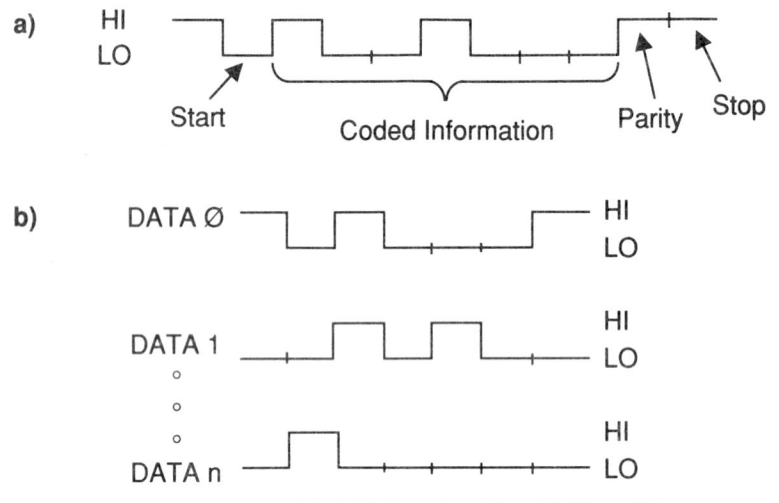

Figure 2.2 Coding of digital information: (a) serial; (b) parallel

on a single conductor. For parallel transmission, a group of conductors are used, each of which contains a portion of the information. Digital devices are the basis for computer systems. Chapter 3 will provide an introduction to some simple digital devices and circuits. Because of the use of only two states (e.g., 0 and 1) in digital devices, binary numbers and arithmetic will be important; this topic is also introduced in Chapter 3.

The operation of any computer is based on two simple yet elegant concepts:

1. The *program*, a sequence of instructions, must be stored internally so that the speed is not limited to that of the operator.
2. The physical realization of a computer requires a device based on a large number of simple *bistable* (two-state) elements.

A computer is a digital device with the following interconnected components:

1. *Arithmetic and logical unit (ALU):* A device capable of doing arithmetic (binary) and logical (Boolean) operations on suitably coded data.
2. *Data memory:* A device for internal storage of suitably coded data that can be altered during operation [e.g., random access read-write memory (RAM)].
3. *Program memory:* A device for internal storage of the sequence of operations to be performed [e.g., read-only memory (ROM)]. The same physical device can be used for both data and program storage.
4. *Control unit:* A device that coordinates operation of the three components described in 1–3. The ALU together with the control unit is called the central processing unit (CPU).

5. *Input devices:* Devices that enable the operator, or various external devices, to provide data and to control operation.
6. *Output devices:* Devices that provide information to the operator, or to various external devices. The devices in 5 and 6 are often referred to as input/output (I/O) devices and constitute the interface between the computer and the external world.

A computer operates by repeating a simple and fixed sequence of operations. The time required to carry out the sequence of operations is termed the *cycle time* and is on the order of 1 microsecond for the current generation of commercially available microprocessors. The sequence of operations performed during one cycle is

1. Fetch an instruction from program memory
2. Decode the instruction
3. If data is required go to 4, otherwise go to 5
4. Fetch data from memory
5. Execute the instruction
6. Go back to 1 for next instruction

Chapters 4 and 5 will consider the operation and interfacing of microcomputers in more detail. In the next section of this chapter the programming of computer systems will be briefly introduced.

2.3 PROGRAMMING MICROCOMPUTER SYSTEMS

In previous sections it was noted that computers can operate at very high speeds by carrying out a sequence of instructions, called a program, that is stored in memory. As explained in later chapters, all information is stored in the computer as collections of binary digits (bits) that take on the values 0 or 1. The microprocessor interprets some of these collections of 0's and 1's as instructions and others as data of various kinds (i.e., integers, real numbers, character strings).

Although binary numbers are the natural language for the computer, it is very cumbersome for humans to work with binary numbers. Programming using binary numbers is referred to as *machine language programming*. Grouping of 3 or 4 bits together leads to the use of *octal* and *hexadecimal* representations, respectively. Although still difficult to work with, these *grouped-bit number systems* are more convenient than programming in binary. The next step in programming convenience is to represent instructions to the computer using *mnemonics*, that is, short symbolic representations of the instructions, which are easier for humans to remember. This is referred to as *assembly language* programming. Assembly language commands correspond directly to the operations carried out by the microprocessor hardware, and not to tasks that programmers may want to perform. This has led to the development of various *high level* programming languages (e.g., FORTRAN, Pascal, Ada) that provide the types of commands and capabilities that are in line with the needs of the human programmer. These offer an application-oriented command structure rather than one corresponding to operations that a computer can perform. Programs are easier to write, maintain, and understand when written in these high-level programming languages. These languages are also independent of the particular computer hardware that is used, so the application programs written in these languages can more easily be used on a variety of different hardware. Table 2.1 shows a segment of a program that counts to 20 in various representations. The program clearly becomes easier for a human to write and understand as we go from machine language to hexadecimal to assembly language to a high-level language.

All programs written in hexadecimal, octal, assembly, or high-level languages must be translated to machine language (binary code) before they can be executed. Most microcomputers come with development software that contains some of or all the following programs:

1. *Monitor:* A program that runs when the system is first started (or "booted"). It accepts commands to execute other available development software.
2. *Filer:* A program that allows the user to perform file-handling functions, such as creating or deleting files and moving files from one device to another.
3. *Editor:* A program that can be used to create and/or modify human readable files such as programs.
4. *Compiler:* A program that translates high-level language code to machine language code. The machine language code is *relocatable*; that is, it can be loaded into different memory locations.
5. *Assembler:* A program that translates assembly language (mnemonic) code into relocatable machine language.
6. *Linker:* A program that links together separately compiled or assembled relocatable machine code into a single machine language program that is then ready to be loaded into memory and executed.
7. *Interpreter:* A program that is similar in function to a compiler and linker, but that can translate, link, load, and execute one line of high-level language code at a time as it is entered by the programmer.
8. *Debugger:* A program that is useful for monitoring the execution of a user program by allowing the user to stop execution and examine the contents of certain memory locations as needed.

The monitor and filer constitute a minimum *operating system* for a particular computer. Various utility programs may be available in addition to the devel-

2.3 Programming Microcomputer Systems

Table 2.1 Program to Count to 20

Z80 Machine Language	Hexadecimal	Z80 Assembler	FORTRAN
0010 0001	21	LD HL,1	I=1
0000 0001	01		
0000 0000	00		
0010 0010	22	LD(I),HL	
0000 0000	00		
0000 0001	01		
0010 1010	2A	LOOP:LD HL,(1)	100 I=I+1
0000 0000	00		
0000 0001	01		
0001 0001	11	LD DE,1	
0000 0001	01		
0000 0000	00		
0001 1001	19	ADD HL, DE	
0010 0010	22	LD(1), HL	
0000 0000	00		
0000 0001	01		
0001 0001	11	LD DE,20	IF(20-I)101,100,100
0001 0100	14		
0000 0000	00		
1110 1011	EB	EX DE,HL	
1010 0111	A7	AND A	
1110 1101	ED	SBC HL,DE	
0101 0010	52		
1111 0010	F2	JP P,LOOP	
0000 1000	08		
0000 0001	01		
0111 0110	76	HALT	101 STOP

opment system described herein. There may, of course, be several different editors, compilers, assemblers, and so forth available on the same computer system. There may also be several different operating systems that can be used on the same computer hardware.

Programming in higher-level languages is much more convenient for the user. However, execution of high-level programs is typically slower than lower-level (assembly language) programming. For compiled high-level languages the execution may be three to five times slower than an equivalent assembly language program. The difference is even greater for interpreted high-level languages. Depending on the features of the high-level language, there may be certain operations that cannot be programmed in that language. Thus, there may be programming situations where assembly language programming is necessary. Better designed high-level languages and more sophisticated processors are continually changing the balance in favor of high-level language programming [AuSa81].

In real-time applications it is often necessary to perform certain time-critical tasks using assembly language programming. Routines for digital input and output and analog-to-digital conversions are hardware dependent and also may require assembly language programming. The case studies that appear in Part IV illustrate how high-level language programming can be used for many tasks, whereas assembly language programming is used for others. The linker is used to combine these routines into a program that can accomplish a particular task. In the case studies we provide examples of FORTRAN programming [Ette83; Gold65] and assembly language programming for Z80 [Zaks80] and 8088 [Inte81] based microcomputers. The particular choice of languages is not an easy one; it may depend on a variety of factors such as availability of compilers, features of the programming language, previous programming

Table 2.2 Some Rules for Programming Style

- Write clearly: Do not be too clever or sacrifice clarity for "efficiency"
- Modularize: Use subroutines, functions, and library functions
- Write and test a big program in small pieces
- Test input for plausibility and validity: Make sure it does not violate the limits of the program; identify bad input and recover if possible; use self-identifying input; allow defaults; echo both default and input data on output
- Make it right before you make it faster; make it fail safe before you make it faster; make it clear before you make it faster
- Check some answers by hand, and test programs at their limiting or boundary values
- Make sure all variables are initialized before use: Initialize constants with DATA statements; initialize variables with executable code
- Use variable names that mean something; use statement labels that mean something; format a program to help the reader understand it
- Make sure comments and code agree; do not just echo the code with comments—make every comment count; do not comment bad code— rewrite it; document your data layouts; do not overcomment

Source: [KePl76]

experience, requirements of the particular programming task, compatibility with other programs, and availability of certain useful routines. Most real-time programming tasks are currently performed using FORTRAN and assembly language, Basic, Forth, or C. However, the more recently introduced languages Ada and Modula-2 are also well suited for such tasks [Barn84; Booc83; Wirt82].

Regardless of the programming language or task, there are some general guidelines for writing good programs. Some rules for programming style, as paraphrased from Kernighan and Plauger [KePl76], are summarized in Table 2.2. We will have occasion to refer to these rules again when we describe software development for the case studies in Chapters 9 through 12. Keep in mind that the design of software should be approached much like the design of any engineering product: we must explicitly state the performance requirements of our software; break it down into its components and identify the relationships between these components; design and evaluate each of the components separately; and carefully document our design for maintainability.

2.4 SUMMARY

This chapter has provided an overview of the role of computers in engineering and focused on microcomputer-based systems for data acquisition, analysis, and control. Such systems find widespread application in various manufacturing tasks and provide flexibility and intelligence.

The structure and operation of such systems has been briefly reviewed. They are digital electrical devices that carry out a sequence of instructions (arithmetic and logical operations) that are stored internally in memory in binary form.

The programming of microcomputer systems can be accomplished at several levels; however, all programs must be translated to machine readable code before they can be executed. High-level programming languages offer many advantages, but may need to be used with assembly language programming for certain time-critical and hardware-dependent tasks. Regardless of the programming task or language, there are rules of good programming that should be followed.

2.5 PROBLEMS

1. For several commercially available computers, prepare a table summarizing their computation speed (e.g., cycle time or time to perform an addition), memory, and cost.

2. For a commercially available microcomputer system, find out and list the power requirements, voltage levels, and clock frequencies used.

Also determine whether parallel and/or serial interfaces are available and whether they conform to any national or international standards, such as ANSI, IEEE, ISO.

3. For a commercially available microcomputer, determine the microprocessor used for the CPU, the amount of available RAM, and the available ROM and its function.

4. For a commercially available microcomputer system, identify and classify some development software programs in each of the categories: (a) monitor, (b) filer, (c) editor, (d) compiler, (e) assembler, (f) linker, (g) interpreter, and (h) others (e.g. debugger).

5. Compare the execution speeds of two or more high-level languages on a particular microcomputer system using benchmarks published in computer magazines (e.g., the sieve in *Byte* magazine).

6. Compare the advantages and disadvantages of at least two of the following high-level languages for real-time programming: (a) BASIC (interpreter), (b) BASIC (compiler), (c) FORTRAN, (d) Ada, (e) Modula-2, (f) Forth, and (g) C.

7. Discuss each of the following types of development software in terms of its purpose, relation to other software (e.g., do you link then edit?), the types of input and output files for each, and so on: (a) editor, (b) FORTRAN compiler, (c) assembler, (d) linker.

8. Write the equivalent assembly language program (Z80 or 8088) for the following FORTRAN program segment:

```
        I = 0
        J = 1
100     I = I + J
        GO TO 100
```

Part II
MICROCOMPUTER OPERATION AND PROGRAMMING

Chapter 3
DIGITAL LOGIC, BOOLEAN ALGEBRA, AND NUMBER SYSTEMS

In this chapter several topics are reviewed as background for understanding the operation and programming of microcomputers. First logic gates and Boolean algebra are introduced. Then some simple logic circuits are considered. Finally, the binary, hexadecimal, and octal number systems are described.

3.1 LOGIC GATES AND BOOLEAN ALGEBRA

Digital signals can represent one of two possible states (e.g., HI or LO, 0 or 1, ON or OFF, TRUE or FALSE). Consider now some digital devices that can be used to construct digital logic circuits. Several logic families or physical technologies are used to construct digital logic devices: for example, transistor-to-transistor logic (TTL) and complementary metal oxide on silicon (CMOS). In general, TTL logic is faster and CMOS consumes less power, but both are used to build up logic gates and combinations of logic gates, called devices, that are analyzed in the same way [Hugh69; ZaLe79; Sloa83; Bart85].

Logic gates or *functions* are digital devices that produce certain predictable output states depending on the states of the inputs. These functions can be described by simply listing the input states and the output state they produce in what are called *truth tables*, as shown in Figure 3.1. This figure also indicates

26 Chapter 3 DIGITAL LOGIC, BOOLEAN ALGEBRA, AND NUMBER SYSTEMS

Name	Symbol	Circuit	Truth Table
AND	·	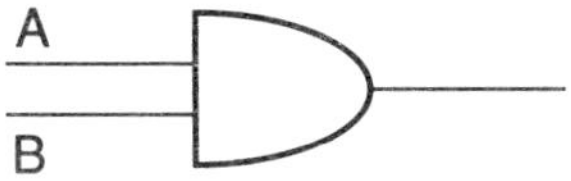 A, B → A·B	A B A·B / HI HI HI / HI LO LO / LO HI LO / LO LO LO
OR / Inclusive or	+	A, B → A+B	A B A+B / HI HI HI / HI LO HI / LO HI HI / LO LO LO
NOT / Negation	( )'	A → A'	A A' / HI LO / LO HI
XOR / Exclusive or	⊕	A, B → A⊕B	A B A⊕B / HI HI LO / HI LO HI / LO HI HI / LO LO LO
EQ / Equivalence	≡	A, B → A≡B	A B A≡B / HI HI HI / HI LO LO / LO HI LO / LO LO HI

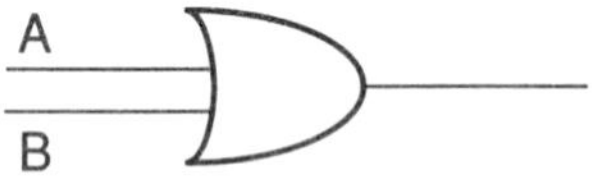

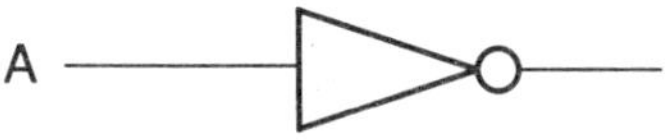

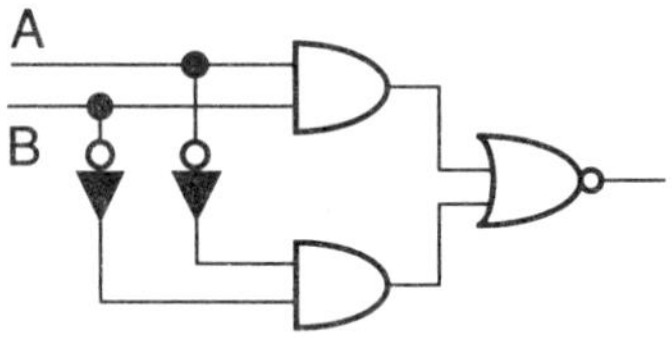

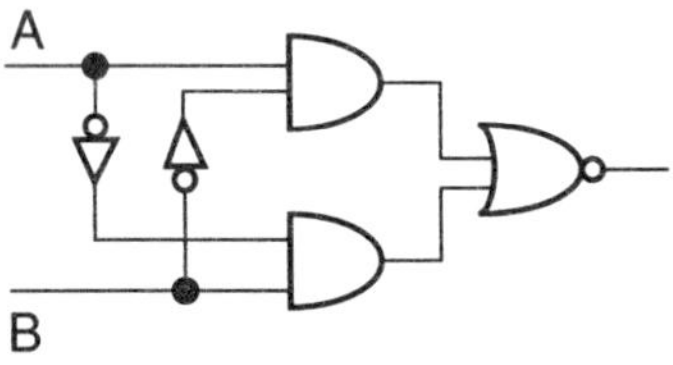

Figure 3.1 *Common Logic Gates or Functions*

the names and symbolic representations of the common logic gates. It is clear that these devices perform logical operations similar to those performed by humans in everyday decision making. Computers and other digital circuits are constructed from such devices. Chips containing one or more logic gates can be purchased and used to construct simple digital circuits like the ones described in the following examples.

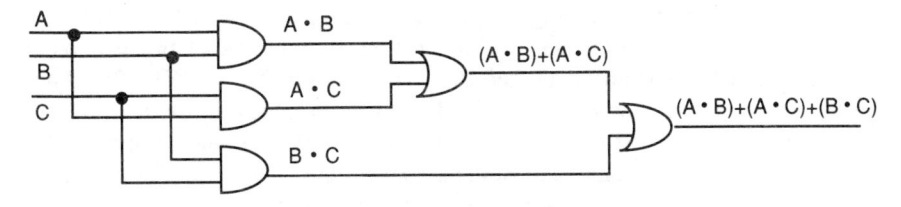

Figure 3.2 Solution to Example 3.1

Example 3.1 Majority Voter Problem

Three pushbuttons (A, B, and C) pressed by three voters are to be the inputs to a digital circuit. We want the output of the circuit to indicate HI if any two or more of the inputs are HI. The circuit shown in Figure 3.2 will do the job. Convince yourself that it will work, and find other possible solutions.

<<>>

Example 3.2 Clamping a Press

We want to design a logic circuit that will set the signal CLAMP to HI when a press is to be clamped. The press is instrumented such that:

1. The signal HAND is HI if a hand is between the dies.
2. The signal PARTIN is HI if a part has not been removed from the dies.
3. When the dies are fully clamped, the signal CLOSED is HI.

Convince yourself that the circuit in Figure 3.3 will work, and also find other solutions.

<<>>

Boolean algebra deals specifically with two-state functions. It offers a means by which digital logic functions can be manipulated algebraically rather than with logic diagrams or truth tables. We noticed from the preceding examples that there may be more than one valid logic circuit to accomplish a particular task. Boolean algebra can be an aid in reducing a particular logic circuit to its simplest form. Economically, this can be important in complex digital circuits with thousands of components.

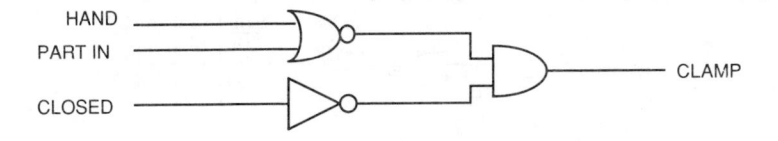

Figure 3.3 Solution to Example 3.2

A Boolean variable (e.g., A, B, C) can have the values "1" or "0" (i.e., HI or LO). A Boolean expression, or function, is composed of Boolean variables, Boolean constants (0 or 1), and Boolean operators (e.g., AND, OR). An entire Boolean expression can be defined using truth tables as we saw earlier with logic gates. The Boolean algebra has certain rules, similar to normal algebra, that allow us to manipulate Boolean expressions. The basic Boolean operators and their symbols were defined in Figure 3.1. Using these as building blocks, more complex functions, such as the NAND gate illustrated in Figure 3.4, can be obtained.

There are three basic laws of Boolean algebra:

1. Commutativity

$$A + B = B + A$$
$$A \cdot B = B \cdot A$$

2. Associativity

$$A + B + C = A + (B + C) = (A + B) + C$$
$$A \cdot B \cdot C = A \cdot (B \cdot C) = (A \cdot B) \cdot C$$

3. Distributivity

$$A \cdot (B + C) = (A \cdot B) + (A \cdot C)$$
$$A + (B \cdot C) = (A + B) \cdot (A + C)$$

Notice that the second part of the distributivity law does not hold in ordinary algebra. Based on these three laws and basic definitions, the first 16 rules in Table 3.1 are obtained. The identities 17 and 18 in Table 3.1 are simple statements of *DeMorgan's theorem* and are very useful because they relate the operators AND, OR, and NOT and permit the reduction of logical functions to simpler forms. DeMorgan's theorem can be stated as: "To form the complement of a Boolean function, all variables are replaced by their complement, all AND's are replaced by OR's and all OR's are replaced by AND's."

FUNCTION	BOOLEAN SYMBOL	LOGIC SYMBOL	TRUTH TABLE
NAND	$(A \cdot B)'$	A, B → $(A \cdot B)'$	A: HI HI LO LO B: HI LO HI LO $(A \cdot B)'$: LO HI HI HI

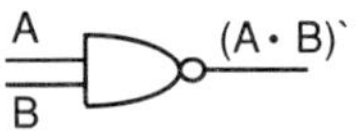

Figure 3.4 NAND function

Table 3.1 Identities for Boolean Variables

1. $A'' = A$
2. $A \cdot 1 = A$
3. $A \cdot 0 = 0$
4. $A \cdot A = A$
5. $A \cdot A' = 0$
6. $A + 1 = 1$
7. $A + 0 = A$
8. $A + A = A$
9. $A + A' = 1$
10. $A + (A \cdot B) = A$
11. $(A \cdot B) + (A \cdot B') = A$
12. $(A + B)(A + B') = A$
13. $A + (A' \cdot B) = A + B$
14. $(A + B') \cdot B = A \cdot B$
15. $(A \cdot C) + (A \cdot B) + (B \cdot C') = (A \cdot C) + (B \cdot C')$
16. $(A + B) \cdot (B + C) \cdot (A' + C) = (A + B) \cdot (A' + C)$
17. $A' \cdot B' = (A + B)'$
18. $A' + B' = (A \cdot B)'$

Example 3.3 Clamping a Press (Revisited)

The result in Example 3.2 can be expressed algebraically as

```
CLAMP = (HAND + PARTIN)' • CLOSED'
```

Using the identities in Table 3.1, this expression can be shown to be reducible to either of the following forms:

```
CLAMP = (HAND'• PARTIN') • CLOSED'
CLAMP = (HAND + PARTIN + CLOSED)'
```
<<>>

3.2 SIMPLE LOGIC CIRCUITS

Using the logic gates introduced in the previous section, a large variety of useful digital circuits can be constructed. In this section flip-flops, registers, and counters are introduced as simple and useful examples of digital devices and logic circuits.

Flip-flops are another type of logic device that, unlike the logic gates discussed previously, have memory. Logic gates are much like a push button; they stay on as long as they are depressed, but go off when released. Flip-flops are more like a light switch; they stay in one state (on or off) until instructed to go to the other state. Flip-flops are important elements of computers, since they can be used as memory devices. Although we will not discuss flip-flops in detail here, we present the R-S flip-flop as an example of how flip-flops are constructed from logic gates. We also present the J-K flip-flop as an element of logic devices such as registers and counters.

Flip-flops have two outputs (1 and 0). If the 1 output is HI, then the flip-flop is considered to be in the "1" state or condition; if the 0 output is HI, then the flip-flop is in the "0" state. Figure 3.5 shows the logic diagram for the

30 Chapter 3 DIGITAL LOGIC, BOOLEAN ALGEBRA, AND NUMBER SYSTEMS

a)

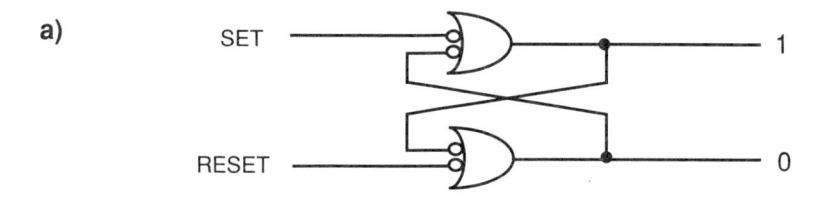

[Logic diagram showing SET and RESET inputs to cross-coupled NOR gates with outputs 1 and 0]

b)

	Initial Conditions		Pulsed Signal Inputs		Output Results	
	1 Output	0 Output	Set Input	Reset Input	1 Output	0 Output
Flip-flop in "1" condition initially	HI	LO	LO	LO	Indeterminate	
	HI	LO	LO	HI	HI	LO
	HI	LO	HI	LO	LO	HI
	HI	LO	HI	HI	HI	LO
Flip-flop in "0" condition initially	LO	HI	LO	LO	Indeterminate	
	LO	HI	LO	HI	HI	LO
	LO	HI	HI	LO	LO	HI
	LO	HI	HI	HI	LO	HI

Figure 3.5 R-S flip-flop: (a) logic diagram, (b) truth table

Reset-Set (R-S) flip-flop and the corresponding truth table. The R-S flip-flop can be constructed from two negated input OR gates and has two inputs called RESET and SET. If the SET input gets a LO pulse the state of the R-S flip-flop is "1", and if the RESET input gets a LO pulse the state is "0". Note that indeterminate states (i.e., either "0" or "1") are possible, as shown in Figure 3.5, so that the R-S flip-flop must be used with caution. When using a flip-flop it is desirable to condition its inputs with the desired levels first, and then allow it to make the desired transition when it receives a clock pulse. This can be accomplished by adding a steering network to the flip-flop circuit. A clocked R-S flip-flop with this type of steering input is shown in Figure 3.6.

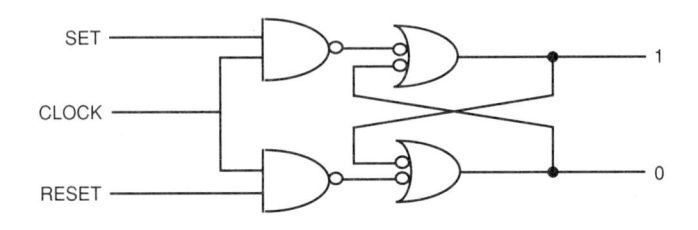

Figure 3.6 Clocked R-S flip-flop logic diagram

The J-K flip-flop, shown in Figure 3.7, is a clocked flip-flop with no indeterminate states. The important features of the J-K flip-flop are

1. If both J and K inputs are disabled (LO), a clock pulse causes no change in flip-flop state.
2. If both J and K inputs are enabled (HI), a clock pulse will complement or change the state.
3. If the J input is LO and the K input is HI, a clock pulse puts the flip-flop into the "0" state.
4. If the J input is HI and the K input is LO, a clock pulse puts the flip-flop into the "1" state.

The following examples illustrate how J-K flip-flops can be used to construct useful digital circuits.

Example 3.4 Four-Bit Shift Register

Figure 3.8 illustrates a four-bit shift register using four J-K flip-flops. This device transmits information from one flip-flop to the next every time it receives a clock pulse. For example, if all flip-flops are initially in the "0" state (this is the purpose of the RESET line) and the SERIAL INPUT signal is HI, when the first clock pulse occurs flip-flop A will go to the "1" state. This is because

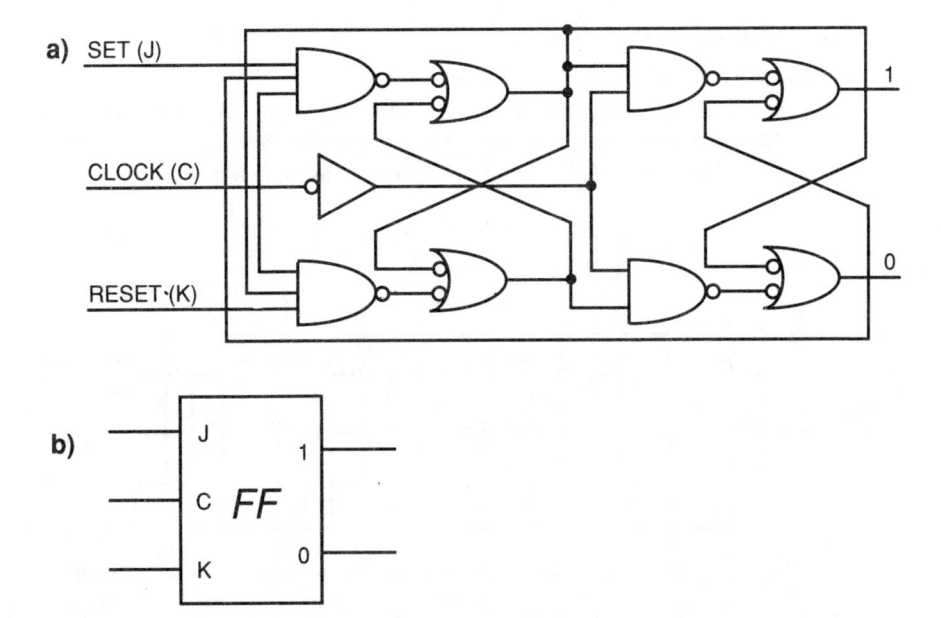

Figure 3.7 J-K flip-flop: (a) logic diagram, (b) symbol

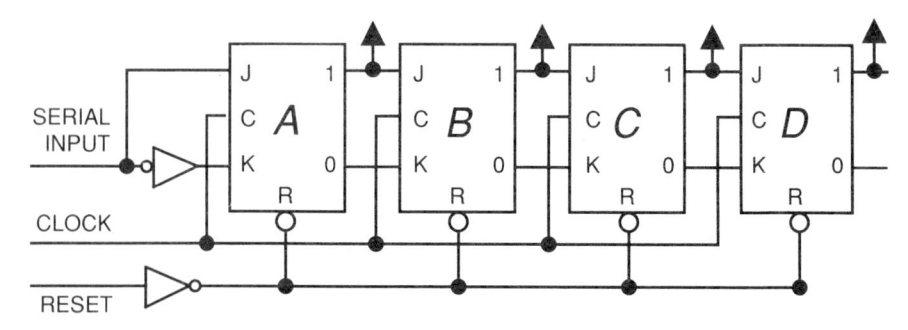

Figure 3.8 Four-bit shift register

its J input will be HI and its K input will be LO. The B, C, and D flip-flops, however, will remain in the "0" state because their J input is LO and K input is HI. If the SERIAL INPUT is LO before the next clock pulse, then flip-flop A will go to state "0", B will go to "1", and C and D remain at "0". Successive clock pulses will shift the "1" next to C then D. After five pulses, all the flip-flops will be back in the "0" state. Shift registers are used in computers as high-speed memory locations and can perform operations like binary division and multiplication.

<<>>

Example 3.5 Four-Bit Binary Up Counter

Although we have not discussed binary numbers, counting in binary is accomplished in much the same way as in the decimal system. If we have four binary digits (bits), we can count from 0 to 15 as shown in Table 3.2. A 4-bit binary counter can be constructed using J-K flip-flops, as shown in Figure 3.9. With

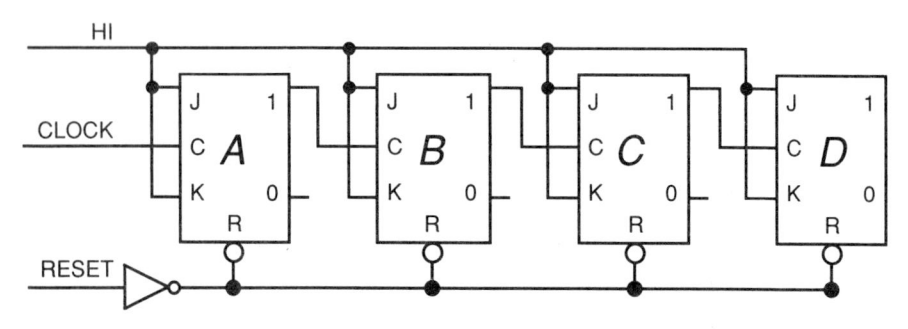

Figure 3.9 Four-bit binary up counter

Table 3.2 Counting from 0 to 15 in Decimal and Binary

Decimal	Binary	Decimal	Binary
00	0000	08	1000
01	0001	09	1001
02	0010	10	1010
03	0011	11	1011
04	0100	12	1100
05	0101	13	1101
06	0110	14	1110
07	0111	15	1111

all four flip-flops initially in the "0" state, the first clock pulse will complement flip-flop A to "1" and the other flip-flops will remain at "0". When the second pulse occurs, flip-flop A is again complemented, going from "1" to "0", and flip-flop B is also complemented from "0" to "1", whereas C and D remain at "0". The counter thus produces the binary code sequence in Table 3.2, where A represents the least significant (rightmost) bit, and D represents the most significant (leftmost) bit of the binary number.

<<>>

3.3 NUMBER SYSTEMS AND ARITHMETIC

As noted earlier, computers can perform arithmetic and logical operations on suitably coded data. Based on our discussions of logic devices, we have some idea of what types of logical operations can be performed and how digital devices can be constructed to perform such operations. Example 3.5 also gave us an indication of how logic devices can perform arithmetic operations. Since computers are based on bistable devices, the natural number system to be employed is *binary* (base 2) rather than the usual *decimal* (base 10). First we will discuss the binary number system and binary arithmetic, then briefly present other number systems (hexadecimal and octal) that are also useful in working with computer systems.

Let us briefly review how the decimal, or base 10, system works. The sequence of decimal numbers has a very specific meaning, for example,

$$204 = (2 \times 10^2) + (0 \times 10^1) + (4 \times 10^0)$$
$$412.05 = (4 \times 10^2) + (1 \times 10^1) + (2 \times 10^0) + (0 \times 10^{-1}) + (5 \times 10^{-2})$$

Notice that we have *positions* representing powers of 10 and carrying *digit position weights* (e.g., 10^0, 10^1) and the *digits* 0,1,2 . . . 9. We multiply the symbol in a particular position by its digit position weight and sum to get the decimal number. In counting or addition we use the 10 digits (0 . . . 9), and when we exceed nine we carry a one to the next position.

In the binary number system there are two digits, 0 and 1, corresponding to the two possible states of the digital signals (e.g., HI and LO). A binary number is, therefore, a sequence of 1's and 0's. Binary numbers can also be represented by multiplying the digit in each position by their digit position weights (powers of 2 in this case), for example,

$$01001101_2 = (1 \times 2^0) + (0 \times 2^1) + (1 \times 2^2) + (1 \times 2^3) + (0 \times 2^4) + (0 \times 2^5) + (1 \times 2^6) + (0 \times 2^7)$$

where the subscript "2" has been used to indicate that this is a binary number. Because 0's and 1's are the digits used in the binary system, they are often referred to as *bits*, an acronym for *bi*nary *dig*its. The leftmost 1 in a binary number is called the most significant bit (MSB), the extreme right digit is called the least significant bit (LSB).

The *conversion from binary to decimal* numbers is readily accomplished by multiplying each symbol by its digit position weight and summing, for example,

$$\begin{aligned}
01001101_2 = 1 \times 2^0 &= 1 \times 1 &&= 1 \\
0 \times 2^1 &= 0 \times 2 &&= 0 \\
1 \times 2^2 &= 1 \times 4 &&= 4_{10} \\
1 \times 2^3 &= 1 \times 8 &&= 8_{10} \\
0 \times 2^4 &= 0 \times 16 &&= 0 \\
0 \times 2^5 &= 0 \times 32 &&= 0 \\
1 \times 2^6 &= 1 \times 64 &&= 64_{10} \\
0 \times 2^7 &= 0 \times 128 &&= \underline{0} \\
& && 77_{10}
\end{aligned}$$

Two methods can be used for *conversion from decimal to binary:*

1. Divide the decimal number by 2 (the binary base) and record the remainder as the LSB of the binary equivalent. Then divide the quotient from the first division by 2 and repeat until the quotient has been reduced to zero. For example, to find the binary equivalent of 29_{10}:

	Quotient	Remainder	
29/2 =	14	1	(LSB)
14/2 =	7	0	
7/2 =	3	1	
3/2 =	0	1	
1/2 =	0	1	(MSB)

Thus, $29_{10} = 11101_2$.

2. Subtract the highest possible power of 2 (the binary base) from the decimal number and place a 1 in the appropriate weighting position, repeat until the decimal number is reduced to zero. If after the first subtraction the next lower power of 2 cannot be subtracted, then place a 0 in the appropriate weighting position. Let us again find the binary equivalent of 29_{10}:

$$29 - 2^4 = \quad 29 - 16 \quad = 13 \quad 1 \text{ (MSB)}$$
$$13 - 2^3 = \quad 13 - 8 \quad = 5 \quad 1$$
$$5 - 2^2 = \quad 5 - 4 \quad = 1 \quad 1$$
$$1 - 2^1 = \text{(cannot be done)} \quad\quad 0$$
$$1 - 2^0 = \quad 1 - 1 \quad = 0 \quad 1 \text{ (LSB)}$$

This confirms our previous result: $29_{10} = 11101_2$.

Binary addition is performed very much like decimal addition. We add two binary symbols and, if the largest symbol is exceeded, we carry a one to the next position or place, for example,

$$\begin{array}{r} 101_2 = 5_{10} \\ +010_2 = +2_{10} \\ \hline 111_2 = 7_{10} \end{array}$$

and,

$$\begin{array}{r} 101_2 = 5_{10} \\ +111_2 = +7_{10} \\ \hline 1100_2 = 12_{10} \end{array}$$

So far we have avoided *negative numbers* and *subtraction*. There are two common ways to handle these in binary arithmetic:

1. *Direct binary subtraction* is like normal decimal subtraction except when we need to borrow we use a two (the binary base), for example,

$$\begin{array}{rr} 110_2 = & 6_{10} \\ -101_2 = & -5_{10} \\ \hline 001_2 = & 1_{10} \end{array}$$

We must always subtract the smaller number from the larger number and then adjust the sign, for example,

$$\begin{array}{rr} 101_2 = & 5_{10} \\ -110_2 = & -6_{10} \\ \hline -001_2 = & -1_{10} \end{array}$$

This is difficult to implement on a computer, and the second approach, which follows, is preferred.

2. *Two's complement arithmetic* is used on computers for negative numbers and subtraction. The two's complement of a binary number is that number which when added to the original number results in a sum of zero, for example,

Binary number: 110110110110_2
Two's complement: $\underline{001001001010_2}$
Carry bit and sum: 1 000000000000_2

Notice that the *carry bit* (1) in the resulting sum is ignored. Thus, we work with a fixed number of bits (12 in this example). This is similar to the odometer of an automobile. For example, if the odometer has five digits and reads 99995_{10} miles, after 5 more miles the reading becomes zero. To obtain the two's complement of a binary number, we first calculate the *one's complement* by setting each bit to the opposite value, then add one to the one's complement, for example,

Binary number: 110_2
One's complement: 001_2
Two's complement: 010_2

Thus, with a fixed number of binary digits the two's complement notation is used to represent negative numbers. For example, with an 8-bit binary number we can represent $2^8 = 256$ possible values (00000000_2 to 11111111_2). Using the two's complement notation, we can represent numbers in the range $+127_{10}$ to -128_{10} (see Fig. 3.10). The leftmost bit serves as a *sign bit*, where 0 implies a positive number and 1 implies a negative number. Note that $0_{10} = 00000000_2$ is considered to be a positive number.

Binary numbers are natural for computers, but very inconvenient for people. It is too easy to make a mistake, and long sequences of binary digits are difficult

3.3 Number Systems and Arithmetic 37

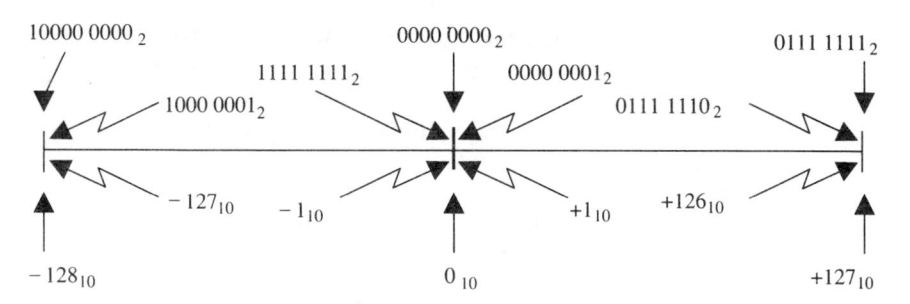

```
10000 0000₂                    0000 0000₂                      0111 1111₂
      \          1111 1111₂    |       0000 0001₂                    |
       \        /1000 0001₂    |      /      0111 1110₂             |
        ▼      ▼               ▼     ▼      /                        ▼
   ─────▲──────────────────────▲▼▲──────────────────▲────────────────
        ▲       ▲              ▲    ▲       ▲                        ▲
        |      −127₁₀  −1₁₀    |   +1₁₀   +126₁₀                    |
     −128₁₀                   0₁₀                                 +127₁₀
```

Figure 3.10 *8-Bit binary numbers in two's complement notation and their decimal equivalents*

to interpret. Other number systems, notably *hexadecimal* (base 16) and *octal* (base 8), are commonly used by computer programmers. These are sometimes referred to as *grouped bit number systems*, since groups of 4 bits are easily converted to hexadecimal, and groups of 3 bits are easily converted to octal. Most microprocessors today use 8 or 16 bits, making groupings of 4 bits more convenient. Thus, the hexadecimal system is generally preferred and will be emphasized here.

Octal numbers use the symbols 0,1,3, . . . 7 and digit position weights that are powers of 8. Similarly, hexadecimal numbers use the symbols 0,1,2, . . . 9 and A,B,C,D,E, and F and digit position weights that are powers of 16. Numbers up to 23_{10} and their hexadecimal, octal, and binary equivalents are shown in Table 3.3. Using this table, we can readily convert binary numbers, in 3-bit groups, to their octal equivalent, for example,

Table 3.3 Decimal Numbers Up to 23 and Their Hexadeximal, Octal, and Binary Equivalents

Decimal	Hexa-decimal	Octal	Binary	Decimal	Hexa-decimal	Octal	Binary
0	0	0	0	12	C	14	1100
1	1	1	1	13	D	15	1101
2	2	2	10	14	E	16	1110
3	3	3	11	15	F	17	1111
4	4	4	100	16	10	20	10000
5	5	5	101	17	11	21	10001
6	6	6	110	18	12	22	10010
7	7	7	111	19	13	23	10011
8	8	10	1000	20	14	24	10100
9	9	11	1001	21	15	25	10101
10	A	12	1010	22	16	26	10110
11	B	13	1011	23	17	27	10111

Binary number: 011   010   111   101
Octal equivalent: 3   2   7   5

Thus, $011010111101_2 = 3275_8$. Similarly, using 4-bit groupings, we can convert to hexadecimal,

Binary number: 0110   1011   1101
Hexadecimal equivalent: 6   B   D

Thus, $011010111101_2 = 6BD_{16}$. The basic methods and concepts described for binary numbers regarding conversion, negative numbers, and arithmetic also apply to octal and hexadecimal numbers. These are illustrated for hexadecimal numbers in the examples that follow.

Example 3.6 Conversion between Decimal and Hexadecimal

First we evaluate the decimal equivalent of $19AF_{16}$.

$$\begin{aligned}19AF_{16} &= (1 \times 16^3) + (9 \times 16^2) + (10 \times 16) + (15 \times 16^0) \\ &= 4096 + 2304 + 160 + 15 \\ &= 6575_{10}\end{aligned}$$

Now we convert the decimal number 29_{10} to hexadecimal using the division method.

	Quotient	Remainder	
29/16	1	$13_{10} = D_{16}$	(LSD)
1/16	0	$1_{10} = 1_{16}$	(MSD)

Thus, $29_{10} = 1D_{16}$. Confirm this result by subtracting powers of the base.

<<>>

Example 3.7 Hexadecimal Addition

Two examples are given here.

$$\begin{array}{r}1D_{16} = 29_{10} \\ +2_{16} = +2_{10} \\ \hline 1F_{16} = 31_{10}\end{array}$$

and,

$$\begin{array}{r}1F_{16} = 31_{10} \\ +5_{16} = +5_{10} \\ \hline 24_{16} = 36_{10}\end{array}$$

<<>>

Example 3.8 Hexadecimal Subtraction

We can handle negative numbers and subtraction by using the 16's complement of a hexadecimal number. For example, to subtract $86B_{16}$ from $A94_{16}$ we add the 16's complement of $86B_{16}$ to $A94_{16}$. The 16's complement of $86B_{16}$ is obtained by getting the 15's complement of $86B_{16}$ then adding one. The 15's complement is obtained by subtracting each digit from 15, for example,

$$15\text{'s complement of } 86B_{16} = 794_{16}$$
$$16\text{'s complement of } 86B_{16} = 795_{16}$$
$$A94_{16} - 86B_{16} = A94_{16} + 795_{16} = 1229_{16}$$

Neglecting the carry, we get the result 229_{16}. Check this result by direct hexadecimal subtraction.

<<>>

3.4 SUMMARY

This chapter has reviewed some background material that will be useful for understanding the operation and programming of microcomputer systems.

First we introduced the basic logic gates (i.e., AND, OR, XOR, NOT, and EQ) and Boolean algebra. These gates were then used to illustrate some simple digital logic circuits (i.e., flip-flops, registers, and counters).

Finally, we introduced the binary, octal, and hexadecimal number systems and their use in simple arithmetic operations. Conversions between these number systems and the two's complement representation of negative binary numbers were also considered.

3.5 PROBLEMS

1. Obtain a solution to the majority voter problem in Example 3.1 using either AND or OR operators but not both.
2. Consider a simple manufacturing cell as shown in Figure 3.P2. Under the control of the cell computer (CC) the robot (R) performs the following tasks:

    Task 0: Wait until requested to perform one of the tasks that follow.
    Task 1: Pick up a part from the input conveyor (IC) and deliver it to the machine tool (MT).
    Task 2: Remove a part from the machine tool (MT) and take it to the inspection station (IS).

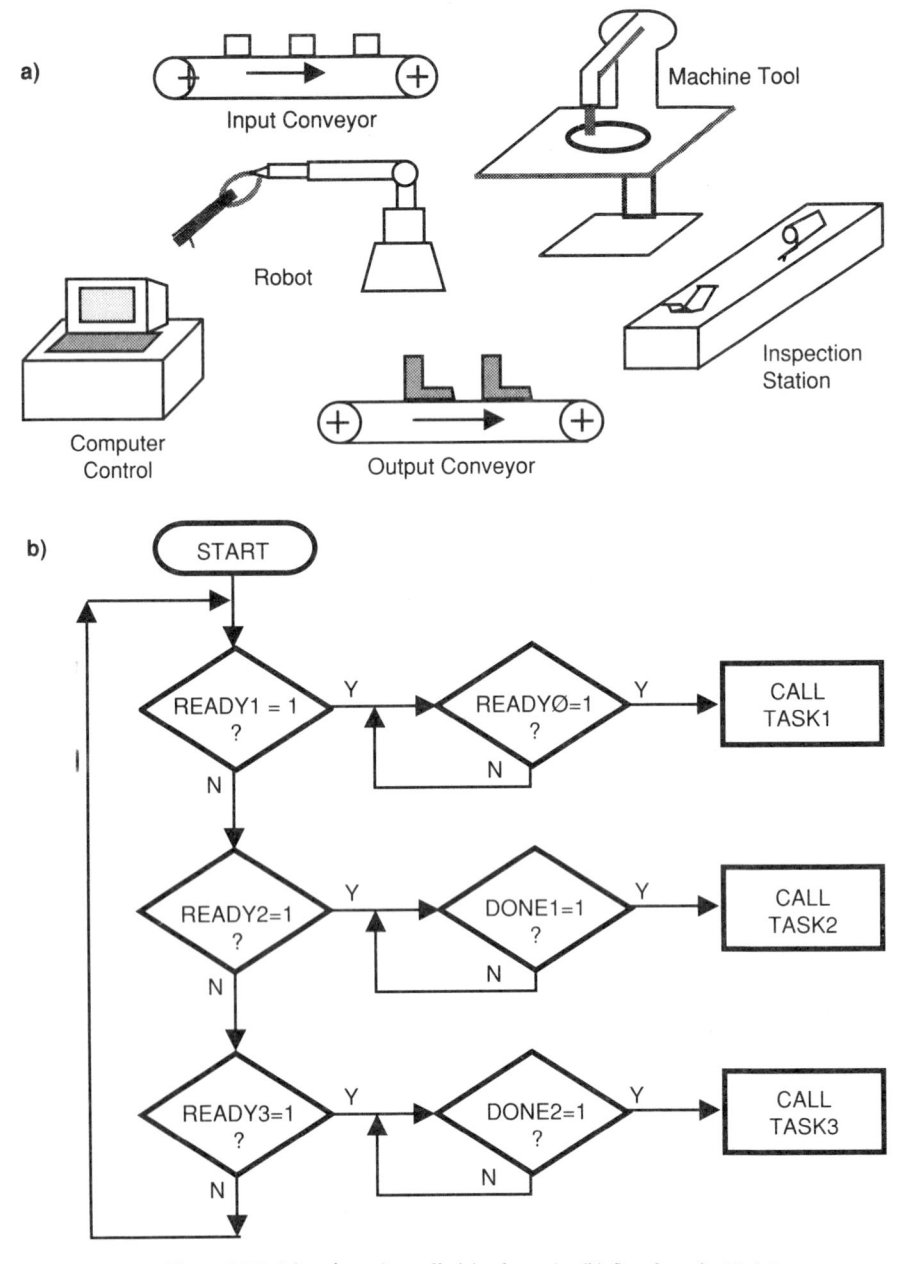

Figure 3.P2 Manufacturing cell: (a) schematic, (b) flowchart for Task 0

Task 3: Remove a part from the inspection station (IS) and place it on the output conveyor (OC).

The CC has a parallel input port where it can read the following digital signals:

```
READY0 = 1 when IC has a new part in position
READY1 = 1 when MT is ready to receive a new part
DONE1  = 1 when MT is ready to have a part removed
READY2 = 1 when IS is ready to receive a new part
DONE2  = 1 when IS is ready to have a part removed
READY3 = 1 when OC is ready to receive a part
```

Thus, the CC can implement Task 0 as shown in the flowchart in Figure 3.P2b. It would be possible to simplify the software for Task 0 if we could replace the 6 signals (Ready0 to Ready3 and Done1, Done2) by the following 3 signals:

```
TSK1R = 1 when Task 1 is ready to be performed
TSK2R = 1 when Task 2 is ready to be performed
TSK3R = 1 when Task 3 is ready to be performed
```

Show two logic circuits to implement evaluation of TSK1R; one that does not use AND gates and another that does not use OR gates.

3. Assume an 8-bit word length and two's complement notation in the following:

   a. Convert -25_{10} to binary.
   b. Convert $+68_{16}$ to decimal.
   c. Compute the binary sum $(01100101_2) + (11111110_2)$ and check the addition with decimal numbers.

4. In the following calculation, assume that all binary values will be stored in 8-bit words using the two's complement sign convention:

   a. Convert -42_{10} to binary.
   b. Convert $B9_{16}$ to decimal.
   c. Compute the binary sum: $(01101101_2) + (01110110_2)$; is the result as stored in the machine correct?

5. For a sawing operation in a lumber mill, assume that sensors have been installed to measure blade vibration, blade temperature, product surface

quality (roughness), and product dimensional accuracy. Also assume that the sensor circuits can send the following digital signals:

EV = "1" when excessive vibration occurs
HT = "1" when the temperature is too high
PS = "1" when the product surface quality is poor
PA = "1" when the dimensional accuracy is poor

These signals are to be sent to a processor that will display the following information on an operator's CRT console:

"Code Yellow" if any one of the signals (EV, HT, PS, PA) are high (="1")

"Code Red" if two or more of the signals (EV, HT, PS, PA) are high (="1")

One approach is to input all four signals to the processor digital I/O port and then implement in software the logical operations required to detect a "code yellow" or "code red" condition. Another approach is to use a hardware solution by designing a digital circuit to detect a "code yellow" or "code red" condition and then to input this information directly to the processor. Show the digital circuit that would be required to implement the second approach.

6. Assume an 8-bit word size and two's complement sign convention in answering the following:

   a. What is the two's complement of $0101\ 1101_2$?
   b. What is the hexadecimal equivalent of $1111\ 1110_2$?
   c. What is the decimal equivalent of $6B_{16}$?
   d. What is the binary equivalent of -25_{10}?
   e. Write the Boolean expression $A + B'$ using only the operators NEG ( )' and AND (·).

7. (a) How will the number -49_{10} normally be represented in an 8-bit processor in two's complement form?

   (b) Supposing both positive and negative numbers can be represented in an 8-bit computer, which of the following additions will produce an erroneous result? Why?

   (i)   11001011   (ii)   11001011   (iii)   11001011
       + 10101101   + 10111001   + 00101101

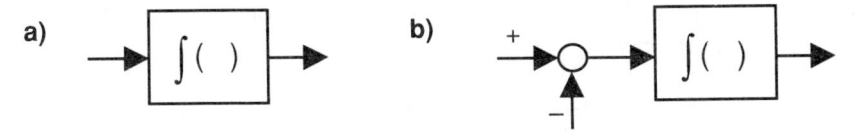

Figure 3.P9 The approximate functions performed by (a) a binary up counter, (b) a binary up–down counter

8. Consider 8-bit binary numbers with two's complement notation and determine how to represent the following:

   a. -112_{10}
   b. AF_{16}
   c. $(73_{10}) - (21_{10}) = (52_{10})$
   d. $(2B_{16}) + (16_{16}) = (?)$

9. The 4-bit binary up counter is the digital equivalent of an integrator. That is, it sums up the number of HI pulses it receives up to the limit of the counter (15). Extend the 4-bit binary up counter in Example 3.5 to an up–down counter, which would be equivalent to a comparator and integrator. See Figure 3.P9.

10. A firm that heat treats steel sheet in process requires a method of determining whether their heat treatment of SAE 1040 cable meets certain hardness specifications. They decide on using an eddy current inspection sensor that provides an analog signal that is a function of the material hardness (microstructure) and integrity (crack detection). The firm devises a logic circuit (using an 8-bit analog-to-digital converter) to provide the outputs shown in Table 3.P10, where RC and RB denote Rockwell C and B hardness measures, respectively.

    Design a circuit that is connected to the three least significant bits and uses a minimum of logic gates (i.e. AND, OR, NOT) that produces a HI signal when the steel sheet is too hard.

Table 3.P10 Digital Output of Eddy Current Sensor Circuit

Measured Hardness	Bit Number								Specification
	7	6	5	4	3	2	1	0	
Cracked	0	0	0	0	0	1	?	?	Cracked
52 RC	0	0	0	0	0	0	1	1	Too hard
26 RC	0	0	0	0	0	0	1	0	Desired hardness
83 RB	0	0	0	0	0	0	0	1	Too soft

11. Assuming an 8-bit word size and two's complement sign convention, determine the following:

   a. The binary equivalent of $6B_{16}$
   b. Decimal equivalent of $6B_{16}$
   c. The octal equivalent of $6B_{16}$
   d. The binary equivalent of $A3_{16}$
   e. The decimal equivalent of $A3_{16}$

Chapter 4
MICROCOMPUTER ARCHITECTURE AND OPERATION

Previous chapters briefly introduced microcomputer systems, digital logic, and binary arithmetic. This chapter will describe the structure and operation of a microcomputer. First, the typical architecture of a microprocessor is introduced, and then its operation is described in terms of typical instructions it can perform. A brief review of assembly language programming, with examples for the Z80 and 8088 microprocessors, is also presented.

4.1 MICROPROCESSOR ARCHITECTURE

As stated earlier, the microprocessor must perform the following sequence of operations:

1. Fetch instruction from program memory
2. Decode the instruction
3. If data is required, go to 4, otherwise go to 5
4. Fetch data from memory
5. Execute the instruction
6. Go back to 1 for next instruction

46 Chapter 4 MICROCOMPUTER ARCHITECTURE AND OPERATION

To better understand how these operations are performed, we will briefly describe the hardware configuration of a typical microprocessor.

Referring to Figure 4.1, we note that a typical microprocessor has on one chip an arithmetic and logical unit (ALU), a control unit, and internal registers. These together are referred to as the central processing unit (CPU). Typically, the microprocessor chip is connected to three *buses* or sets of parallel conductors:

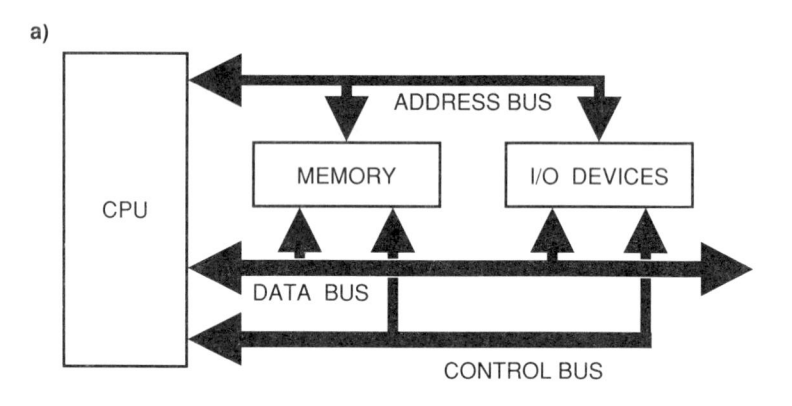

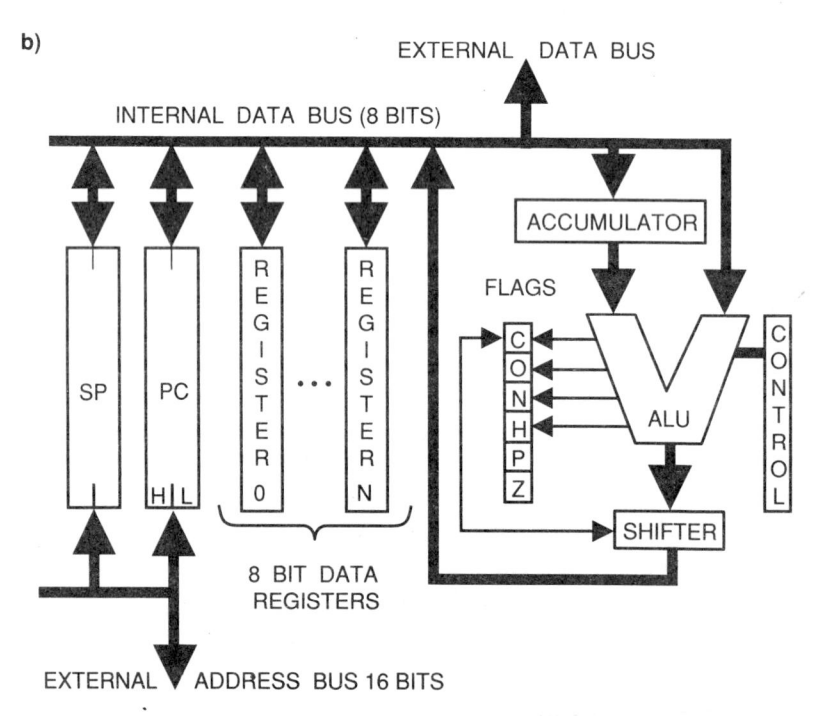

Figure 4.1 Typical 8-bit microcomputer architecture: (a) microcomputer block diagram, (b) microprocessor architecture

1. Data bus
2. Address bus
3. Control bus

Other devices of the microcomputer system (e.g., memory, clock, parallel port) are also connected to these buses. Information (i.e., instructions or data) travels over the data bus between the CPU and memory or I/O devices. The CPU uses the address bus to select a unique location in memory or a particular I/O device. The control unit directs and oversees this sequence of operations, and the control bus carries signals specifying whether data is going in or out of the CPU, and exactly when the data should be transferred.

The ALU performs certain arithmetic and logical operations, most of which are associated with registers. In addition to arithmetic and logical operations, the ALU performs shift and rotate operations, as shown in Figure 4.2. The *registers* are special high-speed memory locations, of which there are several types:

1. *General Purpose Registers:* A number of general purpose registers are connected to the data bus. In some processors these are also grouped in pairs and have a connection to the address bus. Only one register is connected to the bus at a time through a multiplexer.
2. *The Accumulator:* This is a special register provided on one of the inputs to the ALU and as a result is usually the fastest register. All data to be operated on by the ALU must use the accumulator.
3. *Flags or Status Register:* The individual bits of this register are used to store information about conditions, both routine and exceptional. An example of a routine condition would be an indication of whether the last

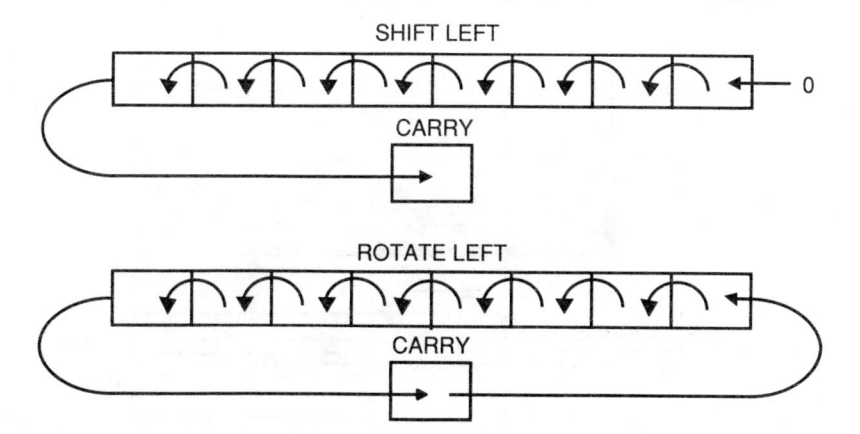

Figure 4.2 *Shift and rotate operations (Note: Some shift and rotate instructions do not include the carry.)*

arithmetic result was positive or negative; an example of an exceptional condition would be if this last result caused an arithmetic overflow.

4. *Address Registers:* These registers (sometimes called data counters or pointers) are for the storage of addresses. They are connected to the address bus.

A program is a sequence of instructions stored in memory that are executed in sequential order, unless an instruction explicitly changes that order. To handle this very important special case, when the CPU finishes executing an instruction it uses the *program counter* (PC), which is a special address register that always contains the address of the next instruction to be executed. The PC is automatically incremented or decremented the correct amount or modified by instructions that change the order of executions so that it will point to the next instruction.

The *stack* is another important concept. It is a set of memory locations with a last-in first-out structure used for temporary storage (see Fig. 4.3). Elements are deposited on the stack by a *push* operation, and removed by a *pop* operation. Stacks are often used to save the contents of the PC when program branching occurs, as in a subroutine call or an interrupt. The *stack pointer* (SP) is an address register that is used to keep track of the top of stack within memory. Thus, it contains the location or address of the top of stack in memory.

Sometimes, special address registers called *index registers* are also provided. They can be readily incremented or decremented to point sequentially to particular elements of a block of data in memory.

In closing this section, to illustrate typical microprocessor architecture, brief overviews of the Zilog Z80A and Intel 8088 microprocessor are provided. For more detailed information on these processors consult [Inte81, Mors82, ReA180, WiKr83, Zaks80]. For information on other popular microprocessors consult [Andr82, Leve79, Leve80, Moto84].

The Z80 microprocessor has an 8-bit data bus and a 16-bit address bus. Thus, directly addressable memory is limited to $2^{16} = 65536$ bytes (i.e., 64k bytes).

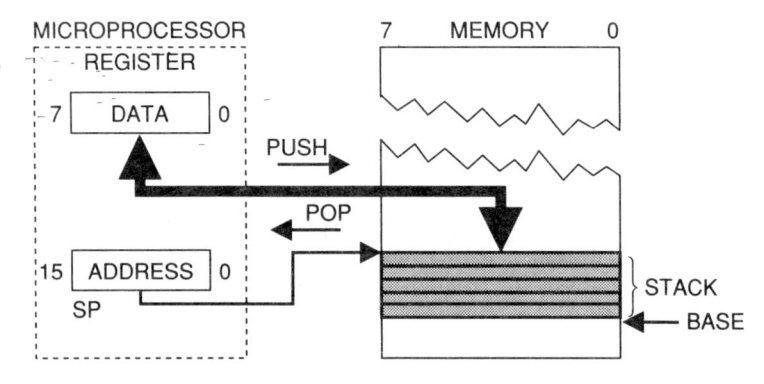

Figure 4.3 The two stack-manipulation instructions: PUSH and POP

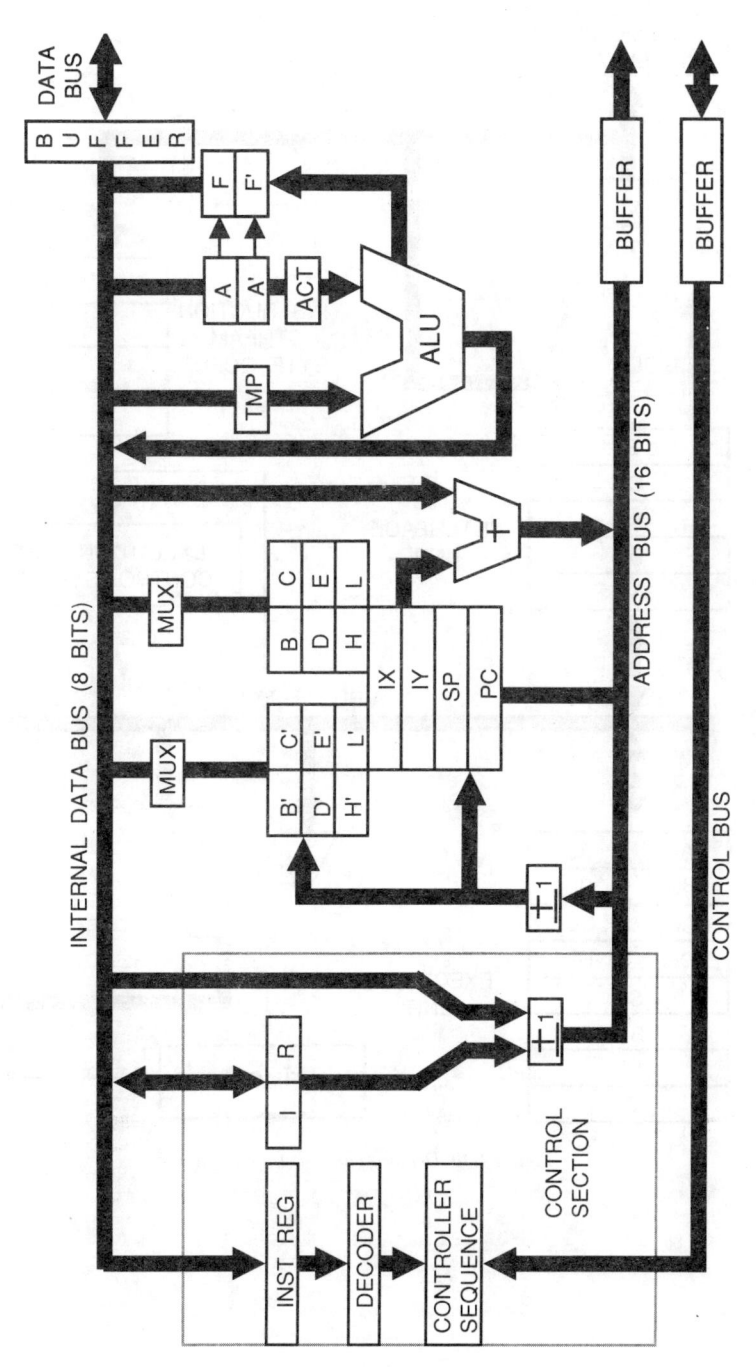

Figure 4.4a Detailed block diagram of the Z80

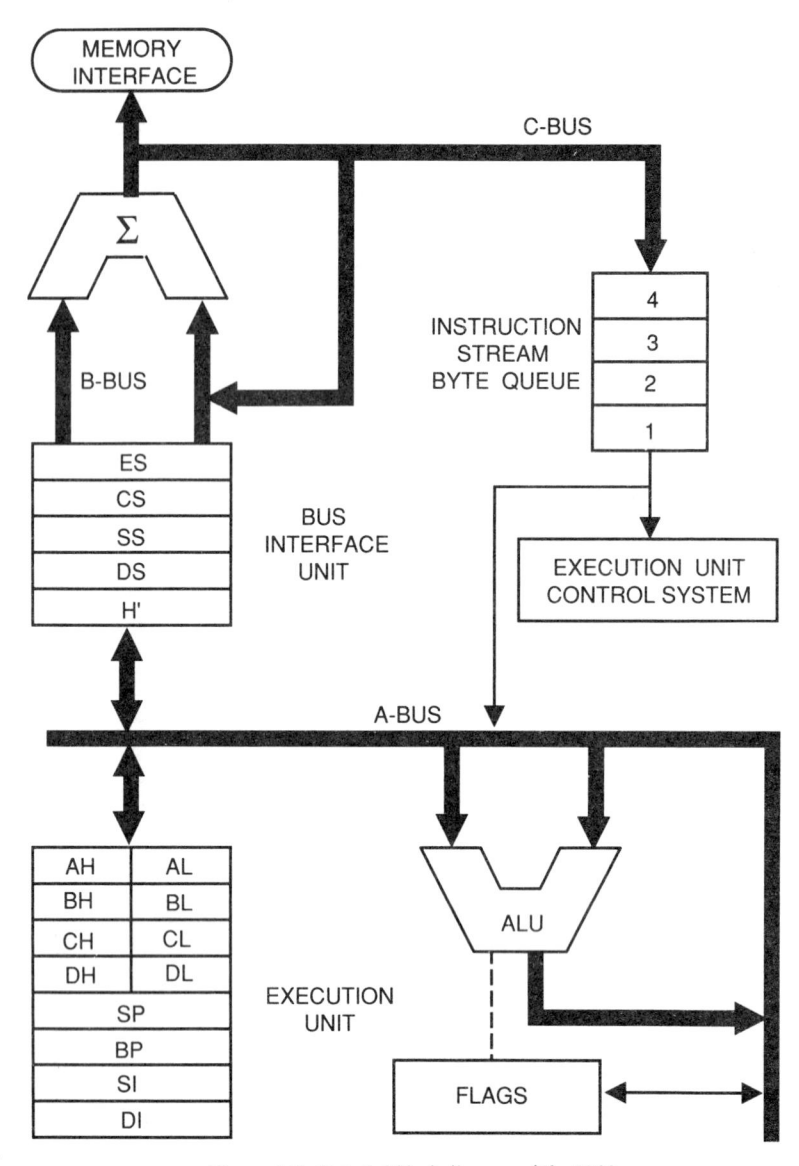

Figure 4.4b Detailed block diagram of the 8088

The Z80 registers, shown in Figure 4.4a, include an 8-bit accumulator (A), an 8-bit flags or status register (F), three pairs of 8-bit general purpose registers (BC,DE, and HL), two 16-bit index registers (IX and IY), a 16-bit stack pointer (SP), and a 16-bit program counter (PC).

The 8088 microprocessor CPU is really two separate functional units termed the bus interface unit (BIU) and the execution unit (EU). The role of the BIU is to fetch instructions from memory and to put them in an instruction stream byte queue, whereas the EU executes a previously fetched instruction (see Fig. 4.4b). The internal A bus in the EU is a 16-bit bus that interfaces only to the BIU. There are 8-bit B and C buses in the BIU, and the C bus provides the interface to external devices (e.g., memory). The 8088 has four general purpose 16-bit registers (AX, BX, CX, and DX). These can also be referenced as pairs of 8-bit registers, for example, the most significant byte of AX can be referenced as AH and the least significant byte as AL. Two string index registers [SI (source index) and DI (destination index)] and two stack pointers [SP (stack pointer) and BP (base pointer)] can also be used as general purpose 16-bit registers. There are also four special purpose 16-bit segment registers [CS (code segment), DS (data segment), SS (stack segment), and ES (extra segment)] that permit addressing of 2^{20} = 1,048,576 bytes (i.e., 1 M byte) of memory using 16 bit "offset" addresses, that can be used with the base addresses stored in the segment registers. Two other special purpose 16-bit registers are the instruction pointer (IP), which is analogous to the program counter described previously, and the flags or status word.

4.2 INSTRUCTIONS AND OPERATIONS

In a computer, information (instructions and data) is represented by a sequence of binary digits called *bits*. A group of 8 bits is called a *byte*, and, depending on the particular computer, one or more bytes form a *word* (e.g., 8-, 16-, or 32-bit words are common). The instructions and data are coded in a particular format dependent on the computer. For example, integers are generally represented in two's complement binary form, alphanumeric data (characters) are generally represented by the American Standard Code for Information Interchange (ASCII), floating-point (real) numbers are represented by a signed integer mantissa and exponent, and instructions (called "opcodes") depend on the particular microprocessor and are typically coded in 1 to 3 words. The ASCII code requires 1 byte, as shown in Table 4.1, and since the most significant byte (MSB) is reserved for error checking, 2^7 = 128 characters can be represented. Figure 4.5 shows a typical format for representation of floating-point data. The coding of instructions on a Z80 and 8088 processor is illustrated in Figure 4.6; each instruction contains space for the opcode, and data or address as required. The Z80 instruction set contains only 1-, 2-, and 3-byte instructions, whereas the 8088 also contains some 4-byte instructions.

Chapter 4 MICROCOMPUTER ARCHITECTURE AND OPERATION

Table 4.1 American Standard Code for Information Interchange (ASCII)

HEX	MSD	0	1	2	3	4	5	6	7
LSD	Bits	000	001	010	011	100	101	110	111
0	0000	NUL	DLE	SP	0	@	P	`	p
1	0001	SOH	DC1	!	1	A	Q	a	q
2	0010	STX	DC2	"	2	SB	R	b	r
3	0011	ETX	DC3	#	3	C	S	c	s
4	0100	EOT	DC4	$	4	D	T	d	t
5	0101	ENQ	NAK	%	5	E	U	e	u
6	0110	ACK	SYN	&	6	F	V	f	v
7	0111	BEL	ETB	'	7	G	W	g	w
8	1000	BS	CAN	(	8	H	X	h	x
9	1001	HT	EM	)	9	I	Y	i	y
A	1010	LF	SUB	*	:	J	Z	j	z
B	1011	VT	ESC	+	;	K	[	k	{
C	1100	FF	FS	,	<	L	\	l	|
D	1101	CR	GS	−	=	M	}	m	}
E	1110	SO	RS	.	>	N	^	n	~
F	1111	SI	US	/	?	O	_	o	DEL

Special ASCII Symbols
NUL Null
SOH Start of heading
STX Start of text
ETX End of text
EOT End of transmission
ENQ Enquiry
ACK Acknowledge
BEL Bell
BS Backspace
HT Horizontal tab
LF Line feed
VT Vertical tabulation
FF Form feed
CR Carriage return
SO Shift out
SI Shift in
DLE Data link escape
DC Device control
NAK Negative acknowledge
SYN Synchronous idle
ETB End of transmission block
CAN Cancel
EM End of medium
SUB Substitute
ESC Escape
FS File separator
GS Group separator
RS Record separator
US Unit separator
SP Space (blank)
DEL Delete

In this section, to illustrate typical microprocessor instructions and operation, an overview of *assembly language* programming is provided for the Z80 and 8088 microprocessors. In assembly language programming, instructions are represented in mnemonic (symbolic) form. Each instruction consists of an *opcode* followed by an optional data or address. The shorter instructions execute more rapidly and are generally preferred. Instructions can be classified according to the type of operation:

```
  31         24 23      16 15              8 7              0
 ┌──┬──────────┬──┬──────────────────────────────────────────┐
 │S │   EXP    │S │              MANTISSA                    │
 └──┴──────────┴──┴──────────────────────────────────────────┘
```

Figure 4.5 Typical floating-point number representation with 32 bits

4.2 Instructions and Operations

```
         7                0
2 Byte  ┌─ GENERALIZED    │ 1 Byte
Instruction │ OPCODE      │ Instruction
        │  ─────────────
        └─ OPTIONAL DATA           ─ 3 Byte
           OR ADDRESS                Instruction
                                                    ─ 4 Byte
           OPTIONAL DATA                              Instruction
           OR ADDRESS

           OPTIONAL DATA
           OR ADDRESS
```

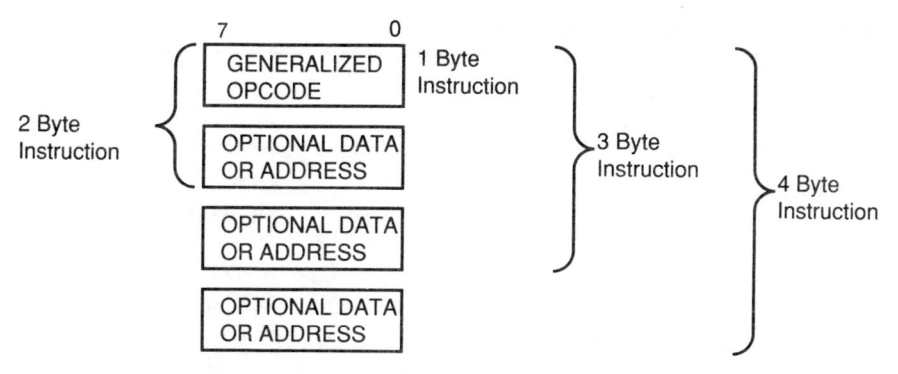

Figure 4.6 *Typical instruction formats for the Z80 and 8088*

1. Data transfer operations
2. Arithmetic operations
3. Logical operations
4. Bit-oriented operations
5. Test and branch type operations
6. Other (e.g., for handling I/O, interrupts)

Table 4.2 lists some Z80 and 8088 assembly language instructions in each category. Complete instruction-set listings for the Z80 can be found in [Zaks80], and for the 8088 in [Inte81].

Addressing refers to the specification, within an instruction, of the location of the operand on which the instruction will operate. Microprocessors typically use 16 bits or 2 bytes to specify an address. Thus, $2^{16} = 65536$ memory locations can be directly addressed, where each memory location is 1 byte. The 8088 processor can directly address $2^{20} = 1,048,576$ bytes (1 M byte) by adding a 4-bit segment address to the 16-bit address within each of the 16 segments. There are numerous methods, or modes, of addressing. Some are described next, and examples of these are given in Table 4.3 for both the Z80 and 8088 microprocessors. For the beginning programmer, addressing is initially one of the more confusing concepts; remember the difference between an address and the contents or data stored at that address. The following list of addressing modes should help to make this clearer:

1. *Implicit or implied addressing:* The instruction does not contain the address of the operand on which it operates. These are instructions that operate exclusively on registers, usually the accumulator.
2. *Immediate addressing:* The 8-bit opcode is immediately followed by an 8-bit or 16-bit constant.

Table 4.2 Some Z80 and 8088 Instructions

Instruction Type	Z80	8088
Data Transfer	LD d, s	MOV d, s
		MOVS d, s
Arithmetic	ADD d, s	ADD d, s
	ADC d, s	ADC d, s
	SUB d, s	SUB d, s
	SBC d, s	SBC d, s
	INC d	INC d
	DEC d	DEC d
	NEG	NEG d
		DIV s
		MUL s
		IDIV s
		IMUL s
Logical	AND s	AND d, s
	OR s	OR d, s
	XOR s	XOR d, s
	CP s	TEST d, s
	CPL	NOT d
Test and Branch	JP d	JMP d
	JR offset	J condition, offset
	JP condition, d	CALL d
	JR condition, offset	RET
	CALL d	where offset is an offset to the instruction pointer within the current segment and the conditions in the flag register that can cause transfer are numerous. Some examples are
	CALL condition, d	
	RET	
	RET condition	
	where offset is an 8-bit offset to the current program counter, and the conditions in the flag register that can cause the transfer are	

Z80 conditions:

Z	Zero bit = 1
NZ	Zero bit = 0
M	Sign bit = 1
P	Sign bit = 0
PE	Parity bit = 1
PO	Parity bit = 0
C	Carry bit = 1
NC	Carry bit = 0

8088 conditions:

JC	Carry bit = 1
JNC	Carry bit = 0
JZ	Zero bit = 1
JNZ	Zero bit = 0
JCXZ	CX register = 0
JNS	Sign bit = 0
JO	Overflow bit = 1
JNP/ JPO	Parity bit = 0

Table 4.2 Some Z80 and 8088 Instructions (*cont.*)

Instruction Type	Z80		8088
Bit-oriented	BIT bit #, d SET bit #, d RES bit #, d Logical (above) and shift and rotate instructions can also be used		These are done with logical instructions (above) and shift: SHL/ SAL d, # of bits SHR d, s SAR d, # of bits and rotate instructions: ROL d, # of bits ROR d, # of bits RCL d, # of bits RCR d, # of bits
Other	IM0 HALT NOP EXX POP d OUT d, s etc.	(interrupt mode 0) (halt the CPU) (no operation) (exchange alternate registers)	INT interrupt type HLT NOP XCHG d, s POP d OUT d, s etc.

d = destination and s = source can take on various forms.

3. *Absolute addressing:* The 8-bit opcode is followed by a 16-bit address that the parentheses around an address refers to the contents of that memory location. Recall that addresses are 16 bit whereas the contents of a memory location are 8 bit.
4. *Relative addressing:* The 8-bit opcode is followed by an 8-bit offset (or displacement). This is used with the "jump relative" instruction and allows branching of up to 127 locations forward or 128 locations backward.
5. *Indexed addressing:* Uses the index registers successively to access the elements of a block of data in memory.
6. *Indirect addressing:* Useful when several routines need to access the same data. An 8-bit opcode is followed by an indirect address, which points to the final address where the data is stored in memory, that is, the address of an address.
7. *Direct addressing:* 8-bit opcode is followed by the address of the memory location where the data is stored.

As with most programming, assembly language programming is not conceptually difficult. However, months or even years of experience may be required to become a proficient assembly language programmer. Some programming tasks, depending on the instruction set of the particular microprocessor, may be

Table 4.3 Some Z80 and 8088 Addressing Modes

Addressing Mode	Z80		8088	
Implicit or implied or register		LD A, B		MOVE AX, BX MOV AL, CL
Immediate		ADD A, 0H		ADD AX, 0H
Absolute		LD (1234H), A		MOV [1234H], AL
Relative	LOOP:	SUB B JR NZ, LOOP	LOOP:	SUB AX, BX JNZ LOOP
Indexed	STATUS	EQU 0FF70H LD IX, STATUS LD A, (IX + 2)	STATUS	DW 0FF70H MOV DI, STATUS MOV AX, [DI + 2]
Indirect		LD A, (BC)		MOV AX, [BX]
Direct	VALUE:	DEFS 1 LD A, (VALUE)	VALUE	DW MOV AX, VALUE

very difficult to carry out in assembly language (e.g., floating-point multiplication). Other tasks (e.g., examining the contents of a memory location) may be conveniently carried out. In this book it is assumed that you are familiar with a high-level programming language such as FORTRAN [Ette83, Gold65], and in the case studies we emphasize the appropriate use of FORTRAN and assembly language as needed to accomplish a particular task. The following example illustrates a simple assembly language program for the Z80 and 8088 microprocessors.

Example 4.1 Assembly Language Program for Addition

The following is a Z80 assembly language program for adding two 8-bit operands (OP1 and OP2) stored at memory addresses ADR1 and ADR2, respectively:

```
ADR1:    DEFS 1
ADR2:    DEFS 1
ADR3:    DEFS 1
         LD A, (ADR1)      ;load OP1 into A
         LD HL, ADR2       ;load ADR2 into HL
         ADD A,(HL)        ;add OP2 to OP1
         LD(ADR3), A       ;save result RES at ADR3
```

The result RES = OP1+OP2 is stored at ADR3 as shown in Figure 4.7. To subtract OP1 from OP2, insert

```
NEG              ; negate A
```

as the second line, or use the SUB instruction.

The same program for the 8088 is given next, where now ADR1 and ADR2 are memory addresses within the data segment (DS):

```
ADR1        DB    ?
ADR2        DB    ?
ADR3        DB    ?
            MOV   SI, OFFSET ADR1    ; or use the equivalent
            MOV   AL, [SI]           ;instructions:
            MOV   BX, OFFSET ADR2    ; MOV AL, ADR1
            ADD   AL, [BX]           ; ADD AL, ADR2
            MOV   DI, OFFSET ADR3    ; MOV ADR3, AL
            MOV   [DI], AL
```

Again the result is stored at the offset ADR3 in the data segment (DS), and to subtract OP1 from OP2 insert

```
NEG AL
```

as the third instruction in the program, or use the SUB instruction.

Notice that the differences in the Z80 and 8088 versions of the program are due to several factors: (1) the differences in the instructions sets (e.g., LD vs. MOV), (2) differences in the architectures (Z80 addresses are 16 bits, whereas 8088 addresses are 20 bits with a 16-bit offset), and (3) notational differences (e.g., ADR1 in the Z80 program corresponds to OFFSET ADR1 for the 8088,

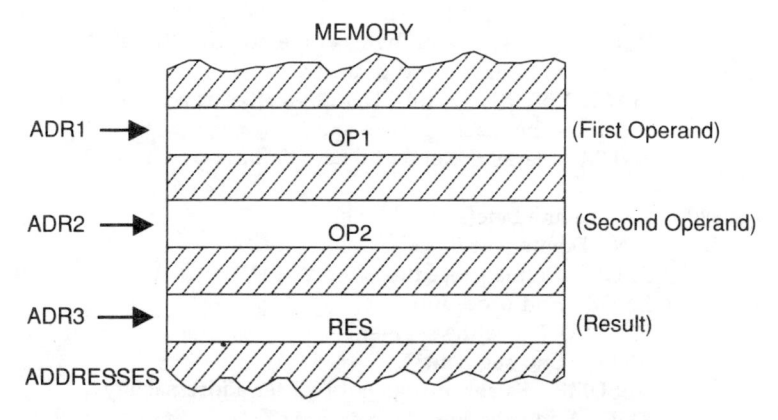

Figure 4.7 *Eight-bit addition, RES = OP1 + OP2*

and (ADR1) for the Z80 corresponds to ADR1 for 8088. Notice also that DEFS, DB, and OFFSET are not actual assembly language instructions, but instructions to the assembler. Such assembler directives or pseudo-opcodes are discussed next.

<<>>

The code given in Example 4.1 is not a complete assembly language program. In addition to the instructions or opcodes for assembly language instructions, there are other assembly language instructions called *assembler directives* or *pseudo-opcodes*. These are commands given by the programmer to the assembler (the program that translates assembly language code). Some examples of assembler directives (e.g., EQU and DW commands) have already been used in Table 4.2. You should refer to the assembler manual for your particular system for a detailed description of these commands. Here we discuss the formal syntax of assembly language programming and provide some typical examples of assembler directives for the Z80 and 8088 in Table 4.4.

Assembly language programs for the Z80 and 8088 typically have several fields: (1) the label field (optional), which may contain a symbolic address for the instruction that follows; (2) the instruction field, which includes the opcode and any operands; (3) the comment field (optional) to the right of the instruction. A label is a symbolic means of referring to a particular location in the program (e.g., LOOP in the example on relative addressing in Table 4.3). The optional label can typically be up to 7 characters long, followed by an optional colon, and must be in the first 8 columns. To the right of the label field is the instruction field, which contains the Z80 or 8088 opcode (or a pseudo-opcode) followed by any operands. A semicolon may be inserted in any line, and all text following the semicolon is treated as a comment. Comments

Table 4.4 Examples of Assembler Directives

Z80	EQU nn	Assigns the value nn to a label
	DEFS nn	Reserves nn bytes of memory for the label
	END	Indicates the end of program
	MACRO p0, p1, ... pn	Defines a label as a macro and defines the macro parameters
	ENDM	Marks the end of macro definition
8088	DB	Define byte(s)
	DW	Define word(s)
	PROC	Define procedure
	ENDP	End procedure
	SEGMENT	Define segment
	ENDS	End segment
	ASSUME	Establish segment register addressability
	END	End program

are used to clarify a program and are ignored by the assembler. In creating an assembly language program, tabs may be effectively used to separate the three fields just described.

Most assemblers also provide a programming feature that is generally referred to as a *macro*. A macro is a collection of assembly language code, which may have formal parameters, which the programmer can refer to by a label. If a group of instructions is referred to several times by a programmer, it may be convenient to define them as a macro instead of repeatedly writing out the instructions. It may appear that a macro is similar to a subroutine, but they are actually somewhat different. A subroutine is also defined only once, then invoked as needed by a CALL statement. The CALL and RET statements of a subroutine actually cause branching in the program. The executing program on encountering a subroutine call jumps to the location in memory where the subroutine code is stored, executes that code, then returns on encountering a RET statement in the subroutine to the instruction following the original subroutine call. All this occurs at execution time. The macro facility, however, is an assembler directive. The assembler simply takes the code associated with the original macro definition and inserts it wherever the macro label is referred to in the program. Use of macro leads to a longer but more efficient code, whereas subroutines result in a more compact but slower code. Example 4.2 presents both a macro and subroutine code for subtracting two 8-bit numbers on a Z80.

Example 4.2 Macro and Subroutine for Subtraction

As in Example 4.1, ADR1 points to OP1, ADR2 points to OP2, and ADR3 points to RES = OP1-OP2 (see Fig. 4.7). The macro to implement the 8-bit subtraction is

```
SUB8:       MACRO A1,A2, A3
            LD A, (A1)          ; load OP1 into A
            LD HL, A2           ; load ADR2 into HL
            SUB (HL)            ; subtract OP2 from OP1
            LD (A3), A          ; put RES in ADR3
            ENDM
```

This macro can be "called" as follows,

```
SUB8 ADR1, ADR2, ADR3
```

A subroutine to implement the same task is,

```
SUB8:       PUSH AF             ; save registers
            PUSH HL
            PUSH BC
            PUSH DE
            LD A, (BC)          ; load OP1 into A
```

```
                SUB  (HL)           ; subtract OP2 from OP1
                LD   (DE), A        ; put RES in ADR3
                POP  DE             ; restore registers
                POP  BC
                POP  HL
                POP  AF
                RET
```

The calling program for this subroutine would look like,

```
LD BC, ADR1             ; load ADR1 into BC
LD HL, ADR2             ; load ADR2 into HL
LD DE, ADR3             ; load ADR3 into DE
CALL SUB8
```

An analogous 8088 program for 16-bit subtraction is given next:

```
SUB 16      MACRO A1, A2, A3
            MOV AX, A1
            SUB AX, A2
            MOV A3, AX
            ENDM
```

This macro can be "called" as follows:

```
SUB 16      ADR1, ADR2, ADR3
```

The subroutine form is,

```
SUB 16      PROC FAR
            PUSHF
            PUSH AX
            MOV AX, [SI]
            SUB AX, [BS]
            MOV [DI], AX
            POP AX
            POPF
            RET
SUB 16      ENDP
```

And the calling program for this subroutine would look like,

```
MOV SI, OFFSET ADR1         ; Remember that ADR's are
MOV BX, OFFSET ADR2         ; defined by "ADRn DW ?"
MOV DI, OFFSET ADR3         ; in contrast to the Z80's
CALL SUB 16                 ; "ADRn: DEFS 1"
```

<<>>

4.3 SUMMARY

A typical microprocessor is a digital device that contains on one chip an arithmetic and logical unit (ALU), a control unit, and internal registers. This microprocessor can be connected to a data bus, address bus, and control bus. The typical operations of such a device, including its registers, stack manipulation, and indexing have been described. As examples, the specific architecture of Z80 and 8088 microprocessors have been briefly introduced.

The operation of a typical microprocessor, and as specific examples the Z80 and 8088 microprocessors, has been discussed by classifying the types of instructions and the addressing modes available. This brief introduction to assembly language programming has been concluded by a description of assembler directives and macros.

4.4 PROBLEMS

1. Suppose that a memory location (8 bits) contains 00100100_2. If this is interpreted as an ASCII code and printed on a terminal, what character will be printed?

2. A peg of diameter D2 is to be inserted in a hole of diameter D1 in a plate. The values of D1 and D2 are measured and are available in the microprocessor at addresses D1LOC and D2LOC, respectively. The following program checks for the proper clearance for assembly by using the following condition,

   ```
   MIN < (D1-D2) < MAX
   ```

   where the minimum and maximum clearance values MIN and MAX are stored in memory at locations MINLOC and MAXLOC, respectively. The program that follows is supposed to check this condition and set bit 7 of the byte stored at RESLOC if the condition is satisfied. Otherwise, bit 7 of the contents of RESLOC should be reset. There are, however, errors in the program. Please identify, explain, and correct any errors.

   ```
   CHKCON:     PUSH AF
               PUSH BC
               PUSH DE
               SET 7,A
               LD (RESLOC), A
               LD A, (D1LOC)
               LD B, (D2LOC)
               SUB A,B
               LD C,A
   ```

```
                    LD   D, (MINLOC)
                    SUB  A,D
                    JP   P, QUIT
                    RES  7,A
                    LD   (RESLOC), A
                    JP   QUIT
NEXT:               LD   A,C
                    LD   D, (MAXLOC)
                    SUB  A,D
                    JP   P, QUIT
                    RES  7,A
                    LD   (RESLOC), A
QUIT:               POP  DE
                    POP  AF
                    POP  BC
                    RET
```

3. Using the given Z80 code as a basis, develop a correct 8088 assembly language program for Problem 4.2.

4. A press is instrumented such that the following signals are HI

```
HAND     when a hand is between the dies
PARTIN   when a part has not been removed from between
         the dies
CLOSED   when the dies are fully clamped
```

Clamping only proceeds when the following signal CLAMP is HI,

```
CLAMP  =  (HAND + PARTIN)'  ·  CLOSED'
```

Assume that the three signals HAND, PARTIN, CLOSED are connected to a microcomputer parallel I/O port. When they are true, they set the most significant bit (MSB) of a single byte word. The convention is that HI is negative (MSB set) and LO is positive (MSB not set)

Given the start of a Z80 program below, complete the instructions to evaluate the Boolean expression and store it in CLAMP. The 'H' following a number indicates the hexadecimal base.

```
              CSEG
HAND          EQU          0FF21H
PARTIN        EQU          0FF22H
CLOSED        EQU          0FF23H
CLAMP         DEFS         1
BOOL:
              PUSH         AF
```

          PUSH   BC

5. Repeat Problem 4.4 using 8088 assembly language.
6. Suppose that a location in memory contains 00100100 and these contents are printed directly on a terminal with an ASCII character set.

   a. If the parity of the terminal is such that a character will be printed, is the parity odd or even?
   b. What character will be printed?

7. If the accumulator contents is 10101101_2, what will happen to its contents after each of the following instructions? The 'H' following a number indicates the hexadecimal base.

   a. OR FFH
   b. XOR 00H
   c. XOR FFH
   d. AND 83H
   e. XOR A

8. Refer to Problem 3.10 on inspection of heat-treated steel. Assume that the three logical signal outputs of the sensor are read into an address location called SERIN. Write a Z80 program that constantly checks to see if the sheet is cracked. If it is cracked, set bit 0 of the digital output

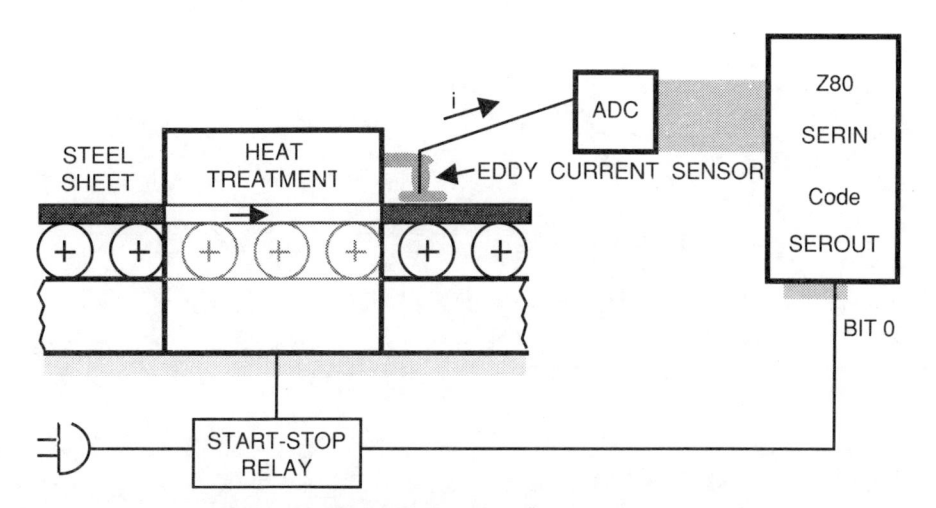

Figure 4.P8 *Eddy current sensing control*

port (address SEROUT) high and send it out (see Fig. 4.P8). The start of such a program is given below, where 'H' following a number indicates the hexadeximal base.

```
              CSEG
SERIN    EQU       OFF21H
SEROUT   EQU       OFF22H
```

9. Repeat the programming assignment in Problem 4.8 using 8088 assembly language.

Chapter 5
INTERFACING

The previous chapter described the structure and operation of microcomputers. In this chapter the interfacing of the microcomputer to external physical devices is discussed. First, a brief introduction to sensors and actuators is provided. Next, the exchange of information with other digital devices is treated. Then, the methods for information exchange with analog devices are described. Finally, the scheduling of service to input and output devices and hardware interrupts are introduced.

5.1 SENSORS AND ACTUATORS

Computers interact with external physical devices through various sensors and actuators, which in turn interact with the physical process or device. Sensors are devices that are used to bring information about the process into the computer, whereas actuators are devices that are used to send commands from the computer to the process. There are a very large variety of sensors and actuators, so we cannot give a complete description of all such devices here [Cons74]. However, we will give an example of a commonly used digital sensor (the optical incremental encoder) and an analog sensor (the tachometer). We will also describe a commonly used digital actuator (the stepping motor) and an analog actuator (the dc motor). For more information on these devices, or other sensors and actuators, refer to [deSi85; Doeb83; KoBe78; Potv85; Ruoc87].

Digital sensors produce signals that can be read directly through a serial or parallel input port on a computer. Common digital sensors include limit switches, digital resolvers, digital tachometers, and optical encoders. Optical encoders are commonly used in machine tool and robotics applications to measure angular position or velocity. *Absolute encoders* directly generate a binary word (or "gray code") corresponding to a particular angular position. A similar, but simpler and less expensive, device is the incremental encoder that generates a pulse for every fixed increment of angular position. Incremental encoders with 1000–2000 pulses per revolution are typical. *Incremental encoders* usually consist of a code disk with alternating transparent and opaque areas, a light source (e.g., a light-emitting diode), and two light sensitive probes (e.g., photocells or phototransistors). The configuration is illustrated in Figure 5.1. The two offset light-sensitive probes are used to determine the direction of rotation. The number of pulses generated in a particular direction is proportional to the angular position, and the pulse frequency gives the angular velocity. Pulses can be read directly by the computer and counted to get angular position, and the number of pulses in a fixed time interval can be calculated to get angular velocity. However, this can take up a lot of processor time, and incremental encoders are often used with external digital circuits that consist of an up–down counter for angular position, and a latch buffer that can be read by the computer while the counter continues to count pulses [ChUl87; deSi85].

Another commonly used sensor for angular velocity measurement is the dc tachometer, or tachogenerator, which is an analog sensing device. This device, much like an electric motor, uses a permanent magnet and a rotating coil (see Fig. 5.2). Relative motion between the magnetic field and coil induces a voltage in the coil proportional to its speed relative to the magnetic field. These can be alternating current (ac) or direct current (dc) devices and produce a voltage (V) proportional to the shaft angular velocity (ω),

$$V = k_t \omega \qquad (5.1.1)$$

where the tachometer gain k_t has units of volts · sec/radian.

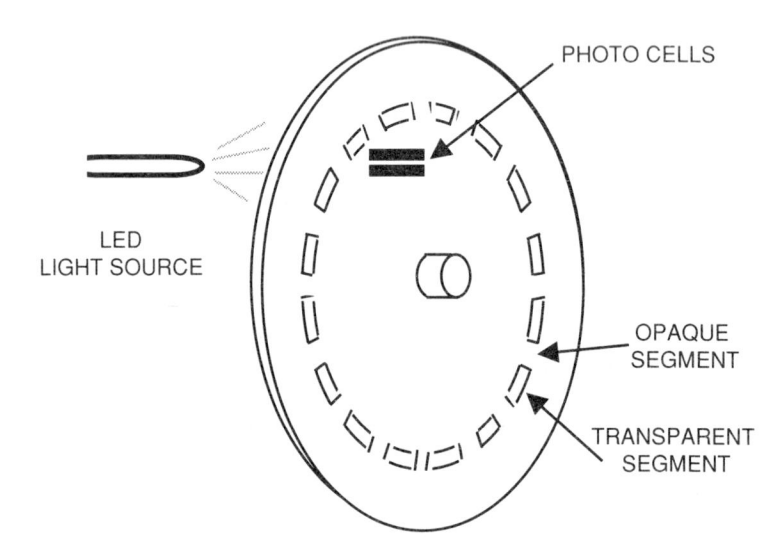

Figure 5.1 Incremental optical encoder

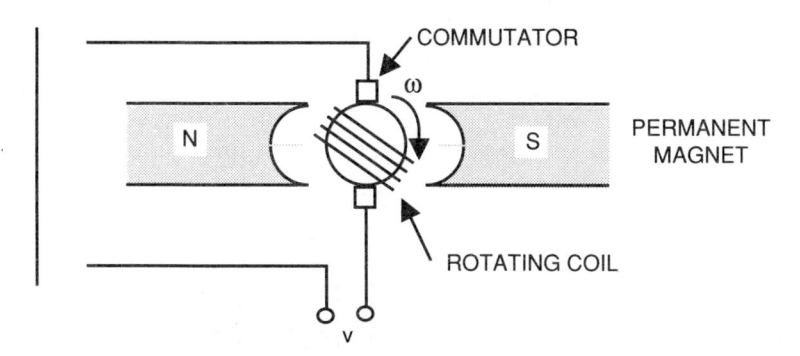

Figure 5.2 A dc generator tachometer

Sensors, like the incremental encoder or tachometer just described, provide information to the computer about the physical process being monitored or controlled. If, as in a control system, the computer is to act on the process, it must be able to send signals to *actuators* that are part of the physical process being controlled. Such actuators take the low-power (information-type) signal being sent by the computer and generate a high-power input to the process. Typical actuators include control valves, electrohydraulic actuators, solenoids, ac or dc motors, and stepping motors. Here we consider the stepping motor as an example of a digital actuator and the dc motor as an example of an analog actuation device.

Stepping motors are the most widely used of the digital actuating devices. The input to a stepping motor is a train of digital pulses, such as can be provided by a digital output port on a microcomputer. A typical application of stepping motors is to produce a desired mechanical motion [AuSa81; SeBo71]; a configuration that will be discussed in detail in Chapter 11 is shown in Figure 5.3.

The input to the stepping motor is a digital pulse train, and each pulse results in one angular increment or step of motion. The pulse frequency determines the rotational speed of the motor. Two important characteristics of a stepping motor are

1. The *steps per revolution* (SPR), which determines the accuracy
2. The *slew rate,* which is the maximum pulse frequency that the motor can handle without missing steps

The slew rate also depends on the motor loading, as shown in Figure 5.4. The fact that this motor characteristic is load dependent points out a disadvantage of stepping motors: they are not suited for variable load applications with high-power requirements (e.g., large industrial robots). The advantages of stepping motors are the simplicity of the system design, low cost, and that there is no need for sensors or digital-to-analog converters.

The digital control of motion with stepping motors requires that we compute the number and frequency of pulses to be sent from the computer to the stepping

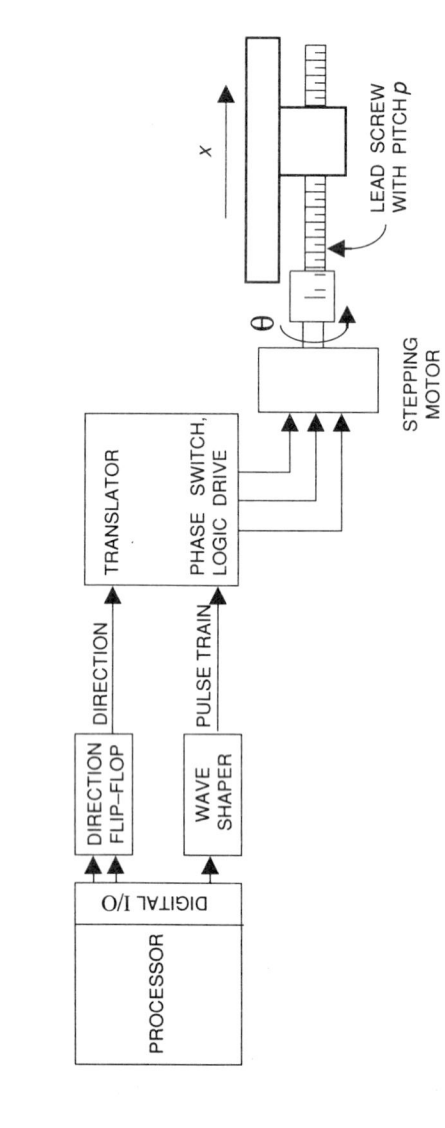

Figure 5.3 Schematic of a stepping-motor drive for linear mechanical motion

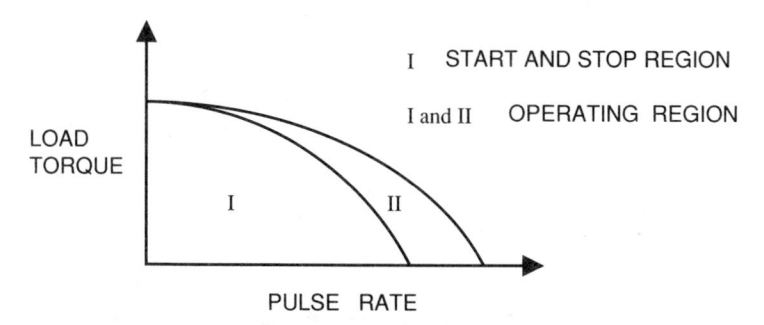

Figure 5.4 *Load versus speed characteristic of a stepping motor*

(Labels in figure: LOAD TORQUE; PULSE RATE; I START AND STOP REGION; I and II OPERATING REGION)

motor to produce a particular desired motion. This is simple in the case of moving a single axis as shown in Figure 5.3. For producing linear motion the stepping motor is coupled to a lead screw. The linear distance moved each time a pulse is sent to the motor is,

$$\Delta x = p/SPR \qquad (5.1.2)$$

where

p = the lead screw pitch (threads/in.)
SPR = steps per revolution of the motor (steps/rev)

Thus, to move a total of x inches, the number of pulses that must be sent to the motor are,

$$N = Round(x/\Delta x) = Round[p/((x)(SPR))] \qquad (5.1.3)$$

The operator *Round* indicates that the results of the calculation are rounded to the nearest integer value. The frequency of the pulses will determine the velocity of the motion. For example, to produce a constant velocity of v (in./sec) along the path,

$$f = Round(v/\Delta x) = Round[p/((v)(SPR))] \qquad (5.1.4)$$

is the required pulse frequency (pulses/sec).

The foregoing observations show that we need to determine the number of pulses N, the frequency of pulses f, and the direction of motion to control linear motion with a stepping motor. The direction of motion is specified either by a separate direction command as shown in Figure 5.3 or by two pulse inputs, one for forward and the other for reverse motion. The pulses sent from the computer to the stepping motor are conditioned by a translator card. The digital logic level may be different between computer and stepping motor, and the motor

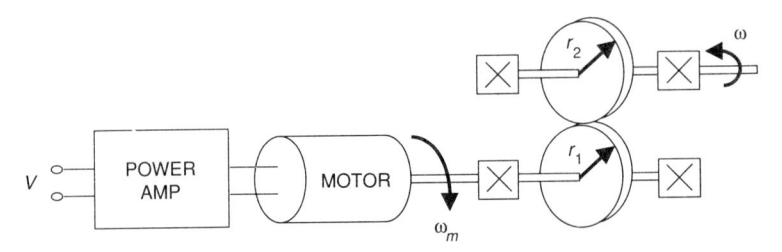

Figure 5.5 A dc motor with power amplifer and gear train

drive circuits may require that the input pulses have a particular waveform; the translator card handles these conversions.

Electrical dc motors are commonly used as actuating devices for variable-load applications where medium power is needed (high-power applications often employ electrohydraulic actuators). Most small to medium-sized robots and machine tools employ dc motor drives. We will consider the mathematical modeling of dc motors in detail in Chapter 7, so we consider here the hardware configuration. To produce mechanical motion, a dc motor is coupled to a transmission device, such as a gear train (see Fig. 5.5). Notice that the output speed ω is related to the speed of the motor shaft ω_m by the gear ratio (r_1/r_2),

$$\omega = (r_1/r_2)\omega_m \qquad (5.1.5)$$

The voltage V applied to the motor is proportional to the motor shaft speed ω_m, and the motor torque T_m is proportional to the current drawn by the dc motor. These relationships are developed in Chapter 7. A power amplifier is needed to produce the electrical power required to overcome loads and generate the desired motion. Sometimes the characteristics of the power amplifier can also influence the actuator characteristics (e.g., owing to nonlinearities or saturation) and several different types of power amplifiers are commonly used (e.g., PWM or SCR type).

5.2 DIGITAL INPUT AND OUTPUT

Digital I/O refers to transfer of information between the computer and other digital devices and is discussed in this section. We have learned how digital information is moved between memory and the registers. Now consider how the microprocessor can exchange digital information with the external world [Andr82; AuSa81; Mone87; MuSh81; ZaLe79]. Input refers to the capture of information from external devices, and output refers to the transfer of information from the microprocessor to the external devices. Common input/output (I/O) devices include the keyboard, cathode ray tube (CRT), printer, and disk drives. For real-time applications we are also interested in sensing devices (e.g., digital encoders) and actuators (e.g., stepping motors).

Many 8-bit microprocessors use *memory-mapped I/O*. That is, input/output devices are connected to the address bus just like memory chips and are addressed like memory locations. Many microprocessors (e.g., the Z80 and 8088) also have specialized I/O instructions, in which case the programmer can use either method. The specialized I/O instructions lead to more compact and faster executing code. Here we do not employ the specialized I/O instructions, but you may want to study these by consulting an assembly language reference manual for your processor (e.g., [Zaks80] and [Inte81]).

As noted previously, digital information can be transferred in either serial or parallel format. For serial transmission the computer must have a serial I/O port to handle devices such as CRT terminals and teletypes. Devices such as printers, and many sensors and actuators, generally employ parallel I/O.

Serial Transfer

In serial data transfer, the bits of information (0's and 1's) are successively sent or received on a single conductor. If the bits are spaced at regular intervals, the transmission is termed synchronous; if they appear in random bursts, it is termed asynchronous.

In a memory-mapped system, the serial I/O port appears as a particular memory location, for example, at address SERIO. One bit of the byte at SERIO will contain the data (e.g., bit 0) and another will be a status bit (e.g., bit 7). The status bit is used to indicate when the data bit is valid. This is sometimes referred to as "handshaking"; that is, the device and the processor use the status bit or register as a means of agreeing when data transfers should occur. In some cases, the status and data bits may be in separate registers. The transfer of words of data from an external device to memory can be performed as indicated in the flowchart shown in Figure 5.6. The following examples show how to use a serial port at location SERIO to turn on a relay and to send out a pulse.

Example 5.1 Set Bit 0 of a Serial I/O Port

The following is a Z80 assembly language program that sets bit 0 of the byte at SERIO; the 'B' following a number indicates the octal base, and 'H' similarly indicates the hexadecimal base:

```
RELON:   PUSH AF          ; save registers
         LD A, (SERIO)    ; load contents of I/O
                          ; port to A
         OR 00000001B     ; force bit 0 to "1"
         LD (SERIO), A    ; load A into (SERIO)
         POP AF           ; restore registers
         RET
```

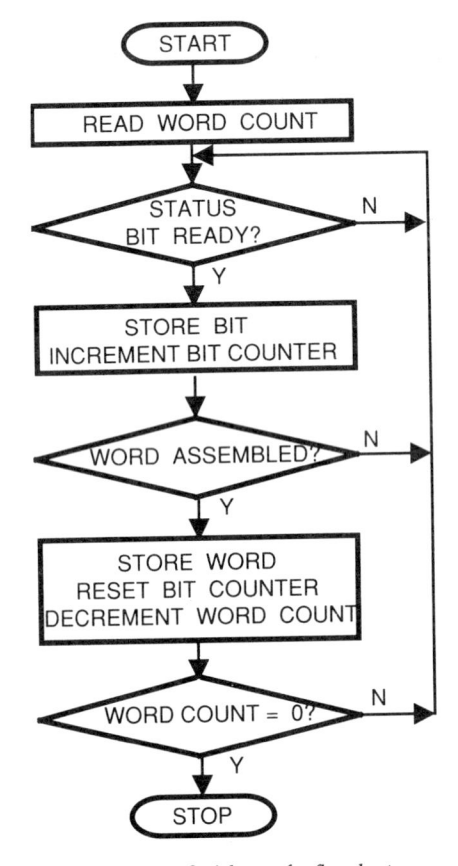

Figure 5.6 Serial transfer flowchart

The 8088 assembly language equivalent is,

```
RELON     PROC FAR          ; SERIO must be defined
          PUSHF             ; and accessible to
          PUSH AX           ; this procedure
          MOV AX, SERIO
          OR AX, 0001H
          MOV SERIO, AX
          POP AX
          POPF
          RET
RELON     ENDP
```

<<>>

Example 5.2 Send a Pulse from a Serial I/O Port

The following is a Z80 assembly language program to send a pulse:

```
SENPLS:   CALL RELON      ; set bit
          CALL DELAY      ; wait
          CALL RELOFF     ; reset bit
          RET
```

where the subroutines DELAY and RELOFF generate a delay and send a 0 to the serial I/O port, respectively. The RELOFF routine is similar to RELON in Example 5.1, except for line 3. Determine how line 3 should be changed. The 8088 assembly language version of SENPLS is,

```
SENPLS    PROC FAR
          PUSH DX
          PUSH AX
          MOV DX, IOPORT     ; DX is needed for indirect
          CALL RELON         ; addressing on the 8088
          MOV AL, SERIO      ; for outside communications
          OUT DX, AL         ; send a high
          CALL DELAY
          CALL RELOFF
          MOV AL, SERIO
          OUT DX, AL         ; send a low
          POP AX
          POP DX
          RET
SENPLS    ENDP
```

A Z80 DELAY routine that generates a delay of 82 clock cycles is shown next, and a basic delay flowchart is given in Figure 5.7.

```
DELAY:    LD A,5
NEXT:     DEC A
          JR NZ, NEXT
          RET
```

The effective delay can be calculated by looking up the number of clock cycles required by each instruction in a Z80 reference manual. Develop an analogous 8088 delay routine.

<<>>

There are several standards for serial I/O; the most widely used is the *RS-232* standard. The *RS-232* standard serial interface provides standard voltage levels (a logical 0 is +3 to +15 V; a logical 1 is −3 to −15 V) and uses

Chapter 5 INTERFACING

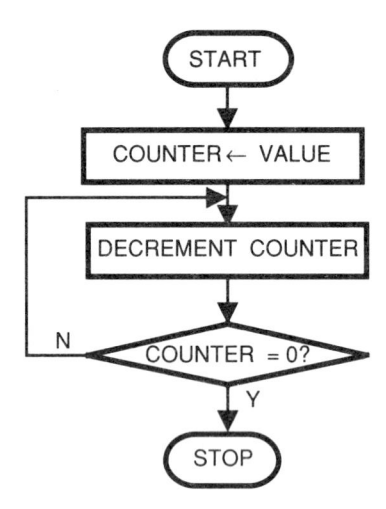

```
        START
          ↓
   COUNTER ← VALUE
          ↓
   ┌──────────────────┐
   │ DECREMENT COUNTER │←┐
   └──────────────────┘ │
          ↓             │
       COUNTER = 0? ──N─┘
          │Y
          ↓
         STOP
```

Figure 5.7 Basic delay flowchart

9 signal lines with a 25-pin connector. The signal lines are described in Table 5.1, along with their standard pin assignments. Typically, many of the functions are only required for use with modems, and in most applications we use only pins 2, 3, and 7. For example, to provide serial communication between two computers we could connect together pin 7 on the RS-232 serial ports of the two computers and connect pin 2 of the first computer to pin 3 of the second computer, and vice versa.

Parallel Transfer

In parallel I/O a complete word of data is transferred at once (for most microprocessors this is an 8-bit word). This provides a faster method for transferring large amounts of information than does serial I/O. For memory mapped I/O, the parallel I/O port is addressed just like a memory location. Again, a status

Table 5.1 The RS-232 Serial Interface Standard

Pin Number	Function
2	Transmitted data, output
3	Received data, input
4	Request to Send (RTS), output
5	Clear to Send (CTS), input
6	Data Set Ready (DSR), input
7	Signal Ground
8	Data Carrier Detect (DCD), input
20	Data Terminal Ready (DTR), output
22	Ring Indicator (RI), input

bit may be assigned to indicate when the data is valid. This status bit may be in a separate memory location. Two examples are presented here.

Example 5.3 Parallel Data Transfer

We will assume that the parallel I/O port address is PARIO, that bit seven of the byte at address STATUS is 1 when the data is valid, and that we want to receive parallel data from the I/O port and store it in memory beginning at location DBLK. The number of words of data to be received is known and is stored in location COUNT. The following Z80 subroutine will perform this task (see Fig. 5.8):

```
PARIN:    PUSH AF                ;save registers
          PUSH BC
          PUSH IX
          LD IX,DBLK             ;load IX with data address
          LD B, (COUNT)          ;load B with number of data
LOOP:     LD A, (STATUS)         ;contents of STATUS into A
          BIT 7, A               ;check bit 7
          JR Z, LOOP             ;if zero go back to loop
          LD A, (PARIO)          ;if one load A with data
```

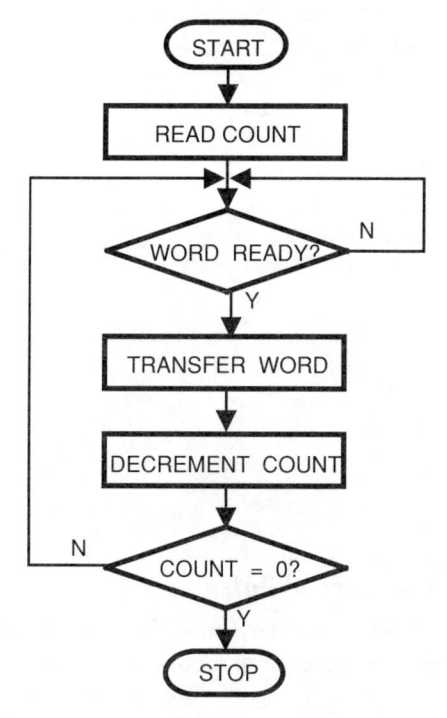

Figure 5.8 Parallel transfer flowchart

```
                LD (IX), A          ;transfer data to memory
                INC IX              ;increment data pointer
                DJNZ LOOP           ;decrement (COUNT)
                                    ;if not zero go back to LOOP,
                POP IX              ; otherwise restore registers
                POP BC
                POP AF
                RET
```

The 8088 assembly language version is,

```
PARIN           PROC FAR
                PUSHF
                PUSH AX
                PUSH BX
                PUSH CX
                PUSH DX
                LEA BX, DBLK        ; this is similar to:
                                    ; MOV BX, OFFSET DBLK
                MOV CX, COUNT
                MOV DX, STATUS      ; STATUS is outside address
LOOP1:          IN AL, DX
                AND AL, 80H
                JZ LOOP1
                MOV DX, PARIO
                IN AL, DX
                MOV [BX], AL
                INC BX
                LOOP LOOP1
                POP DX
                POP CX
                POP BX
                POP AX
                RET
PARIN           ENDP
```

<<>>

Example 5.4 Detect a Pulse

Figure 5.9a is a Z80 assembler listing for a subroutine called TRAIL that is used to detect the trailing edge of a pulse on the parallel I/O port. The routine

```
;----------------------------------------------------------------
;               SUBTTL  DOCUMENTATION
;----------------------------------------------------------------
;       This routine is used to find the trailing edge of a
;       pulse sent to the 1824 digital I/O interface. This
;       routine has no arguments, but could be called from
;       FORTRAN as
;
;               CALL TRAIL
;
;       and would return after finding a trailing edge.
;       Application specific-details are the symbols MASK
;       and IOPORT that indicate the single bit mask
;       that the pulse comes in on and the 1824 slot
;       position on the bus.
;
;       EXTERNAL REFERENCES:  None
;----------------------------------------------------------------
;       W.R. DeVries                    10/80
;----------------------------------------------------------------
;               SUBTTL  CODE
;----------------------------------------------------------------
                CSEG
                PUBLIC TRAIL
;
TRAIL:
SAVE:           PUSH AF         ;SAVE THE REGISTERS
                PUSH BC
                PUSH HL
;       LOAD REGISTERS FOR ADDRESSING THE I/O PORT AND MASKING
                LD B, MASK      ;LOAD REGISTER B W/ MASK
                LD HL, IOPORT           ;LOAD HL PAIR W/ IOPORT
                                        ;ADDRESS
;       WAIT FOR A LOW TO HIGH TRANSITION AND THEN THE
;       HIGH TO LOW TRANSITION ON TRAILING EDGE
HI:             LD A, (HL)      ;INPUT BYTE FROM THE IOPORT TO THE
                                        ;ACCUMULATOR
                AND B           ;LOGICAL .AND. WITH B
                JR Z, HI        ;AND LOOP IF ZERO
LO:             LD A, (HL)      ;AGAIN LOAD A BYTE
                AND B           ;MASK RESULT
                JR NZ, LO       ;AND LOOP IF NOT ZERO
```

Figure 5.9a Listing of the Z80 Routine TRAIL

```
;           RESTORE THE REGISTERS BEFORE RETURNING
RESTOR:         POP  HL
                POP  BC
                POP  AF
                RET
;---------------------------------------------------------------
                SUBTTL USER DEFINED SYMBOLS
;---------------------------------------------------------------
;     DEFINE SYMBOLIC LABELS FOR CONSTANTS
IOPORT          EQU  0FF22H     ;1824 DIGITAL IO PORT
                                ;SETUP TO ADDRESS
                                ;SLOT #3, INPUT PORT #2
MASK            EQU  01H        ;MASK FOR SINGLE BIT SET
                                ;SETUP TO CHECK BIT #0
        END
```

Figure 5.9a *Listing of the Z80 Routine TRAIL (cont.)*

TRAIL is FORTRAN callable and will be useful in the case studies. The 8088 version of TRAIL is given in Figure 5.9b.

<<>>

The most widely adopted standard for parallel I/O is the *IEEE-488* [also known as the general purpose interface bus (GPIB)]. The IEEE-488 parallel interface standard allows for the transfer of one byte of data at a time from an information "source" to one or more information "sinks." This is accomplished using 16 signal lines, as described in Table 5.2. Many laboratory instru-

```
TITLE   TRAIL - PULSE DETECTION ROUTINE
PAGE    ,132             ;Set page width to 132 characters.
;   TRAIL is a routine that returns on the trailing (falling)
;   edge of one complete pulse. Since our interface boxes are
;   built with reverse logic, a high input is a 0 and a low
;   is a 1. We use pin 7 for the signal. Therefore, pin 7 low
;   is 1XXXXXXX (80H), and pin 7 high is 0XXXXXXX (0H) where
;   X= don't care.
;   FORTRAN OR ASSEMBLER ``CALL TRAIL'' calls this procedure.
DATA SEGMENT PUBLIC `DATA'
;   For local assembler program data storage
;   Not used here, but required for linking
```

Figure 5.9b *Listing of the 8088 routine TRAIL*

```
        DATA ENDS
        DGROUP  GROUP  DATA                  ;The DATA segment
                                             ;will be linked into the
                                             ;group called DGROUP, to match
                                             ;Microsoft FORTRAN convention.
        CODE SEGMENT PUBLIC `CODE'
           ASSUME CS:CODE,DS:DGROUP,SS:DGROUP
              PUBLIC TRAIL                   ;Make TRAIL label
        IOPORT          EQU     828          ;available to other segments.
                                             ;Address of Digital
        LO              EQU     80H          ;I/O port in IBM memory
                                             ;Reference pin 7 lo signal,
        HI              EQU     0H           ;reverse logic
                                             ;Reference pin 7 hi signal,
        TRAIL PROC FAR                       ;reverse logic
              PUSH      BP                   ;Save calling frame pointer
              MOV       BP,SP                ;on the stack.
              MOV       AH,HI                ;Put reference HI signal in AH
              MOV       DX,IOPORT            ;Use DX for indirect addressing
                                             ;required by the IN/OUT commands.
        TOP:  IN        AL,DX                ;Get data from port A, input port
              AND       AL,080H              ;Mask off bits 0-6
              CMP       AL,AH                ;Bit 7 hi yet?
              JNE       TOP                  ;If not, loop until it is.
              MOV       AH,LO                ;Put reference LO signal in AH.
        BOTTOM:
              IN        AL,DX                ;Get data from port A
              AND       AL,080H              ;Mask off bits 0-6
              CMP       AL,AH                ;Bit 7 lo yet?
              JNE       BOTTOM               ;If not, loop until it is
              MOV       SP,BP                ;Restore frame pointer
              POP       BP
              RET
        TRAIL ENDP
        CODE ENDS
        END
```

Figure 5.9b Listing of the 8088 routine TRAIL (cont.)

Table 5.2 The IEEE-488 Parallel Interface Standard

Signal Lines	Function
DIO1 to DIO8	Data input/output lines (DIO8 is the MSB)
EOI	End of information
IFC	Interface clear
SRC	Service request
REN	Remote enable
ATN	Attention
DAV	Data valid
NRFD	Not ready for data
NDAC	Not data accepted

ments (e.g., digital oscilloscopes, spectral analyzers, recorders, voltmeters) are designed to communicate with laboratory computer systems using this standard interface.

5.3 CONVERSIONS BETWEEN DIGITAL AND ANALOG DOMAINS

The computer is a digital machine; most of the signals of engineering interest in the external world are analog in nature. Thus, for tasks like data acquisition and control, we must be able to perform conversions between digital and analog signals. In this chapter we consider the devices required for such conversions and their operation and programming.

Figure 5.10 shows how a *digital-to-analog converter* (DAC) and an *analog-to-digital converter* (ADC) can be used to connect the primarily analog world to the digital world of the computer. There are two important considerations in conversions between analog and digital signals:

```
                                          ┌──▶ DAC ──┐
                                          │          ▼
OPERATOR ◀─▶ TERMINAL ◀─▶ COMPUTER                  PROCESS
                                          │          ▲
                                          └──  ADC ◀─┘
```

Figure 5.10 The role of the digital-to-analog converter (DAC) and the analog-to-digital converter (ADC)

5.3 Conversions between Digital and Analog Domains

1. Accuracy: Digital data in the computer is *quantized*, whereas analog signals represent a continuum of values. This quantization process determines the accuracy of the conversion.
2. Conversion rate: To convert an analog signal to digital data, we must *sample* the data at discrete instants in time. We must sample frequently enough to capture the information of interest in the analog signal.

First consider quantization. For example, if we have an 8-bit word we can represent $2^8 = 256$ integer values. If the analog signal is between 0 and 1 V, then the smallest quantity that can be accurately represented is (1/256) V. Thus, the quantization results in an error of 1 part in 256 or ±0.4%. Typically 7 to 12 bits are used for such conversions, and voltage values are in ranges such as ±1 V, ±5 V, ±10 V, 0 to 1 V, 0 to 5 V, 0 to 10 V.

The *sampling theorem* [AsWi84] says that we must sample an analog signal at least twice per period of the highest frequency component of interest, otherwise we will lose information. Typically, sampling rates are chosen to be 5 to 10 times this minimum value from the sampling theorem. A problem with sampling is that high-frequency noise when sampled will lead to spurious low-frequency components. This phenomenon, termed *aliasing*, is illustrated in Figure 5.11. Therefore, high-frequency noise should be filtered before a signal is sampled.

If we denote the sampling period (or interval) by Δt seconds, then $f_s = (1/\Delta t)$ is the sampling frequency in Hz. All signals with frequencies up to and including f Hz can be accurately represented by the sampled data for values of $f \leq (f_s/2)$. However, when $f \geq (f_s/2)$, aliasing can occur and the sampled data contains spurious components at the aliasing frequencies,

$$f_a = (f - kf_s); k = 0, \pm1, \pm2, \pm3, \ldots$$

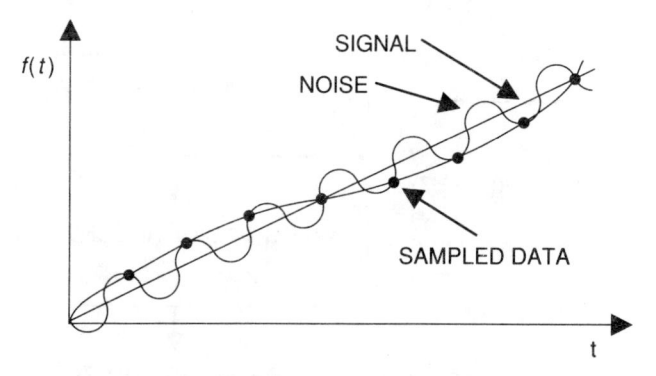

Figure 5.11 Effect of noise on sampling (aliasing)

When $f_a \leq f$, there is no known way to remove these spurious components based on a single signal or without a known underlying structure. As a result, setting up the sensors and signal conditioning is very important to collecting reliable data. One way of handling the aliasing problem is with an *anti-aliasing filter* placed between the ADC and the sensor. This is an analog low-pass filter with a breakpoint at a frequency $f_f \leq (f_s/2)$. The important frequency $(f_s/2)$ is often referred to as the *Nyquist frequency*.

Example 5.5 Force Measurement

Consider the measurement problem depicted schematically in Figure 5.12. A force of 0 to 1000 N, at close frequencies from 0 to 2 Hz, is measured by a force dynamometer with a gain of 0.001 V/N. The dynamometer produces an analog electrical signal in the 0 to 1 V range and can respond fast enough to capture all signal frequencies up to 100 Hz. The dynamometer also introduces a 60-Hz electrical noise into the measurement with amplitudes of up to 0.1 V.

The output of the dynamometer is passed through an analog low-pass (anti-aliasing) filter. The filter break frequency is selected to be 6 Hz, to provide adequate attenuation of the 60-Hz noise signal without significant attenuation of the 0- to 2-Hz signal of interest. A 12-bit ADC with an input voltage range of 0 to 1 V is used with a sampling frequency of $f_s = +20$ Hz. In the digital computer the signal is quantized to one of the 4096 possible levels. Thus, the force resolution of the system is (1/4096)(1/0.001) N. Notice that the effects of

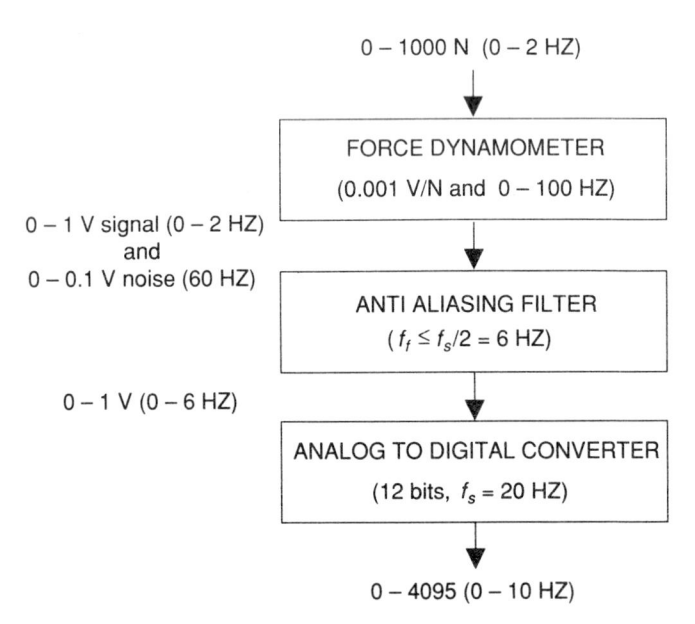

```
                          0 – 1000 N  (0 – 2 HZ)
                                  │
                                  ▼
                    ┌─────────────────────────────┐
                    │    FORCE DYNAMOMETER        │
                    │  (0.001 V/N and  0 – 100 HZ)│
                    └─────────────────────────────┘
 0 – 1 V signal (0 – 2 HZ)
       and                        │
 0 – 0.1 V noise (60 HZ)          ▼
                    ┌─────────────────────────────┐
                    │     ANTI ALIASING FILTER    │
                    │    ( f_f ≤ f_s/2 = 6 HZ)    │
                    └─────────────────────────────┘
  0 – 1 V (0 – 6 HZ)              │
                                  ▼
                    ┌─────────────────────────────┐
                    │  ANALOG TO DIGITAL CONVERTER│
                    │     (12 bits, f_s = 20 HZ)  │
                    └─────────────────────────────┘
                                  │
                                  ▼
                          0 – 4095 (0 – 10 HZ)
```

Figure 5.12 *Force measurement problem in Example 5.5*

all components on both the signal magnitude and frequency must be carefully considered.

<<>>

Digital-to-Analog Conversion

A digital-to-analog converter (DAC), or decoder, is a device that takes an n-bit binary word and converts it to an analog voltage in a specified range. Typically the DAC will output a constant analog value until a new binary word is placed in the DAC register. This is called a zero-order hold; it produces a staircase-type pattern of the output analog voltage versus time.

A DAC is connected to a special DAC register where the binary word to be converted is stored. Each bit of the register causes a switch to open or close, depending on the value of the bit (i.e., 0 or 1). These switches are then connected to a weighted resistor network that produces an output voltage (see Fig. 5.13, where $n = 4$).

$$e_0 = -e_i \sum_{k=1}^{n} (R_f/R_k)$$

There are, of course, numerous variations on the basic DAC circuits described here [Mone87].

Example 5.6 DAC Characteristics

The four bit DAC shown in Figure 5.13 is to be designed such that,

DAC Register		DAC Output
0000_2	=	0.0 vdc
1111_2	=	5.0 vdc

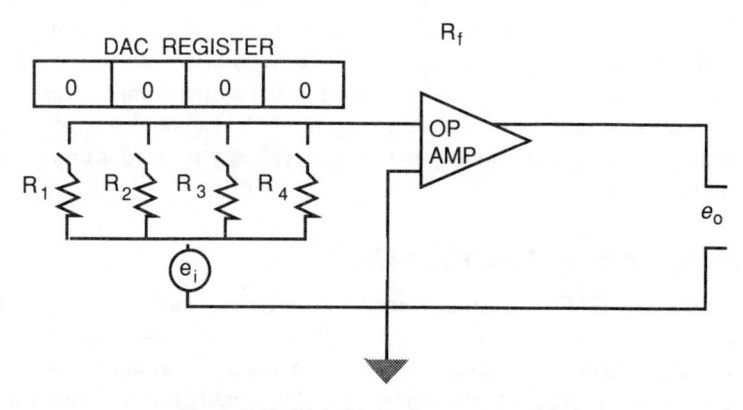

Figure 5.13 Digital-to-analog converter (4 bit)

The output voltage is given by,

$$e_o = -e_i(R_f)[(1/R_1) + (1/R_2) + (1/R_3) + (1/R_4)]$$

When the digital word is 0000_2, $e_o = 0$ since the $R_k = \infty$ for all k. When the digital word is 1111_2, we can select

$$R_1 = 2R_f$$
$$R_2 = 4R_f$$
$$R_3 = 8R_f$$
$$R_4 = 16R_f$$

to obtain

$$e_o = -e_i(15/16)$$

Thus, the input voltage must be

$$e_i = -(5.0)(16/15) = -5.333 \text{ V}$$

What voltage is produced when the DAC register contains 0101_2?

<<>>

Analog-to-Digital Conversion

Analog-to-digital converters (ADCs), or encoders, are more complex (and more expensive) devices than DACs. In fact DACs are typically one of the elements of an ADC [Ciar86; Mone87]. Figure 5.14 shows one way to implement an ADC using a DAC, a switching circuit, a comparator, and an op-amp circuit. In this circuit the input voltage to be converted (e_i) is compared to a voltage from the DAC, and the contents of the DAC register are incremented until the DAC output voltage is greater than or equal to the sampled input voltage e_i. Typically, a multiplexer is used so that the ADC can be shared by several analog input channels. Commonly used boards typically provide 7 to 16 bits and 8 to 16 channels.

Programming ADCs and DACs

ADCs and DACs are programmed by retrieving data from or placing data in special data registers in memory. Additional status and multiplexing registers are also required to indicate when a conversion is complete and to select the appropriate channel for the conversion. The details of programming DACs and ADCs on two typical laboratory systems are described in Appendixes A and B.

Example 5.7 Analog-to-Digital Conversion Routine

The listing in Figure 5.15 shows a FORTRAN callable Z80 assembly program ADC0 that can be used to perform an analog-to-digital conversion on channel zero. Bit 7 of the contents of STATUS is a status bit that is set to 1 when a conversion is complete. The first 4 bits (0 to 3) of the byte at MULTP is used to select one of 16 channels (0 to 15). The two bytes at DATAREG contain the 12-bit data that is converted. The data must be read starting with the most significant byte first from DATAREG, but must be stored in memory with the least significant byte first. The address where the data is to be stored is contained in the HL register pair.

A more general ADC routine for the Z80 is given in the program listings in Appendix B. Similarly, an ADC routine for the 8088 microprocessor can be found in the program listings in Appendix A.

<<>>

5.4 SCHEDULING SERVICE TO I/O DEVICES

Digital I/O devices, DACs, and ADCs have been described. In this section we consider the programming problem of scheduling service for such devices. First, consider the problem of scheduling associated with several external devices connected to the computer. There are two common approaches:

1. *Polling:* Conceptually this is the simplest approach for managing multiple external devices. The processor simply checks each device to see if it needs service; if it does, a service routine for that device is called (see Fig. 5.16).
2. *Interrupts:* As shown in Figure 5.17, this is a hardware solution to the scheduling problem. A special interrupt line is connected to the processor

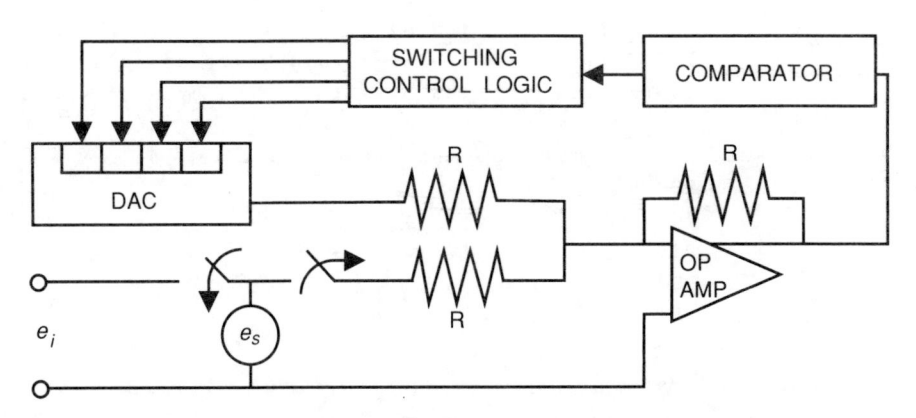

Figure 5.14 Analog-to-digital converter (4 bit)

```
;           ADC0 is a FORTRAN callable Z80 assembly language rou-
;           tine that performs an analog-to-digital conversion on
;           channel zero for the Xycom 3800.  It can be called as,
;                CALL ADC0(ADVAL)
;           where ADVAL is assumed to be passed in the HL
;           register pair.
;                     A.G. Ulsoy   1/82
;           External references:  None.
                      SUBTTL CODE
;

                      CSEG
                      PUBLIC ADC0
;           Define address labels for A/D registers
                      STATUS          EQU     0FF70H
                      MULTP           EQU     STATUS+1
                      DATAREG         EQU     STATUS+2
                      ADONE           EQU     7H     ;A/D complete
                                                     ;test bit
;           Save registers
            ADC0:PUSH IX
                 PUSH AF
                 PUSH BC
                 PUSH HL
;           Conversion is done on channel zero
                 LD IX, STATUS
                 LD (IX+1),0
;           Test bit 7 until it is one;  a 1 implies conversion
;           is completed
            WAIT:BIT ADONE, (IX)
                 JR Z, WAIT
;           Swap the MS and LS bytes of data then store at
;           address in HL register pair.
                 LD A, (IX+2)     ;MS byte of data into A
                 LD B, (IX+3)     ;LS byte of data into B
                 LD (HL), B       ;LS byte into low byte of ADVAL
                 INC HL           ;point to high byte of ADVAL
                 LD (HL),A        ;MS byte into high byte of ADVAL

;           Restore the registers and return
                 POP HL
                 POP BC
                 POP AF
                 POP IX
                 RET
                 END
```

Figure 5.15 Listing of the Z80 routine ADC0 for analog-to-digital conversion on channel zero

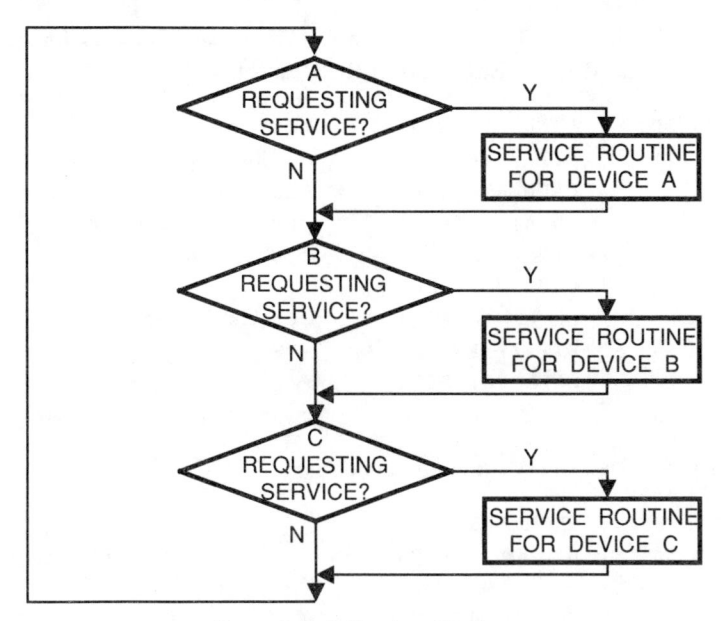

Figure 5.16 Polling loop flowchart

and the external devices, and when a device needs service it sends a signal to the processor. The processor finishes executing the current instruction, then branches to an interrupt service routine.

The advantages of polling are that it is simple and does not require any additional hardware. The disadvantage, of course, is that most of the processor's time is wasted checking to see if devices need service. Also, a request for service might be missed. Example 5.8 gives a simple polling loop program as shown in Figure 5.16.

Example 5.8 Polling Loop

The Z80 Assembly program that follows implements the polling loop in Figure 5.16. The status register location for device A is STATA, for device B it is

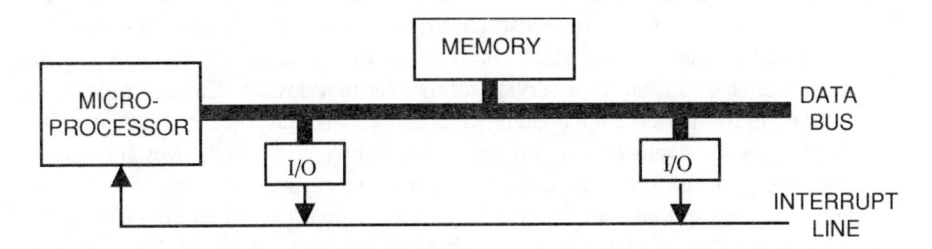

Figure 5.17 Scheduling via hardware interrupts

STATB, and STATC is for device C. The service routines for devices A, B, and C are SERA, SERB, and SERC, respectively.

```
POLER:    LD A, (STATA)        ;get status of device A
          BIT 7, A             ; check bit 7
          CALL NZ, SERA        ;if 1 then call service routine
          LD A, (STATB)        ;check device  B
          BIT 7, A             ; check bit 7
          CALL NZ, SERB
          LD A. (STATC)        ; check device C
          BIT 7, A
          CALL NZ, SERC
          JR POLER             ; no request, try again
```

The 8088 assembly program version is:

```
POLER     PROC FAR
TESTA:    LEA DX, STATA        ; STATA is an outside
                               ; address
          IN AL, DX
          AND AL, 80H
          JZ TEST B
          CALL SERA
TESTB:    MOV AL, STATB        ; STATB is address in RAM
          AND AL, 80H
          JZ TESTC
          CALL SERB
TESTC:    MOV AL, STATC        ; STATC is address in RAM
          AND AL, 80H
          JZ TESTA
          CALL SERC
          JMP TESTA
POLER     ENDP
```

<<>>

Interrupts are a hardware alternative to polling (see Fig. 5.17). When the external device sends an interrupt signal to the processor, the processor must finish executing the current instruction, then save the current contents of the program counter on the stack as well as the contents of the flag register, then branch to an interrupt-service routine for that device. The interrupt-service routine must also save the contents of any registers that it uses. Whereas polling is a scheduled event for the processor, the interrupt (as implied by its name) is like a tap on the shoulder or a ringing telephone. Interrupts are asynchronous events, because they can occur at any time relative to program execution.

The detailed operation of an interrupt depends on the particular processor and the particular hardware used. The Z80 interrupts come in three flavors: (1) the bus request (BUSRQ), (2) the nonmaskable interrupt (NMI), and (3) the ordinary interrupt (INT). Here we will discuss the ordinary, or maskable, interrupt INT. These interrupts are termed maskable because they can be inhibited by the programmer. The instruction EI is used to enable maskable interrupts, and the instruction DI is used to disable them. The maskable interrupt INT can operate in one of three modes:

1. *Interrupt Mode 0:* This mode is selected by the instruction IM0 and is illustrated in Figure 5.18a. If interrupts are enabled (with the EI instruction), then the interrupting device will send an instruction to the processor by placing it on the data bus. The instruction placed on the bus will be one that causes branching to a particular interrupt service routine (e.g., RST, JP, CALL). The processor will then execute the branching instruction. The programmer must supply a RETI instruction to return from the interrupt service routine and must also supply EI and DI instructions as needed. Notice that the branching instructions provide the means for preserving the program counter, but the programmer must explicitly save other registers used by the interrupt service routine.

2. *Interrupt Mode 1:* This mode is selected using the IM1 instruction and is illustrated in Figure 5.18b. This causes an RST 38H instruction to be placed on the data bus and causes branching to the fixed location 38H.

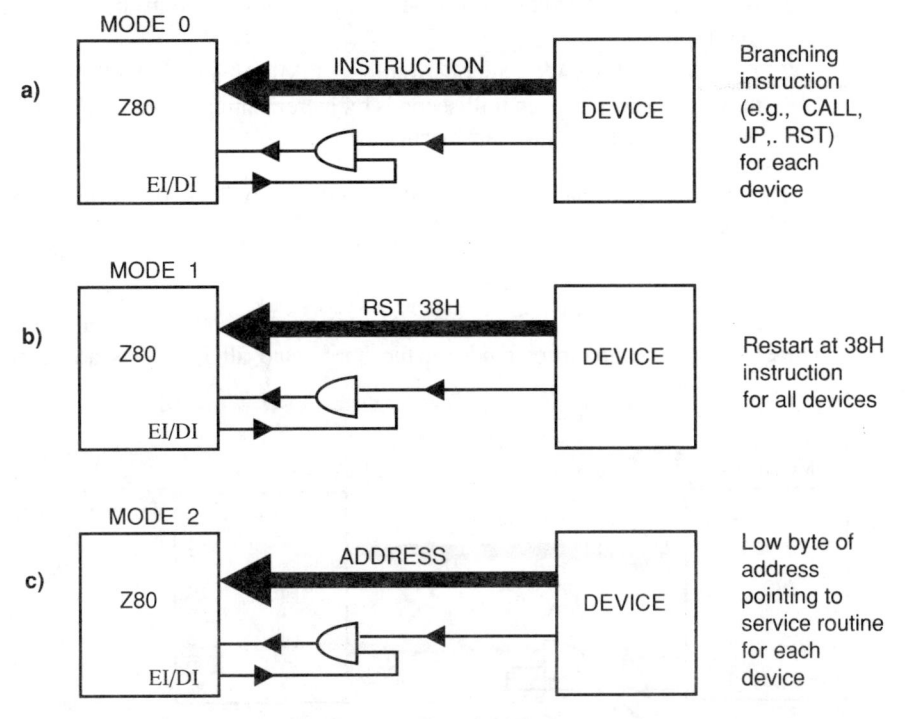

Figure 5.18 *Z80 maskable interrupt modes*

3. *Interrupt Mode 2:* This mode is selected using the IM2 instruction and is sometimes referred to as a vectored interrupt (see Fig. 5.18c). In this mode the interrupting device supplies the least significant byte of an address. The most significant byte of the address is contained in the special I register of the Z80. This resulting 16-bit address is a pointer to the address of the interrupt service routine for that device (i.e., the addressing is indirect). The contents of the PC are automatically pushed onto the stack.

Example 5.9 Z80 Interrupt

Figure 5.19 shows a Z80 processor in interrupt mode zero, together with a programmable interrupt controller (PIC) chip. The PIC can accommodate up to 8 interrupting devices (numbered 0 to 7) and sends an instruction of the form JP ADRn to the processor when device n interrupts. The ADRn is equal to the base address plus $8n$, thus $8 \times 8 = 64$ memory locations are used and 8 bytes are set aside for each device. Since 8 bytes cannot contain a service routine, they contain a branching instruction to the location where the interrupt service routine for that device resides. It is assumed that a 1-kHz clock is connected to the 0 input of the PIC. Figure 5.20a contains the listing of a Z80 assembly language program that performs the following tasks:

1. Initializes the interrupt facility so that the clock and only the clock can interrupt.
2. Provides FORTRAN callable routines TIMEIN and TIMOUT, which allow us to keep time in milliseconds by incrementing a counter (CLKTIC) every time a clock interrupt occurs.

This program and many segments of this program will be useful for several of the case studies.

<<>>

The 8088 in the IBM PC handles interrupts in a slightly different manner. There is only one interrupt mode in the 8088, and all interrupts are treated

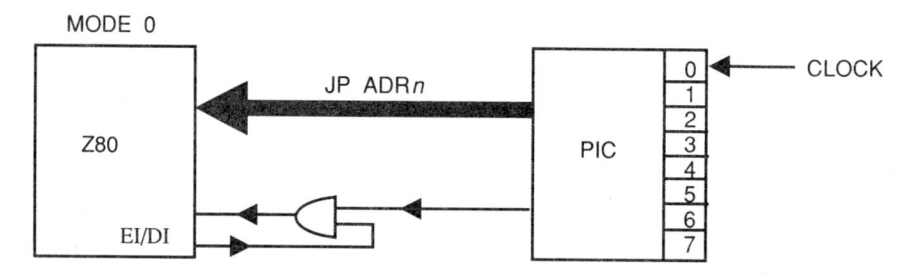

Figure 5.19 *Z80 with a programmable interrupt controller (PIC) as discussed in Example 5.9*

5.4 Scheduling Service to I/O Devices

```
;------------------------------------------------------------
;     INTERRUPT CONTROLLER COMMAND WORDS
;
ICW1   EQU   010H         ;INTERRUPT INITIALIZATION
ICW2   EQU   011H         ;COMMAND WORDS
;
OCW1   EQU   011H         ;INTERRUPT OPERATIONS
OCW2   EQU   010H         ;CONTROL WORDS
;
MASK   EQU   0FEH         ;INTERRUPT MASK
                          ;(EVERYTHING BUT CLOCK)
EOI    EQU   020H         ;END OF INTERRUPT
;------------------------------------------------------------
CLKEOI    MACRO
;     This macro issues the interrupt acknowledge command to the
;     real time clock and is designed to be physically last in an
;     interrupt service routine
          PUSH AF
          LD A, EOI          ;ISSUE AN END OF INTERRUPT
          OUT (OCW2), A      ;TO THE CONTROLLER
          POP AF
          EI                 ;ENABLE INTERRUPTS
          RETI               ;AND RETURN
          ENDM
;------------------------------------------------------------
;     THE REAL TIME CLOCK
          ASEG
          PUBLIC INTVEC
          ORG 04000H         ;MUST START AT A 64 BYTE
                             ;BOUNDARY. HERE IT IS
                             ;LOADED IN HIGH MEMORY
;     INTERRUPT VECTOR
INTVEC:                      ;ONLY FUNCTION IS TO
          JP SERVICE         ;JUMP TO SERVICE ROUTINE
          CSEG
PUBLIC SERVICE, INTINL
INTINL:   PUSH HL
          PUSH AF
;     INTIALIZE THE INTERRUPT FACILITY AND SET THE MASK REGISTER
;     TO ENABLE THE REAL TIME CLOCK (BIT 0 = CLOCK)
          LD HL, INTVEC      ;INTERRUPT VECTOR -> HL
          LD A, L            ;LOWER ROUTINE ADDR -> A
          AND 11100000B      ;MASK OFF BITS 0 - 5
          OR  00010010B      ;MERGE SO VECTOR INTERVAL = 8
                             ;BYTES
```

Figure 5.20a Z80 clock interrupt routines

```
              OUT (ICW1), A      ;INITIALIZE
              LD A, H            ;INTERRUPT
              OUT (ICW2), A      ;ADDRESS
              LD A, MASK         ;INITIALIZE MASK
              OUT (OCW1), A      ;FOR REAL TIME CLOCK
;       SETS INTERRUPT MODE FOR THE CPU
;       THE USER MUST ISSUE AN ENABLE INTERRUPTS
;       COMMAND TO ACTUALLY START THE CLOCK
;       .
;       EI
;       Clock is ticking from this point on
              IM0
;       RETURN FROM SUBROUTINE
              POP AF
              POP HL
              RET
;Service routine that just returns after an interrupt occurred.
;This means that after you enable interrupts and you issue a halt
;command. The processor starts executing again after 1
;millisecond, reloads the clock and executes the next statement
;after halt
SERVICE:
              CLKEOI             ;ACKNOWLEDGE INTERRUPT
                                 ;AND RETURN
              CSEG
              PUBLIC TIMEIN, TIMOUT
;       These routines are used to start and maintain an
;       up-counting clock. TIMEIN sets the clock to zero,
;       CLKINC increments the clock count and is the service
;       routine, and TIMOUT returns the elapsed
;       time as a subroutine argument.
TIMEIN:
              PUSH HL
              PUSH IX
              LD HL,0H
              LD (CLKTIC), HL    ;ZERO THE
                                 ;CLOCK TICK
              LD IX, INTVEC
              LD HL, CLKINC      ;INSTALL THE
              LD (IX+1), L       ;SERVICE ROUTINE
              LD (IX+2), H       ;ADDRESS
              CALL INTINL        ;INITIALIZE THE CONTROLLER
              POP IX
              POP HL
              EI                 ;ENABLE THE INTERRUPT
              RET
```

Figure 5.20a Z80 clock interrupt routines (cont.)

```
CLKTIC:     DS 2
CLKINK:
            PUSH HL
            LD HL, (CLKTIC)
            INC HL
            LD (CLKTIC), HL
            POP HL
            CLKEOI                      ;ACKNOWLEDGE THE INTERRUPT
                                        ;AND RETURN
TIMOUT:
            DI
            PUSH HL
            PUSH DE
            PUSH BC
            EX DE, HL
            LD HL, CLKTIC
            LDI
            LDI
            POP BC
            POP DE
            POP HL
            RET
```

Figure 5.20a Z80 clock interrupt routines (cont.)

the same way. When the 8088 receives an interrupt request, it first checks its interrupt flag (IF). When IF = 1, the 8088 accepts external interrupts. It stops the current program, saves the current code segment and the instruction pointer, saves the content of the flag register, and transfers control to the interrupt service routine. The way it transfers control requires some further explanation.

There is a table in the IBM PC starting from the address $0000_{16}:0000_{16}$. This is a table of the interrupt vectors, which are the addresses of the interrupt service routines. Each of these vectors contains 4 bytes of memory, that is, 2 bytes for the segment address and 2 bytes for the offset address. The table contains a total of 1024 bytes of memory to accommodate 256 different interrupt vectors. Whenever there is an interrupt request, an interrupt type number comes along with it. According to the type number, the 8088 recognizes the request, fetches a specific interrupt vector from the table, and loads the vector into the code segment and the instruction pointer so that it can execute the interrupt service routine.

There are two ways for the user to place interrupt requests. One is in the assembly program by issuing the instruction INT nnn, where nnn is the interrupt type number. The other way is through the interrupt control chip 8259. The 8259 accepts signals from eight different channels, that is, IRQ0 to IRQ7. These channels may connect to the internal clock, the screen, the keyboard, the printer, or other devices. Each channel has its own priority for interrupt requests and

94 Chapter 5 INTERFACING

```
;       TIMEIN   -  CLOCK READING ROUTINE
DATA SEGMENT PUBLIC 'DATA'              ;For local assembler program
DATA ENDS                               ;data storage. Not used but

DGROUP GROUP DATA                       ;required for linking.
CODE SEGMENT PUBLIC 'CODE'
        ASSUME CS:CODE,DS:DGROUP,SS:DGROUP
        PUBLIC TIMEIN
        SUBTTL TIMEIN CLOCK READING ROUTINE
; TIMEIN is a routine that takes a clock reading and stores the
;value in TIME. The clock reading is taken very simply using
;the BIOS routine INT 1AH. If a zero is in AH, INT 1AH will
;read the clock and put the high portion in CX and the low
;portion in DX. Incidentally, the clock is set by putting
;a 1 in AH, the  high portion of the time in CX, the low portion
;in DX, and then issuing an INT 1AH.
TIMEIN PROC FAR
;       FORTRAN "CALL TIMEIN(TIME)" calls this procedure
        PUSH    BP              ;Save calling frame pointer
        MOV     BP,SP
        MOV     AH,0            ;Indicate that we want to READ clock
        INT     1AH             ;Read time from system clock.
;We are only interested in seconds here, so store the low
;portion of the count,DX.
        LES BX,DWORD PTR [BP+6]     ;ES,BX=addr 1st parameter (TIME)
        MOV ES:[BX],DX              ;Put TIME value fróm DX into address
                                    ;in BX
        MOV     SP,BP
        POP     BP              ;Restore frame pointer
        RET     04H             ;Return, pop 4 bytes (address of 1
TIMEIN ENDP                     ;parameter).
CODE ENDS
END
;
;       TIMOUT   -  CLOCK READING ROUTINE
DATA SEGMENT PUBLIC 'DATA'              ;For local assembler program
DATA ENDS                               ;data storage. Not used but
                                        ;required for
DGROUP GROUP DATA                       ;linking.
CODE SEGMENT PUBLIC 'CODE'
        ASSUME CS:CODE,DS:DGROUP,SS:DGROUP
        PUBLIC TIMOUT
        SUBTTL TIMOUT CLOCK READING ROUTINE
;       TIMOUT is a routine that takes a second clock reading,
;       subtracts the first from the second, and returns the
;       difference to the FORTRAN calling program.
```

Figure 5.20b *8088 clock interrupt routines*

```
TIMOUT  PROC  FAR
;  FORTRAN  "CALL  TIMOUT(TIME,CLKTIC)"  calls  this  procedure.  TIME
;  is  the  value  from  FORTRAN  saved  after  TIMEIN  is  called.  CLKTIC
;  is  the  value  returned  to  FORTRAN  from  TIMOUT.
        PUSH    BP                      ;Save calling frame pointer
        MOV     BP,SP
        MOV     AH,0                    ;Indicate that we want to
                                        ;READ clock
        INT     1AH                     ;Read time from system
                                        ;clock into DX
        LES     BX,DWORD PTR [BP+10]    ;ES,BX=addr 1st parameter
                                        ;(TIME)
        MOV     CX.ES:[BX]              ;CX = value of TIME  from
                                        ;TIMEIN.
        SUB     DX,CX                   ;DX  <-- DX - CX ;or
                                        ;Deltatime <-- Newtime -
                                        ;Oldtime
        LES     BX,DWORD PTR [BP+6]     ;ES,BX=addr 2nd param (CLKTIC)
        MOV     ES:[BX],DX              ;Return value of CLKTIC to
                                        ;FORTRAN
        MOV     SP,BP
        POP     BP                      ;Restore frame pointer
        RET     08H                     ;Return, pop 8 bytes (the 2
TIMOUT  ENDP                            ;parameter addresses)
CODE    ENDS
END
```

Figure 5.20b *8088 clock interrupt routines (cont.)*

an assigned interrupt type number starting from 08_{16}. In order for the 8259 to control these different channels, there is an 8-bit register called the Interrupt Mask Register (IMR). Each bit of the IMR controls one channel. If the bit is 1, the associated channel is ignored. When any of the channels gives the 8259 a high signal for the interrupt request, the 8259 based on the IMR and the channel's priority generate one interrupt request, if any, at a time to the 8088; it will not be ready to generate another one unless it receives an End of Interrupt (EOI) from the 8088. Usually the EOI signal is sent at the end of the interrupt service routine.

Example 5.10 8088 Interrupts

The 8088 version of the program in Example 5.9 is given in Figure 5.20*b*. Again, this program (or many segments of it) is used in the case studies. See the program listings included in Appendix A.

<<>>

5.5 SUMMARY

This chapter has treated the various issues associated with interfacing a microcomputer to external physical devices. Some typical sensors and actuators were introduced. These devices may be digital or analog in nature, so digital input and output was considered as well as conversions between analog and digital signals.

Digital input and output devices can be either serial or parallel. Each type of device can be programmed using memory mapped I/O and reading or writing to a memory location, or port, associated with that device. Although the particular details are often device-specific, there are certain standards and protocols that are widely used (e.g., RS-232 for serial ports and IEEE-488 for parallel ports). Some illustrative programs, which will be useful in the case studies, were provided.

This chapter also explained the issues of quantization and sampling, which arise in the conversion of analog signals to digital form. The number of bits used to represent an analog signal in digital form determine the quantization error. The sampling theorem says we must sample more than twice per period of the highest frequency component of interest in the analog signal. To prevent aliasing, it is necessary to employ a low-pass antialias filter before an analog signal is sampled. Conversion devices (DACs and ADCs) were described, and a sample program for performing an ADC was also presented.

Finally, the important notions of polling and interrupts were introduced as two possible approaches to the problem of scheduling service to I/O devices. Interrupts provide a hardware approach and are consequently device-dependent. We discussed the use of clock interrupts and presented some programs that will be of potential use in the case studies.

5.6 PROBLEMS

1. A 4-bit DAC, schematically illustrated in Figure 5.13, has the following resistor values: $R_1 = 2$ MΩ, $R_2 = 4$ MΩ, $R_3 = 8$ MΩ, $R_4 = 16$ MΩ, and $R_f = 1$ MΩ. It is to be designed such that

DAC Register	Output Voltage
0000_2	0.0 vdc
1111_2	5.0 vdc

   a. Determine the required supply voltage e_i.
   b. What voltage will the number 1010_2 produce?

2. Refer back to Problem 3.2 and Figure 3.P2. Assume that the subroutines TASK1, TASK2, and TASK3 are public external routines that are available for your use. Also, assume that the signals TSK1R, TSK2R, and TSK3R are available as bits 1, 2, and 3, respectively, of the byte at address PORT. Implement TASK0 as a Z80 or 8088 assembly language routine. Include comments in your program. Pay some attention to proper assembler syntax and directives, but concentrate on the actual Z80 or 8088 instructions and the program logic.

3. Suppose that an instrument amplifier is to be connected to an ADC so that force measurements can be digitized. The specifications for these two devices are as follows:

ADC	Amplifier
Range: 0 to 10 vdc	Input range: 0 to 500 lbf
Resolution: 12 bits	Output range: 0 to 8 vdc
Conversion rate: 40 kHz	Bandwidth: 10 kHz

   a. If a steady 200 lbf is applied to the system, what would be the digitized decimal value converted.

   b. What is the resolution of the system (i.e., what is the smallest increment of force that can be measured)?

4. An amplifier and ADC are used for force measurement as sketched in Figure 5.P4. The 12-bit ADC has an input range of 0–10 vdc and a conversion rate of 40 kHz. The amplifier has a 0–500 lbf input range, 0–8 vdc output range, and a 100-kHz bandwidth.

   a. What is the required FORTRAN statement(s) to convert the 16-bit INTEGER variable ADCVAL to the corresponding REAL variable F representing the force in lbf?

   b. What is the resolution of the system, that is, what is the smallest change in force that can be measured?

   c. What is the highest frequency component of the force signal that can be detected without violating the sampling theorem?

5. A 4-bit DAC, as in Figure 5.13, has the following resistor values: $R_1 = 2$ MΩ, $R_2 = 4$ MΩ, $R_3 = 8$ MΩ, $R_4 = 16$ MΩ, and $R_f = 1$ MΩ. It is to be designed such that:

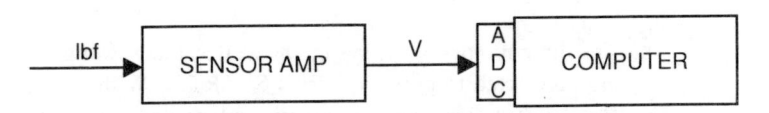

Figure 5.P4 Force measurement system in Problem 5.4

DAC Register	DAC Output
0000_2	0.0 vdc
1011_2	5.5 vdc

What should the supply voltage be to meet this specification?

6. Consider a digital control system implemented on a microcomputer where clock interrupts are used to sample a variable $x(t)$ at discrete times $t = k\Delta t$. For our purposes here assume that the converted value XVAL is available at some specified memory location XLOC as a single byte quantity. You are asked to write an assembly language (Z80 or 8088) subroutine called DELAY3, which when called will introduce a time delay $d = 3\Delta t$ (i.e., a three-step delay). Thus, your subroutine should take the current value XVAL at time $t = k\Delta t$, say $x(k)$, and replace it by the value XVAL from time $t = (k - 3)\Delta t$, that is, $x(k - 3)$. Your subroutine should, of course, make sure that values of XVAL that will be needed later on are not lost. This can be achieved by using a first-in first-out (FIFO) type buffer of the appropriate size for the desired delay. Show clearly the logic of your program (e.g., use a flowchart), use proper assembly language instructions and proper assembler syntax and directives.

7. The automated welding machine shown in Figure 5.P7 has position and velocity control in both the x and y directions. The output velocity for each axis is obtained by counting the pulses generated by an encoder attached to the corresponding lead screw. Timing is achieved with an external clock connected to bit 0 of the digital input port (address is SPORT1) of the control computer. Encoder 1 is connected to bit 1 and Encoder 2 to bit 2. The simplest way to obtain the device outputs (inputs to the computer) is to read the respective bits one at a time. The system is fast enough not to miss any pulses.

   a. Without using the BIT instruction, write a Z80 assembly language routine that reads a pulse from Encoder 1 and starts as follows:

   ```
                   CSEG
                   PUBLIC          ENCOI
   SPORT1          EQU             FF7OH
   ENCOI:
   ```

   b. The three motors of the system as well as the welding current switch are connected to port 2 (address SPORT2). Without using the Z80 Reset instruction, write an assembly language routine that will turn the welding current and motor 3 off simultaneously:

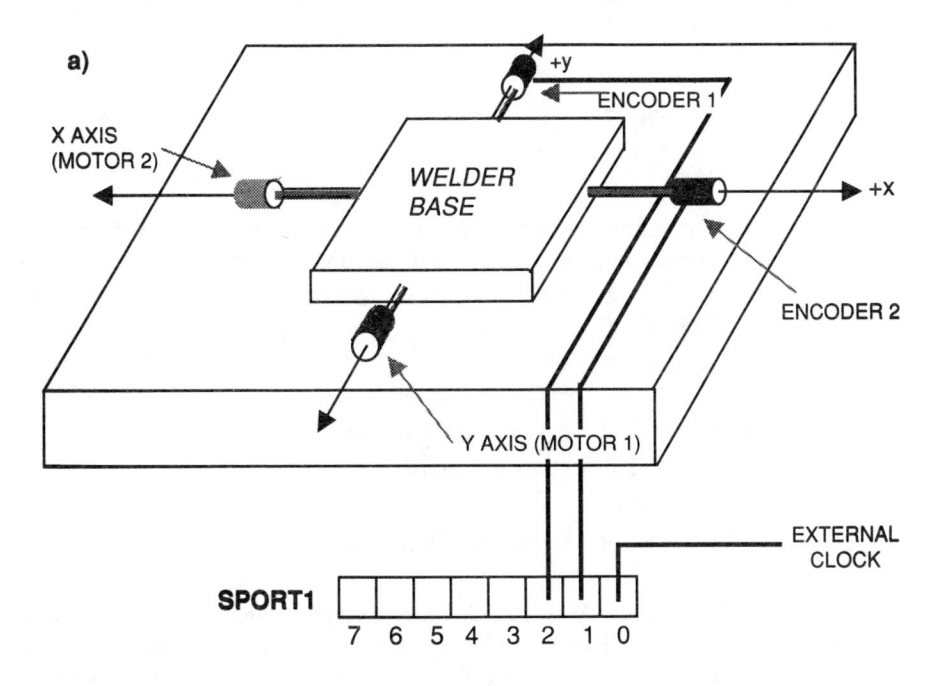

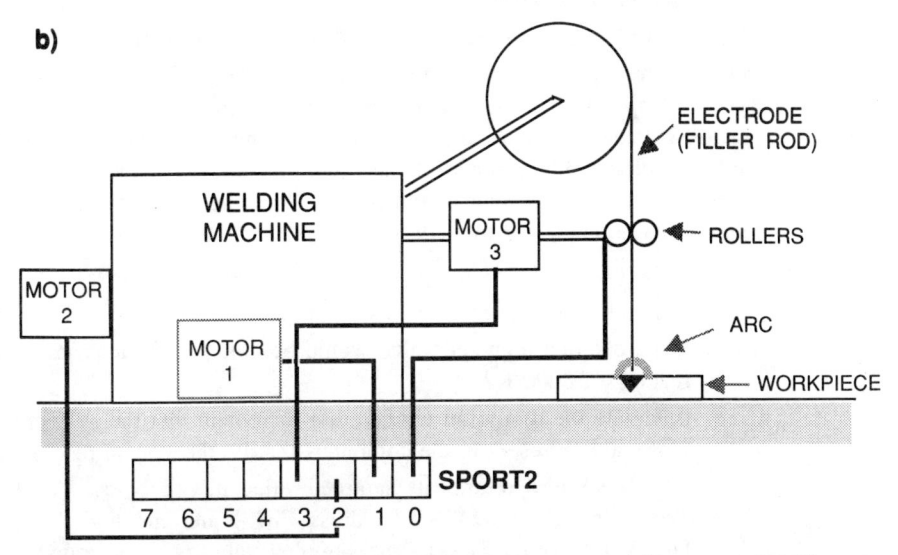

Figure 5.P7 An automated welding machine: (a) clock and encoder, (b) motors and welding current

```
                CSEG
                PUBLIC           MOTOR
        SPORT2  EQU              FF71H
        MOTOR:
```

c. For safety reasons, it should be possible to switch on the welding current only when the electrode does not make contact with the workpiece and the machine is moving along either or both axes. Define:

I = HI or 1 when the current in on
M1 = HI or 1 when Motor 1 is on
M2 = HI or 1 when Motor 2 is on
S = HI or 1 when the start button is pushed
C = HI or 1 when there is contact

The signals M1, M2, C, and S are inputs, and I is an output signal. Design a simple logic circuit (using AND, OR, and NOT gates only) that will permit the current to be turned on only under the appropriate conditions.

8. Repeat Problem 5.7 (a) and (b) using 8088 Assembly language.
9. Suppose that the arc temperature in a welding process is measured with a special temperature probe that converts the temperature into an equivalent analog voltage. The measurable temperature range is 2000 to 10,000°K, and the output voltage range of the probe is 0 to 5 V. For input to the computer, the thermometer output is sampled using a 10-bit ADC whose input voltage range is 0 to 12 V.

  a. What temperature corresponds to an ADC output of 0110100111_2?
  b. What decimal digitized value would be obtained for a temperature reading of 8000°K?
  c. Calculate the minimum temperature increment that the system can sense and represent in the computer, that is, the system resolution.
  d. The ADC used returns its sampled value in two bytes. The low byte is stored in DATA2 (address FF73) and the high byte in DATA 1 (address FF72). The sampled value is to be transferred from these two locations into memory such that the contents of DATA2 is stored in memory location SAMPLE while DATA1's contents are stored in the next higher memory location (see Fig. 5.P9). The following Z80 program is supposed to do that job. Specify any errors or inefficiencies there might be in the program. Or alternatively, write a correct 8088 program.

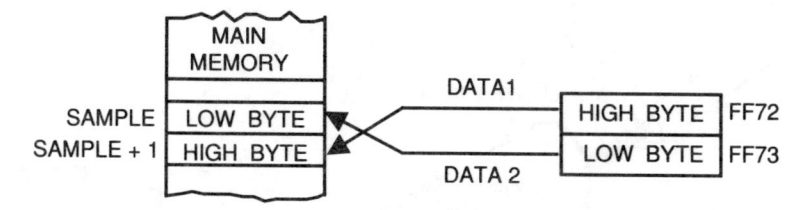

```
                    ┌─────────────┐
                    │    MAIN     │
                    │   MEMORY    │
                    │             │         DATA1
         SAMPLE     │  LOW BYTE   │◄──┐  ┌─────  ┌───────────┐
         SAMPLE + 1 │  HIGH BYTE  │◄─┐└──┼─────► │ HIGH BYTE │ FF72
                    │             │  └───┼─────► │ LOW BYTE  │ FF73
                    └─────────────┘      └─────  └───────────┘
                                           DATA 2
```

Figure 5.P9 Storage of the ADC value in Problem 5.9

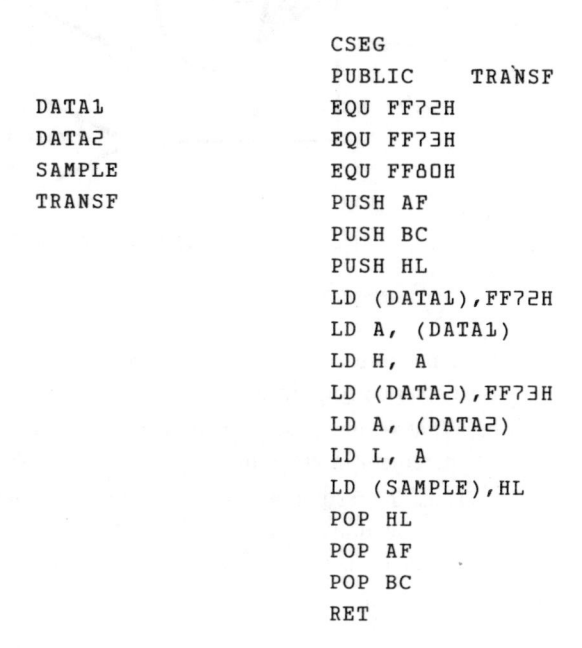

```
                        CSEG
                        PUBLIC      TRANSF
        DATA1           EQU    FF72H
        DATA2           EQU    FF73H
        SAMPLE          EQU    FF80H
        TRANSF          PUSH   AF
                        PUSH   BC
                        PUSH   HL
                        LD     (DATA1),FF72H
                        LD     A, (DATA1)
                        LD     H, A
                        LD     (DATA2),FF73H
                        LD     A, (DATA2)
                        LD     L, A
                        LD     (SAMPLE),HL
                        POP    HL
                        POP    AF
                        POP    BC
                        RET
```

10. Referring to Example 5.6, what is the DAC output voltage when the DAC register contains 0110_2?

11. Select a particular manufacturing automation task, and discuss when the interrupt capability of a microcomputer would be useful for that task.

12. Why are stepping motors not well suited for variable-load applications?

13. Consider a stepping motor with a lead screw driving a single-axis table. The lead screw has $(1/p) = 20$ threads/in., and the stepping motor has SPR = 200 steps/revolution. How many pulses are required to move the table 10 in.? What should the pulse frequency be to produce a constant velocity of 1 in./sec?

14. As shown in Figure 5.P14, we consider the control of the elbow joint (θ) of a robot using a computer. The desired specifications and performance characteristics for the angle sensor, amplifier, and ADC are given in Table 5.P14.

102 Chapter 5 INTERFACING

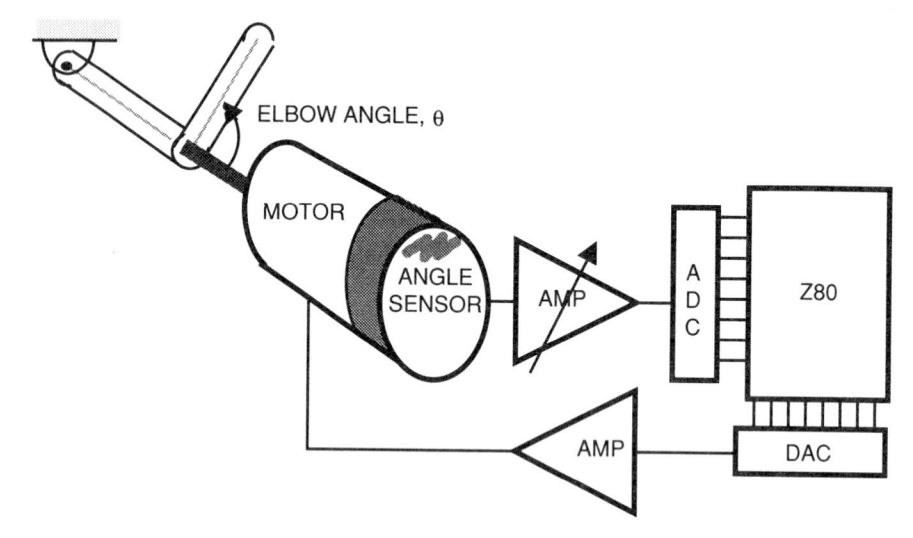

Figure 5.P14 Robot elbow joint for Problem 5.14

a. If the forearm rotates from $\theta = 0°$ to $\theta = 10°$, what would be the converted (digitized) decimal value corresponding to this angle?
b. What is the resolution of the system (i. e., what is the smallest increment in angle that can be measured)?
c. Does the resolution calculated in part b meet the desired performance specifications? If not, explain what hardware change(s) is (are) necessary to meet the 1° system resolution specification.

Table 5.P14 Robot Specifications

Elbow joint	Range = 0–100° Resolution or positioning accuracy = 1° Bandwidth of forearm = 100 Hz
Angle sensor for elbow	Input range = 0–100° Output range = 0–0.1V Bandwidth = 200 Hz
Amplifier with variable gain	Gain is set = 10 V/V Bandwidth = 100 kHz
Analog-to-digital converter	Input = 0–10 V Resolution = 8 bits Bandwidth = 40 kHz

d. A sampling frequency of 100 Hz is used to control the robot elbow joint. In light of the stated desired performance specification on system bandwidth (100 Hz), is this high enough? Why?

15. A robot is used to dip parts into an acid bath to prepare parts for painting (see Fig. 5.P15). Three types of parts, type A, B, and C, are picked up, dipped into the acid bath, and placed onto the output conveyor belt. A type parts are delivered on a constant velocity conveyor belt. The parts are randomly spaced, although a minimum spacing is provided to ensure the robot always has at least enough time to service the next type A item. Part types B and C are delivered by stack loaders that can store parts until the robot "has time" to remove them. All three delivery systems have sensors that indicate when a part is available to be picked up. For type A parts this means that only T seconds remain before the part falls off the conveyor belt, where the time, T, is greater than the time to service one part of type A or B. A computer system is used to control the robot.

    Describe how you would schedule the tasks required of the robot. Include a flowchart(s) of this process. If you wish, you may include real or pseudo program segments. (Hint: What are the important tasks? How important are they with respect to one another?)

16. The cell computer (CC) in Problem 3.2 is instrumented with 8-bit ADCs set up for converting analog voltages in the range of 0 to 5 vdc. Consider a tachometer attached to the shaft of one of the pulleys on the input conveyor (IC). The pulley radius is 0.25 m, the tachometer gain is 1 V/100 rpm, and the belt speed is 100 m/min.

    a. Give the converted value from the ADC as an 8-bit binary number. Also, give the decimal and hexadecimal equivalents.
    b. What is the smallest increment of belt speed that can be detected by this system?

17. Many times a microcomputer system has a time-of-day-clock subprogram that is really an interrupt service routine driven by a clock. To implement such a system, make the following assumptions:

    a. An interrupt will be generated at a rate of 1000 times/sec
    b. A year always contains 365 days
    c. The clock is a 24-hr clock that keeps track of hours, minutes, and seconds. (For your information, 1 sec = 1000 msec, 1 min = 60 sec, 1 hr = 60 min, and 1 day = 24 hr)
    d. A macro, CLKEOI, that does not have any arguments, is available to reenable interrupts and return from a clock interrupt

    Write the Z80 or 8088 assembler code necessary to implement the service routine.

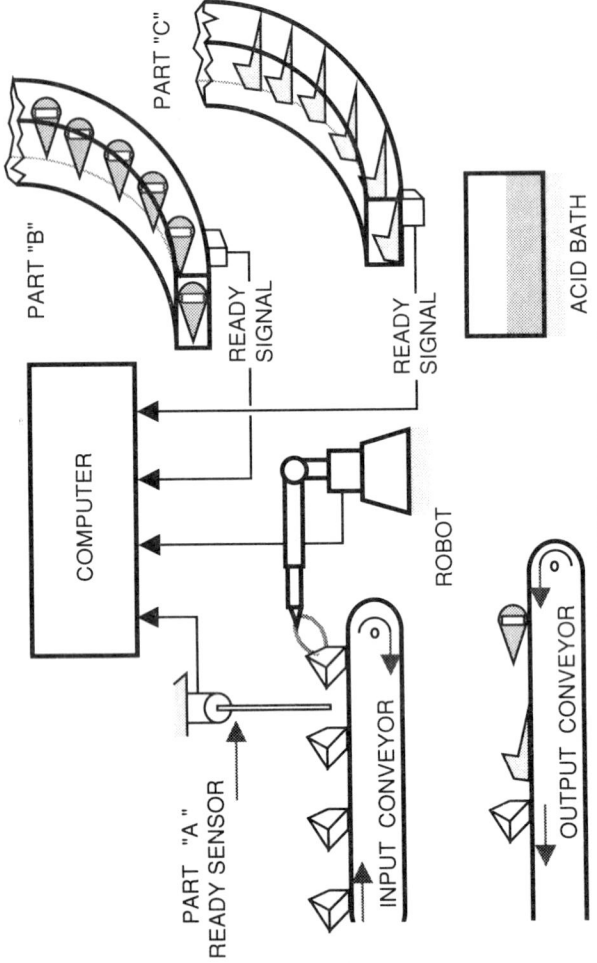

Figure 5.P15 *Part cleaning system*

18. Recall the assembly language routine TRAIL that detects the trailing edge of a pulse. Assume that the parallel I/O port address is IOPORT and write two assembly language subroutines SENDHI and SENDLO that can send, respectively, a HI(="1") and LO(="0") signal to that IOPORT. Use a constant MASK that can select the bits of IOPORT to be effected by these subroutines.

19. Write a program in Z80 and/or 8088 Assembly language called LEAD that detects the leading edge of a pulse sent to the computer digital I/O port and returns.

Part III
METHODS FOR DATA ANALYSIS AND CONTROL

Chapter 6

DATA ACQUISITION AND ANALYSIS

This chapter introduces the basic concepts and methods of data acquisition using a microcomputer system. First, a typical microcomputer configuration for data acquisition is discussed. Next, some statistical techniques are presented as an introduction to signal processing. Finally, the chapter concludes with a discussion of spectral analysis, a very powerful and commonly used signal processing technique.

6.1 BASIC CONCEPTS

A microcomputer system for data acquisition, analysis, and control is schematically illustrated in Figure 6.1. We have already discussed such systems, including microcomputer operation and programming, digital I/O, analog I/O, and scheduling of service to external devices, in previous chapters. Here we provide an overview of how microcomputer systems are configured and how they operate for certain data acquisition, analysis, and control tasks.

First, consider the interactions between the physical system and the microcomputer system as depicted in Figure 6.1. Sensors, which can be either analog or digital in nature, provide information about the physical system. Analog sensors include force dynamometers, tachometers, thermocouples, potentiometers; digital sensors include optical encoders, certain chemical analyzers, and such [deSi85, Potv85]. Some typical analog and digital sensors have been discussed in Chapter 5. Digital and/or analog signal processing is typically required on the sensor output. The signal processing can involve functions such as adjustment of gain and bias (zero offset), filtering, shaping of pulses, and counting.

110 Chapter 6 DATA ACQUISITION AND ANALYSIS

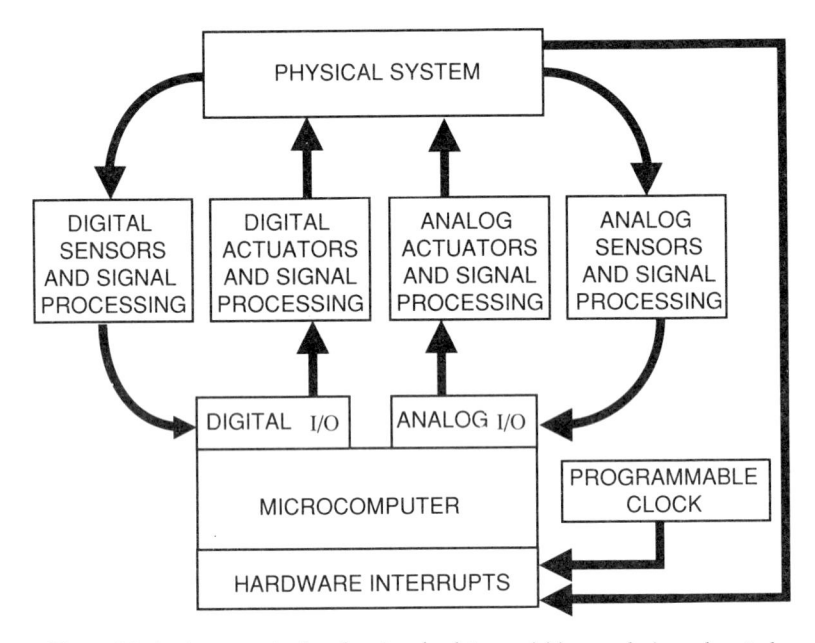

Figure 6.1 *A microcomputer-based system for data acquisition, analysis, and control*

Sensors provide for input to the microcomputer system from the physical system. The microcomputer system can also output signals to affect the physical system for purposes of identification and control. This is done using devices called actuators, which can again be digital (e.g., stepping motors) or analog (e.g., dc motors) in nature. Signal processing is needed here, typically to convert the low-power information-type signal from the microcomputer to a power-type signal that can actually do some work on the physical system. A power amplifier is often used for electronic actuators. Some common analog and digital actuators were discussed in the previous chapter. Additional signal processing may be required to shape, smooth, or limit these actuating signals.

Another point of interaction between the physical system and the microcomputer system may be through the use of external hardware interrupts. A typical example is in rotating machinery where a pulse is generated once per revolution and used to initiate the acquisition of sensor data at each revolution. With this technique, the data collection interval becomes a function of the rotational speed. Another alternative is to collect data at fixed time intervals; this is accomplished using interrupts generated by a programmable clock, as discussed previously in Chapter 5.

The collected data may be stored directly in memory or on a peripheral data storage device (e.g., a disk drive). If the data collection interval is short, there may not be sufficient time to transfer the data to disk. Then, the data collection must be handled in two stages: (1) first collect data at frequent intervals and store it in memory, (2), when the available memory is filled, stop the data acquisition until the data in memory can be transferred to disk. During such data transfers it is extremely important to disable hardware interrupts so that the operation of the disk drive is not disrupted. Devices with direct memory access (DMA) can speed up this type of large data transfer process by bypassing the processor on the microcomputer system.

When the collected data is immediately used in certain computations before the next set of data is collected, the data analysis or control is said to be *on-line* or *recursive*, and the microcomputer system is said to be operating in *real-time*. This is in contrast to *off-line* operation, where the data is collected and stored for subsequent analysis. One of the advantages of microcomputer-based systems is that they can be used for many on-line estimation and control tasks. With current microcomputer technology and without expensive specialized hardware, data acquisition and control tasks requiring sampling intervals as short as 1 msec can be handled.

A microcomputer-based data acquisition and control system is clearly a complex system with many interrelated parts, and it is very important to consider the characteristics of each component of such a system. In addition to certain obvious (and important) considerations such as cost and reliability, one must consider the gain and bias introduced by the component, its bandwidth, and its noise characteristics.

Some of the issues introduced in this section are illustrated in the following example, which is an extension of Example 5.5.

Example 6.1 Force Measurement in Turning

Recall the force measurement system described in Example 5.5 and Figure 5.12, and consider now its use for measuring a component of the cutting force in a turning operation. Although two- or three-component force dynamometers are common, assume that we are measuring a single force component, which we will denote by $F(t)$. There are a number of important issues we must consider:

1. How frequently should we sample the analog force signal?
2. Should we sample at a fixed sampling rate?
3. What type of signal processing is needed before sampling?
4 What signal processing is needed after sampling?
5. What should be the range of the ADC (i.e., 0–5 V, ±5 V)?
6. What do we do with the sampled data?
7. How do we decide when to start and/or stop the data acquisition?

Since the highest frequency signal of interest is 2 Hz, then the sampling theorem requires that,

$$\Delta t \leq (1/2)(1/2) = 0.25 \text{ sec}$$

thus the sampling frequency

$$f_s = (1/\Delta t) \geq 4 \text{ Hz}$$

Sampling at a constant rate f_s seems appropriate in this problem, as long as we sample frequently enough. If our goal had been to sample the $F(t)$ at the same location with regard to the spindle rotation, then we could have selected f_s to be a function of the spindle speed. To prevent aliasing, we need an antialias filter with a filter break frequency of $f_f = 6$ Hz as in Example 5.5. Again, as in that example we select the sampling frequency

$$f_s = 20 \text{ Hz}$$

For a 12-bit ADC with software selectable input ranges of ± 5 V or 0 to 10 V, let us select the latter and introduce an amplifier (before the antialias filter) with no offset (bias) and a gain of 10. Alternatively, we could have selected the ± 5 V range of the ADC and selected an amplifier with a gain of 10 and an offset of -5 V. In either case, the force resolution of the system becomes:

$$\Delta F = (1/4096)(1/0.001) = 0.2244 \text{ N}$$

After sampling the force data we may want to calculate an average force,

$$\bar{F} = \frac{1}{N} \sum_{i=1}^{N} F_i$$

from N sampled force values F_i. To perform this computation of the average force we can collect N data points, then stop data acquisition and calculate $\bar{F}$. Alternatively, we could after each sample form the partial sum

$$S_k = \sum_{i=1}^{k} F_i = S_{k-1} + F_i$$

then calculate

$$\bar{F} = \frac{1}{N} S_N$$

when $i = N$. This latter approach should not be difficult to carry out since $\Delta t = 0.05$ sec. Thus, 50 msec is available for data acquisition and computation of the partial sum. We may want to use clock interrupts at 50-msec intervals, together with an initial user command or external interrupt signal to initiate data acquisition. A counter k could be used to stop data acquisition when $k = N$, and N is specified by the user. Obviously, many of the particular features of our system will depend on the engineering goal of collecting the force data.

<<>>

6.2 SOME STATISTICAL TECHNIQUES

Some signal processing is almost always performed in system such as the one depicted in Figure 6.1[BePi66; Cand88]. Sometimes the signal processing can be very sophisticated, and almost always it incorporates some simple statistical techniques such as averaging. In this section we introduce some of these techniques and the related basic concepts.

Consider two random variables x and y with *probability density functions* $f_x(x)$ and $f_y(y)$, respectively. Thus, the probability that x is between a and b will be given by,

$$P(a \leq x \leq b) = \int_a^b f_x(x)\, dx \qquad (6.2.1)$$

We can now define the *expected values* of the random variables x and y as,

$$E(x) = \int_{-\infty}^{\infty} x f_x(x)\, dx \qquad (6.2.2)$$

$$E(y) = \int_{-\infty}^{\infty} y f_y(y)\, dy \qquad (6.2.3)$$

The following quantities can also be defined,

$\mu_x = E(x)$ is the *mean* of x

$\mu_y = E(y)$ is the mean of y

$\sigma_x^2 = E[(x - \mu_x)^2]$ is the *variance* of x

$\sigma_y^2 = E[(y - \mu_y)^2]$ is the variance of y

$r_{xy} = E[(x - \mu_x)(y - \mu_y)]$ is the *covariance* of (x, y)

$\rho_{xy} = r_{xy}/(\sigma_x \sigma_y)$ is the *correlation* of (x, y)

The square root of the variance is termed the *standard deviation*. Note that if we define a new random variable $z = ax + by$ as a linear combination of x and y, then

$$E(ax + by) = a\mu_x + b\mu_y \tag{6.2.4}$$

and

$$E[((ax + by) - E(ax + by))^2] = a^2\sigma_x^2 + 2abr_{xy} + b^2\sigma_y^2 \tag{6.2.5}$$

When the two random variables x and y are independent, then r_{xy} and ρ_{xy} become zero.

Although the foregoing concepts and definitions are useful, we do not in general know the probability density functions $f_x(x)$ and $f_y(y)$ and cannot compute the various quantities defined. However, results can be obtained by assuming certain probability density functions (e.g., the normal or Gaussian probability distribution) or by conducing experiments to determine these functions empirically. In many engineering applications, we introduce assumptions on the measurement data to make their analysis easier (e.g., independence, normality, stationarity, ergodicity). Discussion of such topics is beyond the scope of our treatment here; refer to [BePi66; PaWu83].

We will consider here, however, the situation where we can obtain N (the sample size) values of x and y denoted by x_i and y_i, where $i = 1, 2, \ldots N$. These may, for example, represent data values obtained by sampling a signal at fixed time intervals. We replace the expectation operation by simple averaging to obtain:

$$\hat{\mu}_x = \bar{x} = \frac{1}{N} \sum_{i=1}^{N} x_i \tag{6.2.6}$$

$$\hat{\sigma}_x^2 = \frac{1}{N} \sum_{i=1}^{N} (x_i - \bar{x})^2 \tag{6.2.7}$$

$$\hat{r}_{xy} = \frac{1}{N} \sum_{i=1}^{N} (x_i - \bar{x})(y_i - \bar{y}) \tag{6.2.8}$$

$$\hat{\rho}_{xy} = \frac{\hat{r}_{xy}}{\hat{\sigma}_x \hat{\sigma}_y} \tag{6.2.9}$$

The estimator $\hat{\mu}_x$ is consistent, so it will tend to converge to the actual mean $\mu_x = E(x)$ as N becomes large, and since $E(\hat{\mu}_x) = \mu_x$, the estimate $\hat{\mu}_x$ is termed *unbiased*. However, the estimates $\hat{\sigma}_x^2$ and $\hat{r}_{xy}$ are known to be biased, and unbiased estimates of σ_x^2 and r_{xy} can be obtained from [PaWu83]:

$$\hat{s}_x^2 = \frac{1}{(N-1)} \sum_{i=1}^{N} (x_i - \bar{x})^2 \tag{6.2.10}$$

$$\hat{\gamma}_{xy} = \frac{1}{N-1} \sum_{i=1}^{N} (x_i - \bar{x})(y_i - \bar{y}) \qquad (6.2.11)$$

Thus,

$$E(\hat{s}_x^2) = \sigma_x^2$$
$$E(\hat{\gamma}_{xy}) = r_{xy}$$

Some additional quantities of interest include the root mean square (rms) value of y:

$$y_{rms} = \sqrt{\frac{1}{N} \sum_{i=1}^{N} y_i^2} \qquad (6.2.12)$$

and the kurtosis:

$$k_y = \left(\frac{1}{N} \sum_{i=1}^{N} (y_i - \bar{y})^4 \right) / (\sigma_y^2)^2 \qquad (6.2.13)$$

Example 6.2 Statistical Properties of Force Measurements in Turning

The normal and radial components of the cutting force as measured in turning under steady cutting conditions are shown in Figures 6.2a and b, respectively. Both components exhibit significant variations about a constant mean value. The force measurements were sampled at 120 Hz and digitized, and the following statistical properties were calculated using $N = 600$ samples:

	Normal Force (N)	Radial Force (N)
$\hat{\mu}$	536.34	231.86
$\hat{\sigma}^2$	36.86	10.93
$\hat{s}^2$	36.92	10.95
$\hat{r}_{xy}$		9.96
$\hat{\rho}_{xy}$		0.50
$\hat{\gamma}_{xy}$		9.98

<<>>

Example 6.3 Tool Wear Estimation from Cutting-Force Data

Colwell Mazur and DeVries [CoMD78] have shown that in turning under constant cutting conditions the cumulative percent deviation (CPD) index calculated

116 Chapter 6 DATA ACQUISITION AND ANALYSIS

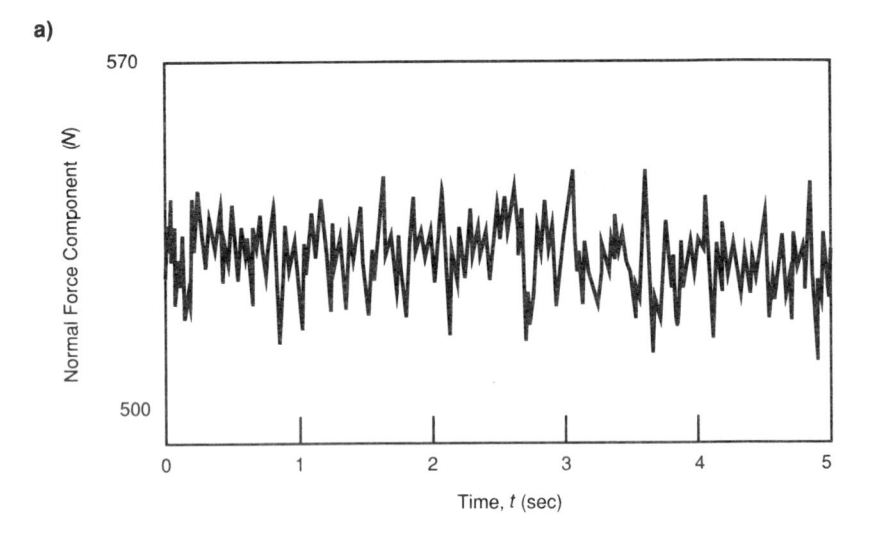

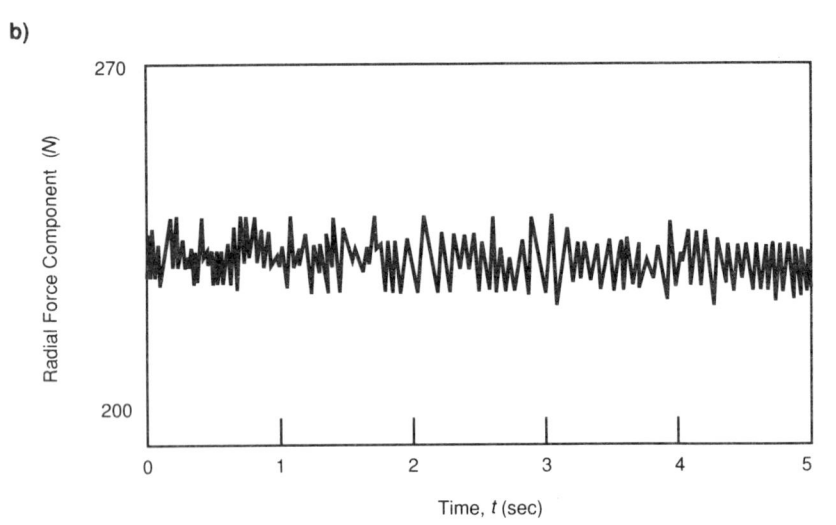

Figure 6.2 (a) Normal and (b) radial force in components in turning

from one of the measured cutting force components is a good indicator of tool wear. The CPD index is calculated from measurements F_k of a force component as follows:

$$S_k = \sum_{i=1}^{k} F_i = S_{k-1} + F_i \qquad \text{(Partial sum)}$$

$$\bar{F}_k = S_k/k \qquad \text{(Average force)}$$

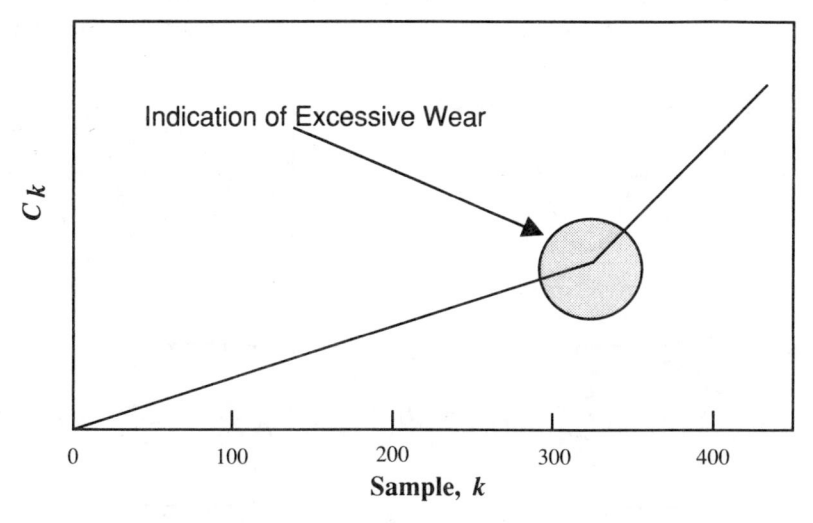

Figure 6.3 The cumulative percent deviation (CPD) as an indicator of tool wear

$$P_i = 100[(\bar{F}_i - F_i)/\bar{F}_i] \quad \text{(Percent deviation)}$$

$$C_k = \sum_{i=1}^{k} P_i \quad \text{(Cumulative percent deviation)}$$

It was determined empirically, from extensive laboratory tests, that a change in the slope of C_k, as shown in Figure 6.3, is a good indicator of excessive tool wear.

<<>>

6.3 SPECTRAL ANALYSIS

In the last 10 to 15 years, a lot of attention has been given to the frequency-domain analysis of time series. Probably this interest is due to the availability of commercial instruments (spectrum analyzers) that can perform a Fourier transform on discrete data in almost real time, and in this way estimate the spectrum of a signal [Berg69; Brig74].

The terms *spectrum, autospectrum, power spectrum,* and *spectral density* are basically synonymous. Each refers to a function that describes how the power or variation in a signal is distributed over a range of frequencies.

Example 6.4 Power Spectrum of Two Signals

Consider the two voltage signals in Figure 6.4. One has a fixed amplitude and frequency (e.g., 60 Hz from a common wall outlet), and the other is a

118 Chapter 6 DATA ACQUISITION AND ANALYSIS

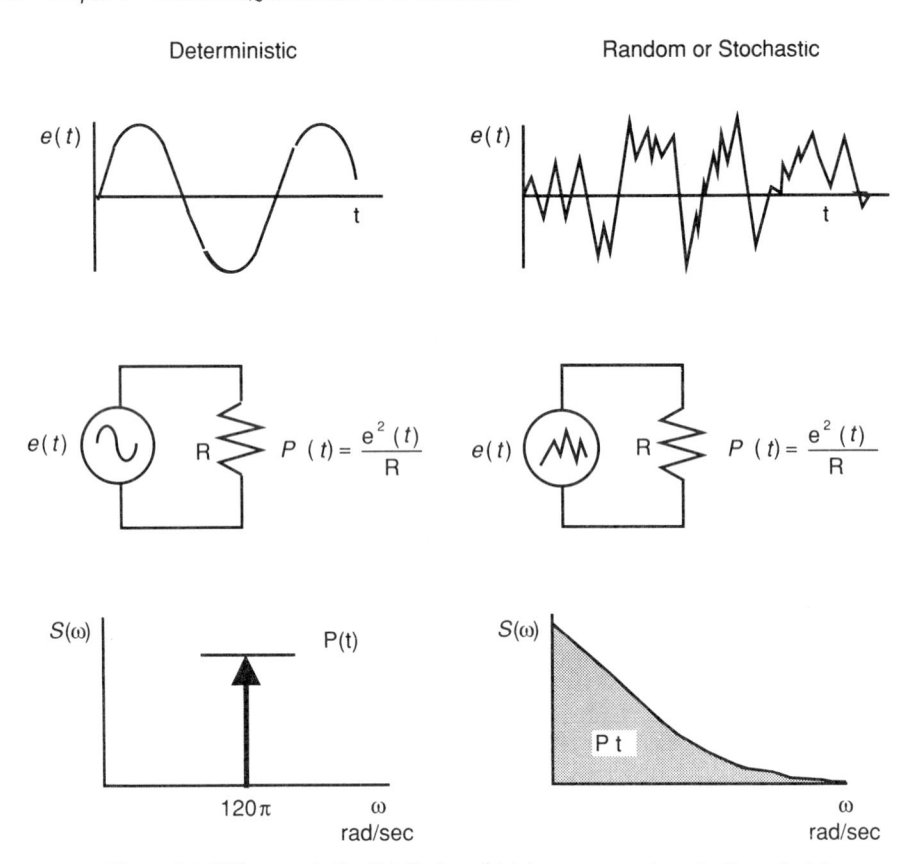

Figure 6.4 *Differences in the distribution of total power or variance in Example 6.4*

random signal without any apparent amplitude and frequency pattern. Although the two signals look different, if they were used as power supplied for a resistive load, experimental measurements would confirm that for either signal the same amount of heat or power would be dissipated in the resistor. The difference would be in the distribution or density of the power at different frequencies. In the first case, the power would be at a single frequency and be represented by an impulse when plotted as a function of the frequency ω. The random signal would have its power distributed over a range of frequencies, ω, and the density would be the spectrum, $S(\omega)$, of this signal. Because the total power provided by both would be the same, it would be true that the areas under the curve $S(\omega)$ for both would be equal.

<<>>

To discuss the power spectrum further, consider the signal $x(t)$ and define its autocovariance as:

6.3 Spectral Analysis

$$\alpha(\tau) = \int_{-\infty}^{\infty} x(t)x(t-\tau)\,dt \quad (6.3.1)$$

Note that $\alpha(0)$ is the variance of $x(t)$. Then the spectrum of $x(t)$ is obtained from the Fourier transform of $\alpha(\tau)$:

$$S(\omega) = \frac{1}{2\pi} \int_{-\infty}^{\infty} \alpha(\tau) \exp(-i\omega\tau)\,d\tau \quad (6.3.2)$$

where $i = \sqrt{-1}$ and ω is the circular frequency in rad/sec. Once $S(\omega)$ is known, $\alpha(\tau)$ can be found from the inverse Fourier Transform of $S(\omega)$,

$$\alpha(\tau) = \int_{-\infty}^{\infty} S(\omega) \exp(i\omega\tau)\,d\omega \quad (6.3.3)$$

Without proof we state some useful properties of $\alpha(\tau)$ and $S(\omega)$ [BePi66]:

1. $\alpha(\tau)$ is an even function [i.e., $\alpha(\tau) = \alpha(-\tau)$]
2. $S(\omega)$ is real
3. $S(\omega)$ is even [i.e., $S(\omega) = S(-\omega)$) and because of this is often plotted as a one-sided function (i.e., $\omega > 0$)
4. $S(\omega)$ is always positive
5. The total area under the curve $S(\omega)$ is $\alpha(0)$

The real difficulty in spectral analysis is estimating $S(\omega)$ from real data, particularly sampled data of a finite length N. This is done by sampling $x(t)$ at a frequency $f_s = (\omega_s/2\pi) = (1/\Delta t)$ to obtain the sampled data x_j for $j = 1, 2, \ldots N$. Then to estimate the sample autocovariance:

$$\hat{\alpha}_k = \frac{1}{(N-k)} \sum_{j=k+1}^{N} x_j x_{j-k} \quad (6.3.4)$$

Then the sample spectrum can be calculated from the Discrete Fourier Transform (DFT):

$$S_x(\omega_k = (k\pi/N\,\Delta t)) = \sum_{j=-(N/2)}^{(N/2)-1} \hat{\alpha}_j \exp\left(\frac{-i\pi j k}{N}\right) \quad (6.3.5)$$

```
              SUBROUTINE FFT                              FFT  1
C M. UHRICH   4/20/68    FFT/V1                           FFT  2
C             4/23/68                                     FFT  3
C FAST FOURIER TRANSFORM SUBROUTINE. REFERENCE:" THE      FFT  4
C FAST FOURIER TRANSFORM" BY M. UHRICH, RAYTHEON          FFT  5
C COMPANY,BEDFORD, MASSACHUSETTS
C             JANUARY 5, 1968.                            FFT  6
C                                                         FFT  7
C             COMMON VARIABLES                            FFT  8
C X(2,2084): DATA; COMPLEX. INPUT IN COLUMN 1, OUTPUT FOR FFT  9
C INVERSE  TRANSFORM IN COLUMN 1. NORMALIZED OUTPUT FOR   FFT 10
C FORWARD TRANSFORM IN COLUMN 2.UNNORMALIZED FORWARD      FFT 11
C TRANSFORM OUTPUT IN COLUMN 1 N STAGE: NUMBER OF         FFT 12
C STAGES AND POWER OF TWO WHICH N IS:
C               N=2**NSTAGE                               FFT 13
C       SIGN: IDENTIFIES DIRECTION OF TRANSFORM           FFT 14
C              SIGN=-1.: FORWARD TRANSFORM.               FFT 15
C              SIGN=+1.: INVERSE TRANSFORM.               FFT 16
C                                                         FFT 17
C                                                         FFT 18
            COMMON/COMFFT/X(2,2084),NSTAGE,SIGN           FFT 19
            COMPLEX X,W                                   FFT 20
            INTEGER R                                     FFT 21
            N=2**NSTAGE                                   FFT 22
            N2=N/2                                        FFT 23
            FLTN=N                                        FFT 24
            PHI2N=6.2831853/FLTN                          FFT 25
            DO 3 J=1,NSTAGE                               FFT 26
            N2J=N/(2**J)                                  FFT 27
             NR=N2J                                       FFT 28
            NI=(2**J)/2                                   FFT 29
            DO 2 I=1, NI                                  FFT 30
            IN2J=(I-1)*N2J                                FFT 31
            FLIN2J=IN2J                                   FFT 32
            TEMP=FLIN2J*PHI2N*SIGN                        FFT 33
            W=CMPLX(COS(TEMP),SIN(TEMP))                  FFT 34
            DO 2 R=1,NR                                   FFT 35
            ISUB=R+IN2J                                   FFT 36
            ISUB1=R+IN2J*2                                FFT 37
            ISUB2=ISUB1+N2J                               FFT 38
```

Figure 6.5 *FFT Algorithm by M. Uhrich in IEEE Transactions on Audio and Electroacoustics, Au-17, No. 2 June 1969*

```
        ISUB3=ISUB+N2                                FFT 39
        X(2,ISUB)=X(1,ISUB1)+W*X(1,ISUB2)            FFT 40
        X(2,ISUB3)=X(1,ISUB1)-W*X(1,ISUB2)           FFT 41
2       CONTINUE                                     FFT 42
        DO 3 R=1,N                                   FFT 43
3       X(I,R)=X(2,R)                                FFT 45
C                                                    FFT 45
        IF (SIGN.GT. 0.) RETURN                      FFT 46
        DO 4 R=1,N                                   FFT 47
4       X(2,R)=X(1*R)/FLTN                           FFT 48
        RETURN                                       FFT 49
        END                                          FFT 50
```

Figure 6.5 *FFT Algorithm by M. Uhrich in IEEE Transactions on Audio and Electroacoustics, Au-17, No. 2 June 1969 (cont.)*

For $k = 0, \pm 1, \pm 2, \ldots, \pm(N/2) - 1, (N/2)$. There are errors introduced by these estimates, and smoothing "windows" or weighting functions are often used.

When digital frequency domain analysis is discussed, the term Fast Fourier Transform (FFT) usually comes up [Berg69; Brig74]. The FFT is an efficient algorithm for computing the DFT. It is a computational "trick" suited for digital computers; results using the FFT are no different than those obtained using the DFT. The FFT requires a sample size N that is a power of 2, for example, $N = 2^p$, where values of $p = 8, 9,$ and 10 are commonly used. Of special importance to engineers and scientists is that this fast algorithm makes "real-time" spectral analysis possible. The speed of the algorithm is related to the details of how a digital computer works. Exponentiation and evaluating special functions such as sine and cosine take much longer to do on a computer than addition. In computing the DFT, multiplication and addition of complex numbers are required, and if this is termed an operation, for N data points (usually N is a power of 2) it can be shown that using the DFT would require about N^2 operations, but using the FFT algorithm this would be reduced to about $N \log_2 N$ operations. As an example, if $N = 1024$, then using the DFT would require 1,048,578 complex multiplications and additions, but with the FFT algorithm this would be reduced by a factor of 100 to only 10,240 operations. Figure 6.5 shows a FORTRAN FFT algorithm.

Example 6.5 Spectrum of the Force Signal in Turning

A component $F(t)$ of the cutting force in a turning operation is measured and its power spectrum computed using an FFT algorithm as implemented on a commercial (digital) spectrum analyzer. The result illustrated in Figure 6.6, shows a large peak at a frequency of approximately 4000 Hz.

This particular turning experiment employed a short workpiece with no tailstock, and the 4000 Hz was identified as the fundamental flexural natural frequency of the workpiece. Thus, a component of $F(t)$ near the 4000-

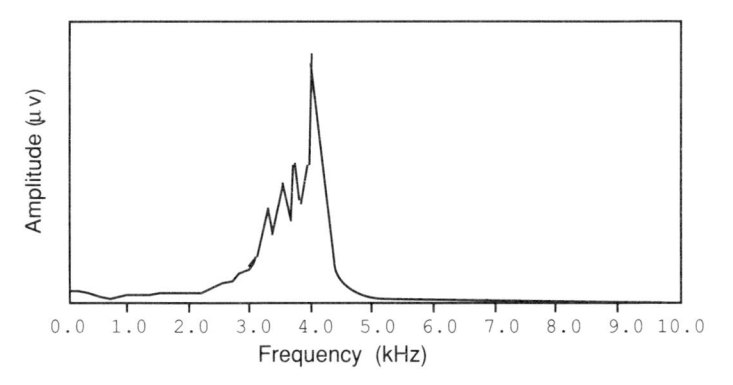

Figure 6.6 *Power spectrum of the cutting force F(t)*

Hz frequency range can be attributed to the workpiece transverse vibration. [DaUl87a,b].

<<>>

6.4 SUMMARY

A microcomputer-based system for data acquisition and control has been described in terms of how it interacts with the environment (a physical system of interest). We have considered issues such as on-line versus off-line computations, real-time operation, analog and/or digital signal processing, and data transfer.

Some basic statistical concepts and techniques have also been introduced (e.g., mean, variance, covariance, correlation). These are often used in signal processing tasks, either directly or as part of a more sophisticated signal processing method.

One of the most widely used signal processing techniques is to determine the power content of a signal as a function of frequency. This is termed spectral analysis, and the power spectrum is obtained from the Fourier Transform of the autocovariance of the signal of interest. The Discrete Fourier Transform (DFT) estimates the power spectrum from a finite set N of sampled values of the signal. The Fast Fourier Transform (FFT) provides an efficient method for computing the DFT.

6.5 PROBLEMS

1. A robot is used for loading and unloading two machine tools as shown in Figure 6.P1. The robot is capable of performing the following tasks when the following subroutines are called:

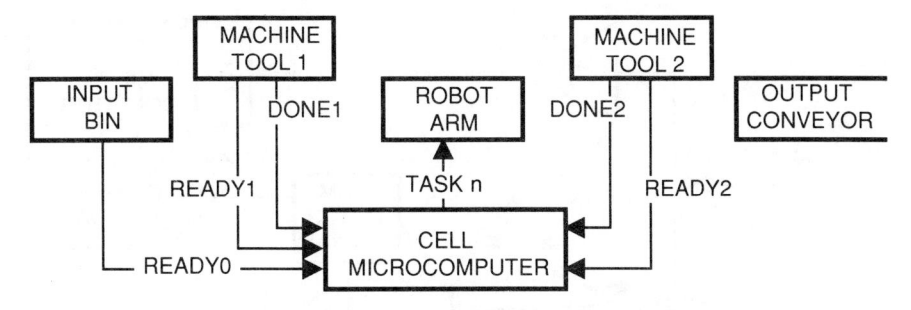

```
          MACHINE                      MACHINE
          TOOL 1                       TOOL 2
INPUT              DONE1    ROBOT              OUTPUT
                            ARM      DONE2
BIN                                            CONVEYOR
             READY1         TASK n    READY2
                            CELL
       READY0            MICROCOMPUTER
```

Figure 6.P1 A robot-handling system for a mechanical cell

Subroutine Name	Task Performed
Task 0:	Wait (do nothing)
Task 1:	Take part from input bin and load machine tool 1
Task 2:	Unload machine tool 1; load machine tool 2 with part removed from machine tool 1
Task 3:	Unload machine tool 2; place the part on the output conveyor

Assuming that the following signals are available,

READY0 = HI when parts are available in input bin
READY1 = HI when part is removed from machine tool 1
READY2 = HI when part is removed from machine tool 2
DONE1 = HI when machining on tool 1 is complete
DONE2 = HI when machining on tool 2 is complete

a. Describe the logic (using flowcharts) of a program to control the sequence of operations in the cell.

b. Describe how you would implement such a microprocessor control for the cell. In particular what hardware would you need, what software would have to be developed, and how would you approach this problem?

2. Consider the inspection of backplane assemblies (essentially a junction box interconnecting all the printed circuit boards on a computer). Each backplane has to be tested for electrical integrity between each pin to make sure each has been wired correctly. It has been proposed to use a robot to automate this inspection process. The set up is shown in Figure 6.P2.

  a. What is the digitized value corresponding to the arm being extended 1 ft?

  b. What is the smallest linear displacement (x direction) the robot can make that can be measured by the computer? (Please state your answer in inches/digitized unit).

124 Chapter 6 DATA ACQUISITION AND ANALYSIS

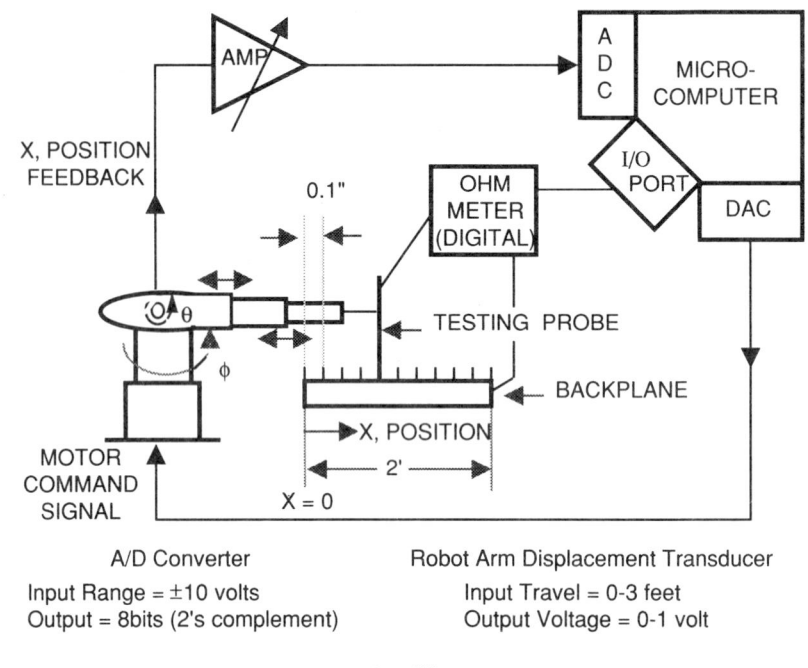

Figure 6.P2 Backplane inspection setup

A/D Converter
Input Range = ±10 volts
Output = 8bits (2's complement)

Robot Arm Displacement Transducer
Input Travel = 0-3 feet
Output Voltage = 0-1 volt

Amplifier
Input Range = ±15 volts
Adjustable Gain (0-100) volts/volt currently set at 2 volts/volt
Adjustable Bias or offset, set at 0volts

    c. Does the system resolution computed in part b allow the robot to test every pin? If not what are the problem(s) preventing the robot from testing every pin? What corrective action would you take?

3. The spindle or chuck speed of a lathe used in a flexibly automated factory is under closed loop control. The workpiece speed is automatically adjusted according to the conditions (cutting force, temperature, tool wear, etc.) at the tool workpiece interface. The conditions are measured by sensors (dynamometers, thermocouples, etc.) and provide feedback through a Z80 processor. An adaptive control algorithm is implemented as an interrupt service routine to adjust the lathe speed constantly so that optimum part quality and production rates are obtained (see Figure 6.P3a).

    The plant production manager asks you to write an inspection-monitoring program. This program will allow, at random, a user to request the next 10 sec of data read in on the ADCs to be saved in a file on the disk. A hard copy of the disk data file can then be obtained by a request from any of the many computers connected to that disk (e.g., the one in the manager's office). The main requirements are that:

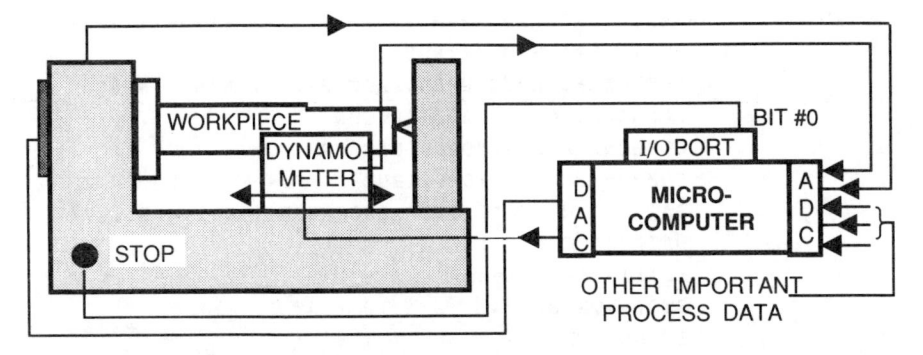

Figure 6.P3a Lathe under adaptive control

a. The normal operation of the lathe not be interfered with
b. A file be written out to disk with the successive contents of the ADCs for a period of 10 sec

The current program that needs to be modified is shown in Figure 6.P3b. The interrupt clock frequency is 10 Hz. The program can be terminated by pushing the stop button on the lathe, which is connected to Bit 0 of the parallel I/O port.

- a. A "user select" switch is installed on Bit 3 of the parallel I/O port. When the switch is pushed, Bit 3 is set. Expand the Interrupt Service Routine so that it will check the I/O port for Bit 3 and put the status of Bit 3 into a variable such as SAVE and provide it to the other parts of the program. Remember to include any additional assembler directives that might be needed and take care of the stack appropriately.
- b. Now expand the main program and/or the control subroutine to interrupt the SAVE variable that the service routine manipulates. On the basis of the status of the variable save, store the next 10 sec (assume a sampling rate 10 Hz) of data by writing it out to a *high-speed* disk (the slow floppy disk drives have been replaced). This process should not interfere with the normal control program. Please state all your assumptions clearly.
- c. Provide a FORTRAN subroutine for calculating the sample mean and standard deviation for any of the data saved on the file.

4. A manufacturer of a computer numerically controlled (CNC) machine wants to design a two-axis table that holds the workpiece. The table position should be repeatable to within ± 0.0003 in. The axes of the table are positioned using two servo motors with closed-loop position control. An optical sensor ("encoder") will be used to indicate the table's position. This device reads and counts notches carefully etched into a glass strip attached to the bottom of the table as shown in Figure 6.P4a. There is one such sensor for each axis.

```
C-----------MAIN PROGRAM--------------------------------
          LOGICAL*1 STOP,ITRUE
          LOGICAL*1 ADCCHN(8), DACCHN(4), NADC, NDAC
          COMMON/ADCBLK/NADC,ADCCHN
          COMMON/DACBLK/NDAC, DACCHN
          COMMON/FLAGS/STOP,ITRUE
          DATA NADC, NDAC/8,2/,ADCCHN/0,1,2,3,4,5,6,7/
          DATA DACCHN/0,1/
C         Initialize flags
          STOP=.TRUE.
          ITRUE=.TRUE.
C         Other relevant code should be inserted here ....
C         Initialize interrupts
          CALL INTINL
C         Enable interrupts
          CALL ENABLE
10        IF(STOP.EQ..TRUE.)GO TO 100
          GO TO 10
100       CALL DISABLE
          STOP
          END
C-----------CONTROL SUBROUTINE --------------------------
          SUBROUTINE CONTRL
          DIMENSION IADVAL(8)
          DIMENSION IDAVAL(2)
          COMMON/ADCBLK/IADVAL
          COMMON/DACBLK/IDAVAL
C         Get data from ADC
          CALL ADC(IADVAL)
C         Adaptive Control Algorithm : calculations go here
C
C         Send out command signal on the DAC
          CALL DAC(IDAVAL)
          RETURN
          END
C--------------------------------------------------------
;         INTERRUPT SERVICE ROUTINE
IOPORT            EQU       0FF2H
                  COMMON/FLAGS/
STOP:             DEFS 1
ITRUE:            DEFS 1
                  CSEG
                  PUBLIC SERVICE
                  EXT CONTRL
```

Figure 6.P3b *Program for Problem 6.3*

```
SERVICE:            DI
                    PUSH AF
                    LD A,(IOPROT)
                    BIT 0, A
                    JP Z, NOSTOP
                    LD A, (ITRUE)
                    LD (STOP), A
NOSTOP:             CALL CONTRL
                    POP AF
                    CLKEOI              ;assume this macro (CLKEOI)
                                        ;is available
```

Figure 6.P3b Program for Problem 6.3 (cont.)

The voltage shown is the output of the encoder's light-receiving device. The voltage pattern looks very similar to the side view of the notches in the glass. The angled notches will enable us to see which direction the table is moving. The output of the encoder's light-receiving device is digitized and processed to produce a digital number based on the number of notches counted. The digital number is incremented at the beginning of each rising level, and at the beginning of each high level (see Figure 6.P4b), for example,

Position	Voltage	Output
0.0	1.0	000
0.0001	2.0	000
0.00022	3.0	001
0.0005	2.0	010
0.00075	3.0	001
0.0009	2.0	100

a. What is the smallest change in distance we can measure?
b. What is the resolution of the optical sensor in inches/digital count?
c. Will we be able to meet desired specification of ± 0.0003 in.
d. Given a desired positioning range $-6 \leq x \leq 6$ in. $0 \leq y \leq 10$ in., what is the minimum number of bits needed to find the x-axis position? (Assume 1 bit for the sign)
e. What would the binary output of the encoder be for $x = 5.1363$ in.?
f. What about for $x = -5.1362$ in.?

Also write an assembly language procedure called ENCODR to process the output of the ADC (see encoder diagram). This procedure will run on a chip that is dedicated to the encoder and will therefore run in a continuous

128 Chapter 6 DATA ACQUISITION AND ANALYSIS

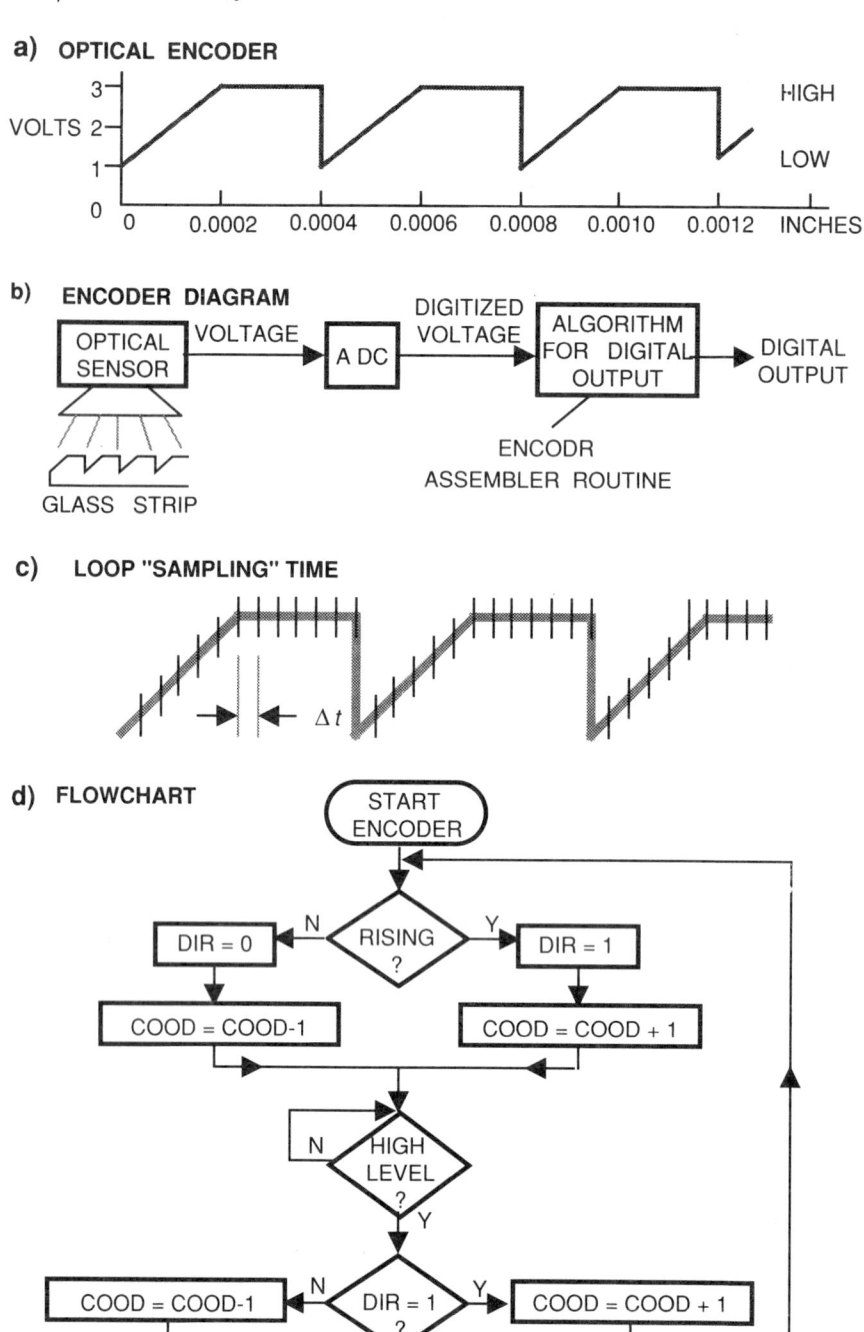

a) OPTICAL ENCODER

b) ENCODER DIAGRAM

c) LOOP "SAMPLING" TIME

d) FLOWCHART

Figure 6.P4 CNC machine tool system in Problem 6.4

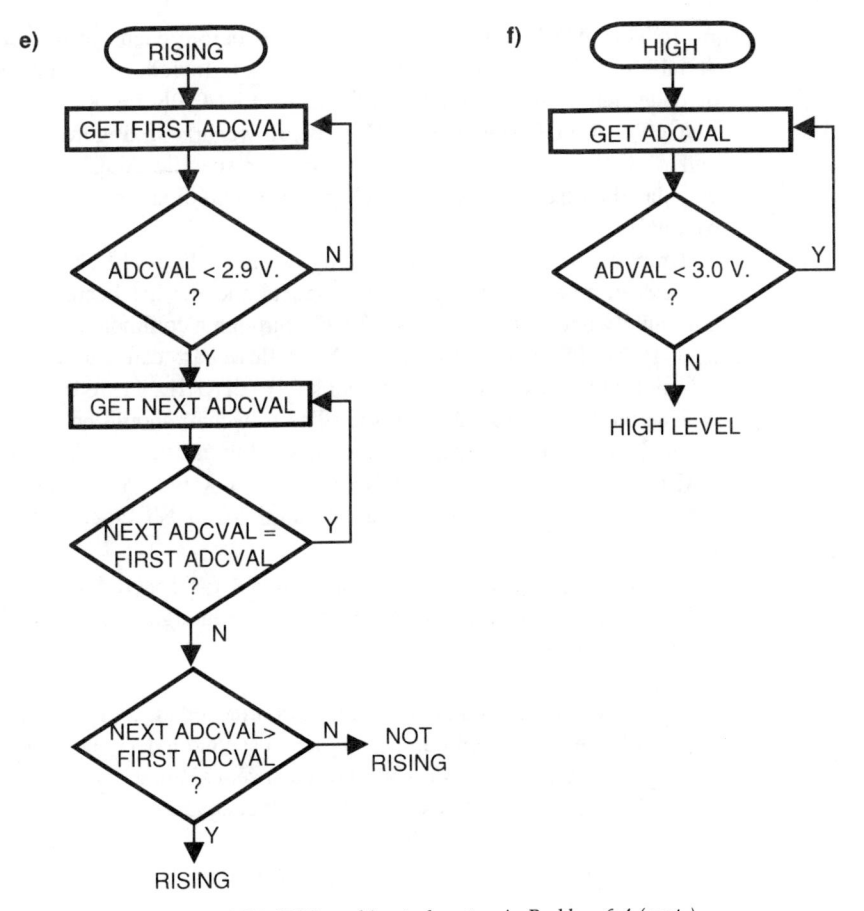

e)
```
        ┌─────────┐
        │ RISING  │
        └────┬────┘
             ▼
   ┌──────────────────┐◄──┐
   │ GET FIRST ADCVAL │   │
   └────────┬─────────┘   │
            ▼             │
       ╱ ADCVAL   ╲   N   │
      ╱  < 2.9 V.  ╲──────┘
      ╲     ?     ╱
       ╲         ╱
            │Y
            ▼
   ┌─────────────────┐◄──┐
   │ GET NEXT ADCVAL │   │
   └────────┬────────┘   │
            ▼            │
       ╱ NEXT ADCVAL= ╲ Y│
      ╱  FIRST ADCVAL  ╲─┘
      ╲      ?         ╱
       ╲              ╱
            │N
            ▼
       ╱ NEXT ADCVAL> ╲  N     NOT
      ╱  FIRST ADCVAL  ╲──►   RISING
      ╲      ?         ╱
       ╲              ╱
            │Y
            ▼
         RISING
```

f)
```
        ┌────────┐
        │  HIGH  │
        └────┬───┘
             ▼
    ┌─────────────┐◄──┐
    │ GET ADCVAL  │   │
    └──────┬──────┘   │
           ▼          │
      ╱ ADVAL     ╲ Y │
     ╱  < 3.0 V.   ╲──┘
     ╲     ?      ╱
      ╲          ╱
           │N
           ▼
      HIGH LEVEL
```

Figure 6.P4 CNC machine tool system in Problem 6.4 (cont.)

loop so we can read the encoder at any time through a common block without having to call the procedure to update the encoder output value for the coordinate axis, COOD. An IBM PC (8088 microprocessor) will control the CNC table, working in conjunction with the dedicated chip and sharing memory and programs.

Assume that the assembly procedure will loop as fast as possible, and this will allow us to read in at least five ADC samples before the table moves 0.0002 in. when the table is moving at its maximum speed. DIR is the direction flag with 0 = negative and 1 = positive. Note also that a rising signal will be increasing from 1.0 V to 2.9 V, and that the encoder signal is always greater than or equal to 1.0 V. Any encoder output value greater than or equal to 3.0 V is a high signal. See Figure 6.P4c and d.

Assume that your procedure can CALL ADC to find the digitized voltage from the optical sensor. The ADC will be ±5 V, 8-bit. Assume that only one axis is being measured and that data is digitized on ADCHN

= 1. This ADC channel must be selected before each conversion. All variables will be passed in common blocks. Include only the portion of code that would normally go from the ENCODR PROC FAR to the ENCODR ENDP statements. Use the flowcharts in Figure 6.P4d, e, and f. Include comments. The values for specific ADCVALs can be calculated by hand; do not use multiplication and division in the assembly routine.

Draw a flowchart for a FORTRAN routine called VELOC to find the velocity of an axis of the table. The optical encoder has been designed to continually update the position, COOD, through a common block. COOD is a 16-bit INTEGER variable. Your flowchart can call any of the assembler routines used in the textbook (e.g., TRAIL, TIMOUT). SPEED will be the table velocity in in./sec. The flowchart should be general; do not include specific numbers. The command to use the subroutine will be CALL VELOC(SPEED,NCOUNT,FREQCK). (Hint: You may want to CALL TIMOUT(CLKTIC) repeatedly until NCOUNT clock counts have passed.)

5. Many computers provide system subroutines for generation of pseudorandom numbers, or they can be generated using algorithms in textbooks such as [Sedg83].

   a. Use a pseudorandom number subroutine or algorithm to generate 100 pseudorandom numbers x_i and a second independent set of 100 pseudorandom numbers y_i. For each set calculate the sample mean and variance and also determine the covariance and correlation between the two sets. Recalculate the same statistics using only the first 50 numbers in each sample and compare these to the values calculated using all 100.

   b. Use the pseudorandom number generator to generate 1024 values x_i, then use the FFT subroutine in Figure 6.5 to determine and plot the spectral density of the pseudorandom variable x_i.

   c. Consider the variable $x(t) = 0.9 \sin 2\pi t + 0.1 \sin 5\pi t$, sampled at $\Delta t = 0.1$ sec. Collect 1024 samples x_i and determine the spectral density using the FFT subroutine in Figure 6.5.

Chapter 7

SYSTEM MODELING AND IDENTIFICATION

This chapter considers the mathematical modeling of dynamic systems because such models are essential for monitoring, signal processing, and control. First, a general discussion introduces the concept of dynamic systems, the types of models that will be employed, and the basic approaches to modeling. Solution of the mathematical model for determining the system stability and response is presented. The solution methods are then used to introduce discrete time systems, difference equations, and Z transforms. The determination of model structure and order for system identification are discussed from both physical and empirical viewpoints, as well as the on-line estimation of model parameters from input and output measurements.

7.1 MODELING DYNAMIC SYSTEMS

A *dynamic system* is one that has memory; that is, its current state $x(t_1)$ depends on its current inputs $u(t_1)$ and its state $x(t_0)$ before those inputs were applied. A *static system*, on the other hand has a state $x(t_1)$ that depends only on current inputs $u(t_1)$. In reality, all physical systems are dynamic; however, if they respond fast enough, we may be able to model them as static systems. This concept is illustrated schematically in Figure 7.1, where for a small value of the time constant τ in Figure 7.1a, the relationship in Figure 7.1b may be a suitable approximation. The relationship between x and u in Figure 7.1a must be described by a differential or difference equation, whereas the one in Figure 7.1b requires an algebraic equation.

Models of dynamic systems, whenever possible, are developed from basic physical laws such as Newton's second law, conservation of energy, and Kirch-

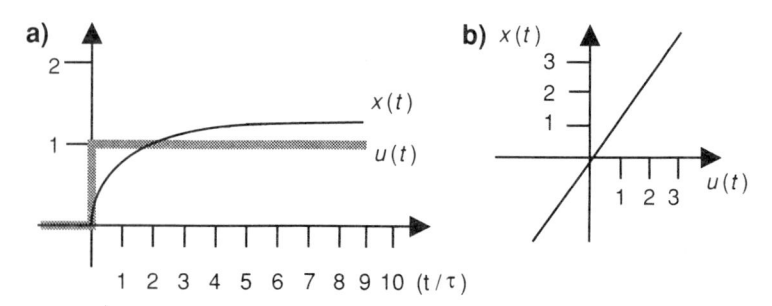

Figure 7.1 (a) Response $x(t)$ of a dynamic system to an input $u(t)$, (b) a static input-output relationship between $u(t)$ and $x(t)$

hoff's laws. Such models are said to be derived from first principles [Cann67]. However, it may not always be possible to derive models based on such fundamental relationships. In such cases, one can employ an empirical approach based on measurement of system inputs and outputs [Ljun87; PaWu83]. Regardless of how they were developed, mathematical models are always limited in their ability to simulate the performance of physical systems. Despite such limitations, they can be very powerful tools for engineering analysis and design. For our purposes, mathematical models will be useful in processing and interpreting data and for control.

We will concentrate here on linear systems, because many physical systems can be approximated as linear and because linear systems are much easier to handle analytically owing to the following properties:

1. $y(t) = L[u_1(t) + u_2(t)] = L[u_1(t)] + L[u_2(t)] = y_1(t) + y_2(t)$
2. $y(t) = L[au_1(t)] = aL[u_1(t)] = ay_1(t)$

where the linear system is represented by the operator L, its inputs by u, and its outputs by y. The first property is termed *linear superposition*. The symbol a in the second property represents a scalar constant.

We will mainly restrict our discussion to systems with a single input variable $u(t)$ and a single output variable $y(t)$. Systems with multiple inputs and outputs can often be treated as several systems with a single input and output using a procedure called *decoupling*. Decoupling is discussed in Section 8.3 in the context of control-system design.

Linear single-input single-output (SISO) dynamic systems can be represented in the form of an nth order differential equation:

$$\left(\frac{d^n y}{dt^n}\right) + \alpha_{n-1}\left(\frac{d^{n-1} y}{dt^{n-1}}\right) + \cdots + \alpha_1\left(\frac{dy}{dt}\right) + \alpha_0 = \beta_m\left(\frac{d^m u}{dt^m}\right) + \cdots$$

$$+ \beta_1\left(\frac{du}{dt}\right) + \beta_0 \qquad (7.1.1)$$

where $m \leq n$ owing to the requirement of *causality* (i.e., physical systems cannot be pure differentiators). The coefficients a_i and b_i in Eq. 7.1.1 will be assumed to be constant in our discussions here. Equation 7.1.1 can always be rewritten in the form of n first-order ordinary differential equations called the *state variable formulation*:

$$dx/dt = \dot{x} = Ax + bu \qquad (7.1.2)$$

is termed the *state equation*, where:

$$x = \begin{pmatrix} x_1 \\ x_2 \\ \cdot \\ \cdot \\ x_n \end{pmatrix}, \quad A = \begin{pmatrix} a_{11} & a_{12} & \cdots & a_{1n} \\ \cdot & \cdot & \cdot & \cdot \\ a_{n1} & \cdots & \cdots & a_{nn} \end{pmatrix}, \quad b = \begin{pmatrix} b_1 \\ b_2 \\ \cdot \\ \cdot \\ b_n \end{pmatrix}$$

The algebraic equation:

$$y = c^T x + du \qquad (7.1.3)$$

where $c^T = [c_1 c_2 \ldots c_n]$, is termed the *output equation*. It should be noted that there is not a unique state representation for the system in Eq. 7.1.1; uniqueness depends on the definitions of the state variables $x(t)$ used to derive Eqs. 7.1.2 and 7.1.3. Figure 7.2 shows a state variable representation of a general dynamic system. Linear dynamical systems can be modeled using three types of ideal elements: *resistance* (R) elements, *capacitance* (C) elements, and *inductance* (L) elements. Although the terminology comes from electrical circuits, these ideal elements define the relationships between the *potential* (e.g., voltage, velocity, pressure, temperature) and *flow* (e.g., current, force, fluid and heat flow) variables in all types of dynamical systems, as summarized in Table 7.1 [Cann67; Palm86].

Linear SISO dynamic systems as in Eq. 7.1.1 can also be represented by a transfer function:

$$Y(s)/U(s) = G(s)$$
$$= (\beta_m s^m + \cdots + \beta_1 s + \beta_0)/(s^n + \alpha_{n-1} s^{n-1} + \cdots$$
$$+ \alpha_1 s + \alpha_0) \qquad (7.1.4)$$

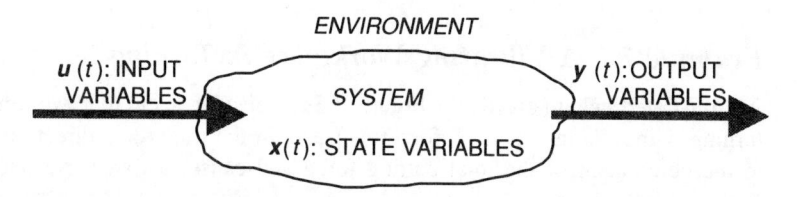

Figure 7.2 *Schematic representation of a dynamic system showing the inputs u(t), outputs y(t), and states x(t)*

134 Chapter 7 SYSTEM MODELING AND IDENTIFICATION

Table 7.1 Summary of Relations for Linear Ideal Elements.

	p(Potential)	f(Flow)	R Element	C Element	L Element
Mechanical (translational)	Velocity v	Force F	$F = cv$	$m\frac{dv}{dt} = \frac{1}{m}F$	$\frac{dF}{dt} = kv$
Mechanical (rotational)	Angular velocity ω	Torque T	$T = b\omega$	$\frac{d\omega}{dt} = \frac{1}{I}T$	$\frac{dT}{dt} = G\omega$
Electrical	Voltage e	Current i	$e = Ri$	$\frac{de}{dt} = \frac{1}{C}i$	$\frac{di}{dt} = \frac{1}{L}e$
Fluid	Pressure p	Volumetric flow rate q	$p = R_f q$	$\frac{dp}{dt} = (\frac{pq}{A})q$	$\frac{dq}{dt} = (\frac{A}{pL})_l$
Thermal	Temperature T	Heat flow q	$T = \frac{1}{H}q$	$\frac{dT}{dt} = \frac{1}{C}q$	—

Notes: 1. The product (pf) is power in all cases except thermal systems.
2. For mechanical systems the potential-force analogy is also valid.
3. The fluid relations given are for incompressible fluids.
4. All the element relations presented linearized versions.

where $Y(s)$ and $U(s)$ represent the Laplace transform of the variables $y(t)$ and $u(t)$, respectively. The ratio $Y(s)/U(s) = G(s)$ is termed the *transfer function* and can also be obtained from the state variable formulation in Eqs. 7.1.2 and 7.1.3:

$$Y(s)/U(s) = G(s) = \mathbf{c}^T(s\mathbf{I} - \mathbf{A})^{-1}\mathbf{b} \qquad (7.1.5)$$

Transfer functions are particularly useful for representation of dynamic systems using block diagrams as illustrated in Example 7.1.

Example 7.1 A Vibrating Workpiece in Turning

As illustrated schematically in Figure 7.3a, consider a circular workpiece in turning without a tailstock. A force $u(t)$ acts in the transverse direction owing to the component of the total cutting force and causes a transverse deflection $y(t)$ under the load.

A simple model can be formulated as in Figure 7.3b, and application of Newton's second law gives:

$$m(d^2y/dt^2) + c(dy/dt) + ky = u \tag{7.1.6}$$

which is of the form of Eq. 7.1.1. Note that if the influence of m and c are small, we could use the static relationship:

$$y = (1/k)u$$

However, if we are interested in the vibration $y(t)$ for changing $u(t)$, and not just the steady state deflection, we must use the dynamic model in Eq. 7.1.6. This can also be written as:

$$\begin{Bmatrix} \dot{x}_1 \\ \dot{x}_2 \end{Bmatrix} = \begin{bmatrix} 0 & 1 \\ -(k/m) & -(c/m) \end{bmatrix} \begin{Bmatrix} x_1 \\ x_2 \end{Bmatrix} + \begin{Bmatrix} 0 \\ 1/m \end{Bmatrix} u$$

and

$$y = x_1$$

where, x_1 = displacement $y(t)$, and x_2 = velocity $\dot{y}(t)$. Convince yourself that state equations equivalent to Eq. 7.1.6 could also be written using x_1 = spring force and x_2 = momentum. The transfer function can be obtained using Eqs. 7.1.4 or 7.1.5:

$$Y(s)/U(s) = G(s) = (1/m)/[s^2 + (c/m)s + (k/m)]$$

Thus, we can represent the system graphically as in Figure 7.3c.

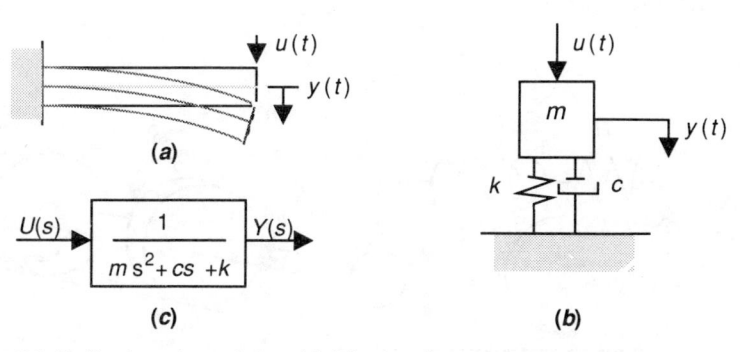

<<>>

Figure 7.3 *Idealizations of a workpiece: (a) deflection of a workpiece in turning (no tailstock), (b) simple model, (c) block diagram*

Example 7.2 Armature-Controlled dc Motor

An armature controlled dc motor, illustrated schematically in Figure 7.4, operates by manipulation of the armature voltage (V) to control the motor shaft rotational speed (ω), or manipulation of current (i) to control the torque (T). Two characteristics of a particular motor are the voltage constant (K_v) and the torque constant (K_t). Referring to Figure 7.4, the links between the electrical and mechanical part of the motor are:

$$V_m = K_v \omega \qquad (7.1.7)$$

$$T_m = K_t i \qquad (7.1.8)$$

where ω is the rotation speed at the motor shaft, V_m is the back EMF (electromotive force), T_m is the motor torque, and i is the armature current. Then, using Ohm's and Kirchhoff's laws, the voltage drop in the motor is

$$V = Ri + L(di/dt) + V_m \qquad (7.1.9)$$

where R is the armature resistance, L is the armature inductance, and V is the voltage at the armature terminals.

Considering the mechanical part of the system as shown in Figure 7.4, a dynamic torque balance gives the following relationships:

$$J(d\omega/dt) + B\omega = T_m \qquad (7.1.10)$$

where J is the sum of the moment of inertia of the rotor (J_1) and the load (J_2) and B is a viscous damping coefficient. Notice that we have neglected any flexibility in the shaft and assumed that the rotation speed ω is the same for both J_1 and J_2. If we make the further assumptions that inductance and damping can be neglected (i.e., $L = B = 0$), then Eqs. 7.1.7–7.1.10 can be combined to yield:

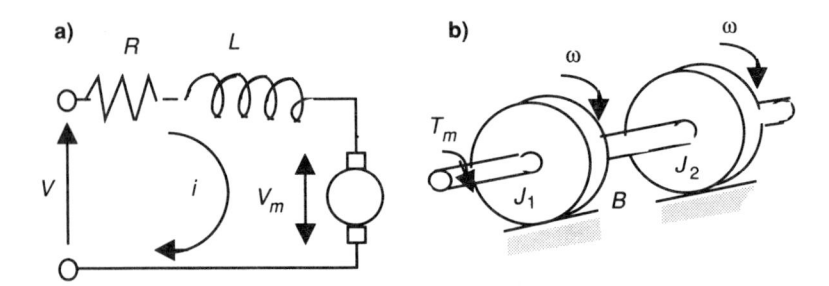

Figure 7.4 Schematic of a dc motor: (a) electrical, (b) mechanical

$$d\omega/dt = \dot{\omega} = -(K_v K_t/JR)\omega + (K_t/JR)V \qquad (7.1.11)$$

which gives the dynamic relationship between the applied voltage V and the motor speed ω.

As a numerical example, consider a dc motor with $J = 5 \times 10^{-4}$ (Nm/sec^2), $K_v = K_t = 0.1$ (V/rad/sec), and $R = 25$ ohms (Note that $K_v = K_t$ for a particular motor when SI units are employed). Substituting these values into Eq. 7.1.11 gives:

$$\dot{\omega} = -0.8\,\omega + 8.0V$$

as the differential equation of motion for the motor speed $\omega(t)$.

<<>>

7.2 SOLUTION OF LINEAR DIFFERENTIAL EQUATIONS

Equation 7.1.11 for the dc motor is a linear first-order ordinary differential equation, which can be written in the form:

$$\dot{x} = ax + bu \qquad (7.2.1)$$

The solution to this equation for a given initial condition x_0 and input function $u(t)$, can be written as:

$$x(t) = \exp[at]x_0 + \int_0^t \exp[a(t-\tau)]b\,u(\tau)\,d\tau \qquad (7.2.2)$$

The first term on the right side of Eq. 7.2.2 is called the free or homogeneous solution, and the second term is the forced or particular solution. Since the system is linear, the total solution is obtained through a linear superposition of the free $[x(0) = x_0$ and $u(t) = 0]$ and forced $[x(0) = 0$ and $u(t) \neq 0]$ solutions.

If we consider systems of equations of the form:

$$\dot{\mathbf{x}} = \mathbf{A}\mathbf{x} + \mathbf{B}\mathbf{u} \qquad (7.2.3)$$

where $\mathbf{x}$ and $\mathbf{u}$ are vectors, and $\mathbf{A}$ and $\mathbf{B}$ are matrices, then Eq. 7.2.2 can be generalized to:

$$\mathbf{x}(t) = \mathbf{\Phi}(t)\mathbf{x}_0 + \int_0^t \mathbf{\Phi}(t-\tau)\mathbf{B}\,\mathbf{u}(\tau)d\tau \qquad (7.2.4)$$

where $\mathbf{\Phi}(t)$ is termed the *state transition* or *solution matrix*, and is given by:

$$\mathbf{\Phi}(t) = \exp[\mathbf{A}t] = \mathbf{I} + \mathbf{A}t + \mathbf{A}^2(t^2/2!) + \mathbf{A}^3(t^3/3!) + \cdots \quad (7.2.5)$$

Laplace transform methods can be used to write Eq. 7.2.1 in the form

$$sX(s) - x_0 = aX(s) + bU(s) \quad (7.2.6)$$

and the solution in the Laplace domain is

$$X(s) = [x_0 + bU(s)]/(s + a) \quad (7.2.7)$$

Similarly, the Laplace domain solution to the system of equations 7.2.3 is:

$$\mathbf{X}(s) = [s\mathbf{I} - \mathbf{A}]^{-1}[\mathbf{x}_0 + \mathbf{B}U(s)] \quad (7.2.8)$$

For zero initial conditions, the relationships in Eqs. 7.2.7 and 7.2.8 give the transfer function between input and output.

Sometimes we are interested not in the system response $x(t)$ to initial conditions x_0 and inputs $u(t)$, but rather the characteristics of the system itself. Insight into the characteristics of the free system can be obtained by solving the *eigenvalue problem*. Consider first Eq. 7.2.1 with $u(t) \equiv 0$, and let $x(t) = v \exp(\lambda t)$ to get:

$$(\lambda - a)v \exp[\lambda t] = 0 \quad (7.2.9)$$

Since $\exp[\lambda t] \neq 0$ for all t, we obtain the eigenvalue problem:

$$(\lambda - a)v = 0 \quad (7.2.10)$$

The case when $v \equiv 0$, termed the trivial solution, is not particularly interesting. Rather, we require that $\lambda = a$, which means that $x(t) = v \exp[at]$ for $u(t) = 0$ and v must be determined from initial conditions x_0. We note, however, that regardless of the initial condition x_0 the stability of this system is completely determined by the *eigenvalue* $\lambda = a$. The response will either grow, remain constant, or decay with time, depending on whether a is positive, zero, or negative (see Fig. 7.5).

These notions can be generalized to the system in Eq. 7.2.3, where with $\mathbf{u}(t) = \mathbf{0}$ and $\mathbf{x}(t) = \mathbf{v} \exp[\lambda t]$ we obtain the eigenvalue problem:

$$(\lambda \mathbf{I} - \mathbf{A})\mathbf{v} = \mathbf{0} \quad (7.2.11)$$

For $\mathbf{v} \neq \mathbf{0}$, the eigenvalues are found from the *characteristic equation*:

$$\det[\lambda \mathbf{I} - \mathbf{A}] = |\lambda \mathbf{I} - \mathbf{A}| = 0 \quad (7.2.12)$$

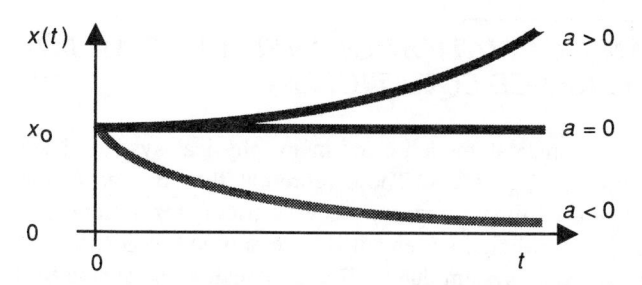

Figure 7.5 Stability of a first-order system

The characteristic equation is an nth-order polynomial in λ, and its n roots λ_j ($j = 1, 2, \ldots n$) are the eigenvalues of the system. These λ_j will in general be complex:

$$\lambda_j = \sigma_j \pm i\omega_j \qquad (7.2.13)$$

where $i = \sqrt{-1}$ and if any of the σ_j are positive there will be an exponentially growing term in the solution, and the system will be unstable. We will see in Section 8.2 that a similar stability condition can be defined in terms of the eigenvalues of linear discrete time systems described by difference equations. Note also that for a given λ_j we can use Eq. 7.2.11 to determine a vector $\mathbf{v}^{(i)}$ called the *eigenvector* corresponding to the eigenvalue λ_j. The eigenvectors can be useful in developing a modal transformation technique that transforms the original state equations in Eq. 7.2.3 into modal state equations that are decoupled. Such decoupled equations are useful in computation of the state transition matrix and in controller design for multiinput, multioutput systems (see Section 8.3).

Example 7.3 Step Response of a dc Motor

In Example 7.2 we had obtained the following equation for a dc motor:

$$\dot{\omega} = -0.8\,\omega + 8.0\,V$$

Given $\omega(0) = 60$ rad/sec and $V(t)$ a step function of magnitude 5 V, we can use Eq. 7.2.2 to obtain,

$$\omega(t) = (e^{-0.8t})(60) + \int_0^t e^{-0.8(t-\tau)}(8.0)(5.0)d\tau = 50 + 10e^{-0.8t}$$

Note also that the eigenvalue of this system is $\lambda = -0.8$, so the system is stable.

<<>>

7.3 TRANSFORMATION OF DIFFERENTIAL TO DIFFERENCE EQUATIONS

The mathematical modeling of many physical systems leads to equations of the form of Eq. 7.2.3. These represent the changes in the system state as a function of time t and can be very useful for system analysis and design. However, for digital systems it is necessary to consider the discrete time nature of the physical system due to digital-to-analog and analog-to-digital conversion [CaMa70; DeAs81]. Thus, we wish to transform differential equations of the form of Eq. 7.2.3, to difference equations of the form:

$$\mathbf{x}(k+1) = \mathbf{P}\,\mathbf{x}(k) + \mathbf{Q}\,\mathbf{u}(k) \tag{7.3.1}$$

where $\mathbf{x}(k) \equiv \mathbf{x}(t = k\Delta t)$, and Δt is termed the sampling period or interval. Thus, Eq. 7.3.1 allows us to calculate the values of $\mathbf{x}$ at the $(k+1)$ time step given the values of $\mathbf{x}$ and $\mathbf{u}$ at the kth time step. To obtain a difference equation as in Eq. 7.3.1 from a differential equation as in Eq. 7.2.3, we assume that the sampling period Δt is known and constant and that the value of $\mathbf{u}(t)$ is constant during each sampling interval (as it would be from a zero-order hold-type DAC). Under these assumptions, we can apply the general solution form in Eq. 7.2.4:

$$\mathbf{x}(t = (k+1)\Delta t) = \mathbf{\Phi}(\Delta t)\,\mathbf{x}(t = k\Delta t) + \int_{k\Delta t}^{(k+1)\Delta t} \mathbf{\Phi}[(k+1)\Delta t - \tau]\mathbf{b}\,\mathbf{u}(\tau)d\tau \tag{7.3.2}$$

which gives:

$$\mathbf{x}(k+1) = \mathbf{\Phi}(\Delta t)\,\mathbf{x}(k) + [\mathbf{\Phi}(\Delta t) - \mathbf{I}]\,\mathbf{A}^{-1}\,\mathbf{B}\,\mathbf{u}(k) \tag{7.3.3}$$

Therefore,

$$\mathbf{P} = \mathbf{\Phi}(\Delta t) = \exp[\mathbf{A}\Delta t] \tag{7.3.4}$$

$$\mathbf{Q} = (\mathbf{\Phi}(\Delta t) - \mathbf{I})\mathbf{A}^{-1}\mathbf{B} = (\mathbf{P} - \mathbf{I})\mathbf{A}^{-1}\mathbf{B} \tag{7.3.5}$$

Thus, for a system described by Eq. 7.2.3 with a constant sampling period Δt, and $\mathbf{u}$ held constant during each sampling period, we can obtain difference equations as in Eq. 7.3.1 by calculating $\mathbf{P}$ and $\mathbf{Q}$ from $\mathbf{A}$ and $\mathbf{B}$ using Eqs. 7.3.4 and 7.3.5.

Example 7.4 Difference Equation for dc Motor Speed

In Example 7.3, we had

$$\dot{\omega} = -0.8\,\omega + 8.0\ \text{V}$$

using Eqs. 7.3.1, 7.3.3, and 7.3.4 we find the corresponding difference equation for $\Delta t = 0.1$ sec:

$$p = \exp[(-0.8)(0.1)] = 0.92312$$
$$q = (8.0)[(0.92312) - 1](-1.25) = 0.7688$$

Thus,

$$\omega(k + 1) = (0.923)\, \omega(k) + (0.769)\, V(k)$$

is the difference equation for the motor speed ω. If the values of the voltage $V(k)$ are constant during each sampling interval, this equation yields the exact values of the motor speed at the sampling instants; it is not an approximation.

<<>>

Another approach for transforming the continuous time system to a discrete time system is to use transform methods. Just as Laplace transforms are used in continuous time systems, and operationally:

$$sX(s) \rightarrow dx/dt$$
$$X(s)/s \rightarrow \int x(t)\,dt$$

For discrete time systems we employ the Z transform [Jury64], and operationally:

$$z\,X(z) \rightarrow x(k + 1)$$
$$z^{-1} X(z) \rightarrow x(k - 1)$$

Thus, the z corresponds to a forward time-shift operator of one step Δt.

If the input for a discrete time system is constant over each time interval Δt (as for the output of a DAC), then the input–output relationship is termed the pulse transfer function. For example, the pulse transfer function corresponding to Eq. 7.3.1 is given by:

$$\mathbf{G}(z) = (z\mathbf{I} - \mathbf{P})^{-1} [\mathbf{x}_0\, z + \mathbf{Q}\, \mathbf{U}(z)] \qquad (7.3.6)$$

It is also possible to obtain the pulse transfer function $\mathbf{G}(z)$ from the transfer function $\mathbf{G}(s)$ of the continuous-time system (see Fig. 7.6). To do this, note that the transfer function of a zero-order hold is [TaRA74]:

$$G_{ZH}(s) = (1 - \exp[-s\Delta t])/s \qquad (7.3.7)$$

and that $z^{-1} = e^{-s\Delta t}$ is a one time step delay. Thus, we can write:

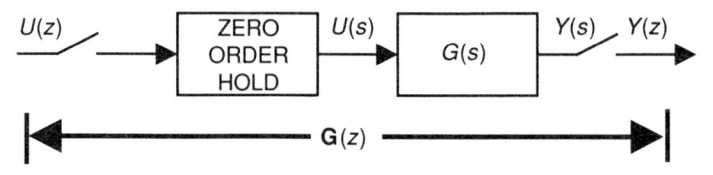

142 Chapter 7 SYSTEM MODELING AND IDENTIFICATION

Figure 7.6 The Pulse Transfer Function G(z)

$$G(z) = (1 - z^{-1})Z[L^{-1}(G(s)/s)] \qquad (7.3.8)$$

where Z and L^{-1} denote the Z transform and the inverse Laplace transform, respectively.

Example 7.5 Pulse Transfer Function for the dc Motor Speed

From the results of Example 7.4 and Eq. 7.3.6, we can write:

$$G(z) = \omega(z)/V(z) = (0.769)/(z - 0.923)$$

We can also derive this result from Eq. 7.3.8. First note that:

$$G(s) = \omega(s)/V(s) = (8.0)/(s + 0.8)$$

and,

$$L^{-1}(G(s)/s) = 10(1 - e^{-0.8t})$$

Then, from Eq. 7.3.8 we get

$$G(z) = ((z-1)/z)[(10(1-e^{-0.8\Delta t})z)/(z-1)(z-e^{-0.8\Delta t})] = (0.769)/(z-0.923)$$

The block diagrams for both the continuous-time and discrete-time models of the dc motor are shown in Figure 7.7.

<<>>

In concluding this section, note that SISO discrete-time dynamic systems can be represented by a single high-order difference equation:

Figure 7.7 Block diagram of the dc motor in Examples 7.2 through 7.5: (a) continuous time, (b) discrete time

$$y(k) + \rho_1 y(k-1) + \cdots + \rho_n y(k-n)$$
$$= \eta_0 u(k-d) + \eta_1 u(k-d-1) + \cdots + \eta_m u(k-d-m) \quad (7.3.9)$$

where $d = (n - m) \geq 0$ and represents the system delay. Equation 7.3.9 is analogous to Eq. 7.1.1 for continuous-time systems. For a system described by Eq. 7.3.9, the pulse transfer function is

$$Y(z)/U(z) = G(z) = \frac{z^{-d}(\eta_0 + \eta_1 z^{-1} + \cdots + \eta_m z^{-m})}{1 + \rho_1 z^{-1} + \cdots + \rho_n z^{-n}}$$
$$= \frac{\eta_0 z^m + \eta_1 z^{m-1} + \cdots + \eta_m}{z^n + \rho_1 z^{n-1} + \cdots + \rho_n} \quad (7.3.10)$$

<<>>

7.4 MODEL STRUCTURE AND ORDER

Mathematical models of linear SISO dynamical systems are represented by Eq. 7.1.1. Causality considerations require $m \leq n$, but we have not discussed how m and n, or how the values of the α_i and β_i are to be determined. We will consider the determination of the parameters α_i and β_i in Section 7.5; in this section we discuss the determination of m and n.

In Example 7.1 the values of m and n came naturally from our simple model in Figure 7.3b. It is well known, however, that the simple single degree of freedom model is only an approximation to the vibration of the beamlike workpiece in Figure 7.3a, which in fact has an infinite number of degrees of freedom. Thus, in that example the values of n and m were determined by our modeling assumptions. One way to determine how good this assumption is might be to evaluate the power-spectral density of $y(t)$ from measurements during cutting of the workpiece transverse vibration (see Fig. 7.8a). If a one degree of freedom model is adequate, we will see a single dominant peak in the spectrum at the damped natural frequency:

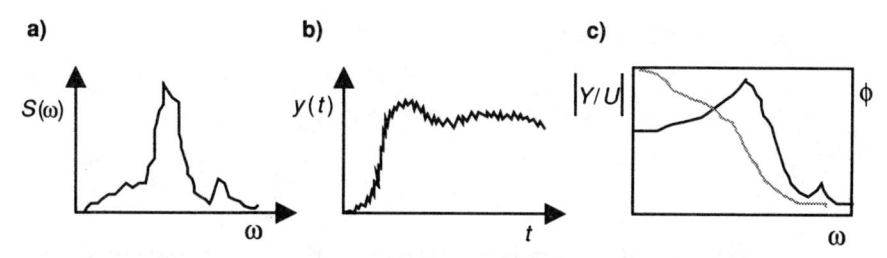

Figure 7.8 Workpiece vibration in turning (see Example 7.1): (a) spectral density, (b) step response, (c) frequency response magnitude and phase

$$\omega_d = \omega_n \sqrt{1 - \zeta^2} \qquad (7.4.1)$$

where $\omega_n = \sqrt{k/m}$ and $\zeta = (c/m)/(2\omega_n)$. If the contribution of motion at higher natural frequencies to the power spectrum is small, then our simple model will be adequate.

Another approach to this problem might be to use $y(t)$ as measured during cutting with special test inputs $u(t)$, such as a step function. A step increase in $u(t)$ could be achieved in turning at the beginning of the cut and might lead to a response like the one sketched in Figure 7.8b. The higher frequency component of the signal represents the contribution due to higher natural frequencies and can be neglected if it is small enough. Another useful test signal is the sinusoid at varying frequencies. Although this would be difficult to generate for our turning example, it can be useful in many other situations. A frequency response curve, such as the one sketched in Figure 7.8c, can be obtained by exciting the system at various frequencies ω and recording the amplitude Y and phase ϕ of the response $y(t)$. Such a frequency response curve can also be useful in selecting an appropriate model order.

Finally, as a method for model order determination consider an input–output approach using the discrete time model given in Eq. 7.3.9. Given values of m and n, in the next section we will discuss methods for determining ρ_i and η_i from input-output measurements. If we repeat such a *parameter estimation* procedure for different values of m and n, and determine a method for selecting the case that gives a best fit to the data, then we can determine the model order [PaWu83]. There are many such procedures that use this basic approach. Note that owing to causality we must have $d \geq 0$ (i.e, $m \leq n$), and typically $m < n$, so we can reduce the number of cases to be tried by selecting $m = n - 1$. If m is less than $n - 1$, then some of the η_i (e.g., η_0) will turn out to be close to zero. So one starts with the case $n = 1$ (or larger if some prior knowledge is available) and increases by increments of one until a stopping criteria is satisfied. For example, one could use

$$J = \sum_{i=1}^{k} e(i)^2 \qquad (7.4.2)$$

where the estimation error is

$$e(i) = y(i) - \hat{y}(i) \qquad (7.4.3)$$

and

$$\hat{y}(i) = -\hat{\rho}_1(i)\, y(i-1) - \hat{\rho}_2(i) y(i-2) - \cdots - \hat{\rho}_n(i)\, y(i-n)$$
$$\hat{\eta}_0(i)\, u(i-1) + \cdots + \hat{\eta}_{n-1}(i)\, u[i-(n-1)] \qquad (7.4.4)$$

is an estimate of $y(i)$ calculated using the parameter estimates $\hat{\rho}_1$ and $\hat{\eta}_i$. Thus, one could stop when J is decreased below a certain absolute level, or when J is minimized with respect to the value of n [PaWu83].

The methods just cited for model order determination can all be useful in certain situations, and there are also numerous other methods and extensions to the ones described here. Note, however, that modeling and model order determination are necessarily approximate. The usefulness of a particular model or modeling scheme depends on its utility to the engineer in a specific application.

7.5 PARAMETER ESTIMATION

As discussed in the previous section, the determination of the parameters $\hat{\rho}_i$ and $\hat{\eta}_i$ of Eq. 7.3.9 from measurements of y and u is termed parameter estimation [GoSi84; LjSo83; Ljun87]. This is accomplished by first selecting an appropriate model structure, as discussed in Sections 7.1 and 7.4, then using input and output measurement to determine the best set of model parameters for that model structure and data.

First consider the simple problem of parameter estimation for a static model where y is a linear function of an independent variable x:

$$y(x) = ax + b \qquad (7.5.1)$$

Given two measurements $y_1(x_1)$ and $y_2(x_2)$ that should fit this structure, we can write

$$y_1 = ax_1 + b$$
$$y_2 = ax_2 + b \qquad (7.5.2)$$

or

$$\begin{bmatrix} x_1 & 1 \\ x_2 & 1 \end{bmatrix} \begin{Bmatrix} a \\ b \end{Bmatrix} = \begin{Bmatrix} y_1 \\ y_2 \end{Bmatrix} \qquad (7.5.3)$$

which has the solution

$$\begin{Bmatrix} a \\ b \end{Bmatrix} = \begin{Bmatrix} (y_1 - y_2)/(x_1 - x_2) \\ (x_1 y_2 - x_2 y_1)/(x_1 - x_2) \end{Bmatrix} \qquad (7.5.4)$$

This solution is valid as long as x_1 and x_2 are different. This restriction is only a simple example of what is termed *poor experimental design* for static systems or *lack of signal richness* for dynamic systems. However, if the number of measurements N is larger than the number of unknown parameters (2 in this example), in addition to Eq. 7.5.2, we get the data $y_3 = ax_3 + b, \ldots, y_{N-1} = ax_{N-1} + b, y_N = ax_N + b$. Since there are more equations than unknowns, we can obtain a number of nonunique solutions $[a\ b]^T$. This problem is solved by the well-known least squares method [Ljun87]. First define

$$J = \sum_{i=1}^{N} e(i)^2 \qquad (7.5.5)$$

where

$$e(i) = y(i) - \hat{a}x(i) - \hat{b} \qquad (7.5.6)$$

then minimize J over all possible values of the parameter estimates $\hat{a}$ and $\hat{b}$. The least squares solution is

$$\begin{Bmatrix} \hat{a} \\ \hat{b} \end{Bmatrix} = \left\{ \begin{bmatrix} x_1\ x_2 \ldots x_N \\ 1\ \ 1\ \ \ldots\ 1 \end{bmatrix} \begin{bmatrix} x_1 & 1 \\ x_2 & 1 \\ \vdots & \vdots \\ x_N & 1 \end{bmatrix} \right\}^{-1} \begin{bmatrix} x_1\ x_2 \ldots x_N \\ 1\ \ 1\ \ \ldots\ 1 \end{bmatrix} \begin{Bmatrix} y_1 \\ y_2 \\ \vdots \\ y_N \end{Bmatrix} \qquad (7.5.7)$$

and gives a "best" fit to all the data points (y_i, x_i) in the sense defined by minimization of J. (Note that if you set $N = 2$ in Eq. 7.5.7, you get the same solution as in Eq. 7.5.4.

Example 7.6 Estimating the Time Constant

A system described by a first-order differential equation

$$\dot{z} = az + bu$$

has a free response (for $u = 0$) of

$$z(t) = \exp[at]z(0) = \exp[-t/\tau]z(0)$$

where $\tau = -1/a$ is the time constant and $z(0)$ the initial condition.

The parameter estimation method outlined above can be used to determine τ and $z(0)$ from measurement of the free response $z(t)$. First take the natural logarithm of both sides to get

$$\ln(z(t)) = \ln(\exp[-t/\tau]) + \ln(z(0))$$

or

$$\ln(z(t)) = (-1/\tau)t + \ln(z(0))$$

If we define

$$\begin{aligned} y &= \ln(z(t)) \\ x &= t \\ a &= (-1/\tau) \\ b &= \ln(z(0)) \end{aligned}$$

we can use the result in Eq. 7.5.7 to get the values of τ and $z(0)$ from measurements $z_i = z(t_i)$ of the free response [Palm86].

<<>>

Now consider the more general parameter estimation problem associated with Eq. 7.3.9. Using the criteria J as defined by Eqs. 7.4.2 through 7.4.4, a least squares solution can be obtained by minimizing J with respect to the parameter estimates a_i and b_i. This is most conveniently expressed by using the following notation for Eq. 7.3.9:

$$y(k) = \boldsymbol{\theta}^T \boldsymbol{\phi}(k-1) \tag{7.5.8}$$

where

$$\boldsymbol{\theta}^T = [\rho_1 \rho_2 \ldots \rho_n \eta_0 \eta_1 \ldots \eta_m] \tag{7.5.9}$$

is termed the parameter vector, and

$\boldsymbol{\phi}^T(k-1)$

$$= [-y(k) \; -y(k-1) \ldots -y(k-n) \; u(k-d) \; u(k-d-1) \ldots u(k-d-m)] \tag{7.5.10}$$

is the measurement vector. Then we can also write Eq. 7.4.4 as

$$\hat{y}(k) = \hat{\boldsymbol{\theta}}(k)^T \boldsymbol{\phi}(k-1) \tag{7.5.11}$$

where

$$\hat{\boldsymbol{\theta}}(k)^T = [\hat{\rho}_1(k) \; \hat{\rho}_2(k) \ldots \hat{\rho}_n(k) \; \hat{\eta}_0(k) \; \eta_1(k) \ldots \eta_m(k)] \tag{7.5.12}$$

is a parameter estimate vector containing the (n + m) parameters to be estimated. The minimization for $k \geq (n + m)$ of

$$J = \sum_{i=1}^{k} [y(i) - \hat{\boldsymbol{\theta}}(k)^T \boldsymbol{\phi}(i-1)]^2 \tag{7.5.13}$$

gives the solution:

$$\hat{\theta}(k) = \mathbf{P}(k) \left[\sum_{i=1}^{k} y(i) \, \phi(i-1) \right] \quad (7.5.14)$$

where

$$\mathbf{P}(k) = \left[\sum_{i=1}^{k} \phi(i-1) \, \phi(i-1)^T \right]^{-1} \quad (7.5.15)$$

Notice that the least squares parameter estimate as given by Eqs. 7.5.14 and 7.5.15 is only suitable for off-line computation. That is, we must first collect the input $u(i)$ and output $y(i)$ measurements for all $i = 1, 2, \ldots k$. Then we must calculate the terms in Eq. 7.5.15 and invert the matrix to get $\mathbf{P}(k)$. Finally, we use Eq. 7.5.14 to get $\hat{\theta}(k)$.

There will, of course, be situations where we would like to obtain an estimate of $\theta(k)$ as we collect data and continue to update it as new data becomes available. Such an on-line version of the least squares parameter estimation algorithm can be obtained by writing Eq. 7.5.14 at k and at $k + 1$, then writing an expression for $\hat{\theta}(k+1)$ in terms of $\hat{\theta}(k)$. The matrix inversion implied by Eq. 7.5.15 can also be avoided by using the matrix inversion lemma [GoSi84]. This on-line version of the algorithm, usually termed *recursive least squares* (RLS), is given by:

$$\hat{\theta}(k+1) = \hat{\theta}(k) + \frac{\mathbf{P}(k) \, \phi(k) \, [y(k+1) - \hat{\theta}(k+1)^T \phi(k)]}{1 + \phi(k)^T \mathbf{P}(k) \phi(k)} \quad (7.5.16)$$

and

$$\mathbf{P}(k+1) = \mathbf{P}(k) - \frac{\mathbf{P}(k) \phi(k) \phi(k)^T \mathbf{P}(k)}{1 + \phi(k)^T \mathbf{P}(k) \phi(k)} \quad (7.5.17)$$

where initial guesses $\hat{\theta}(0)$ and $\mathbf{P}(0)$ are required to start the algorithm. Best initial guesses of the parameters, or simply the zero vector, can be used for $\hat{\theta}(0)$. For $\mathbf{P}(0)$, one often selects the matrix

$$\mathbf{P}(0) = \delta \mathbf{I} \quad (7.5.18)$$

where $\mathbf{I}$ is the identity matrix and δ is a design parameter selected to be a number greater than zero (e.g., 10, 100, or 1000) and then varied until good parameter convergence is obtained. Many variants of the RLS algorithm have been designed to handle problems such as measurement noise, slowly time-varying parameters, and numerical problems [LjSo83]. Note also that the input $u(k)$ must satisfy certain conditions for the parameter estimates $\theta(k)$ to converge to the actual values of θ [GoSi84; LjSo83]. Basically, the input must be selected so that it will excite the system completely and persistently. Inputs with sharp corners such as pulse trains or inputs with a random perturbation signal superimposed are often used.

<<>>

Example 7.7 Flank Wear from Force Measurements in Turning

As described in [DaUl87a,b], the flank wear w in a turning process can be related to the feed-rate f and a component F of the cutting force by a model of the form:

$$w(k+1) = aw(k) + bf(k)$$
$$F(k) = cw(k) + df(k)$$

For a sharp tool $w(0) = 0$, and if we measure $F(0)$ and $f(0)$ before any significant wear develops, then we can determine

$$\hat{d} = F(0)/f(0)$$

and if we perform off-line tests where both $F(i)$ and $w(i)$ are measured for $i = 1, 2, \ldots, k$, then we can determine a least squares estimate of the parameter c:

$$\hat{c} = \left[\sum_{i=1}^{k} w(i)(F(i) - \hat{d}f(i))\right] / \left[\sum_{i=1}^{k} w(i)^2\right]$$

Now using these values for $\hat{d}$ and $\hat{c}$, the parameters a and b in the model can be estimated by RLS:

$$\rho = -a$$
$$\eta = bc$$

where

$$y(k+1) = -\rho y(k) + \eta u(k)$$
$$y(k) = [F(k) - \hat{d}f(k)]$$
$$u(k) = f(k)$$

and the parameters ρ and η are estimated by Eq. 7.3.16 with

$$\hat{\boldsymbol{\theta}}(k)^T = [\hat{\rho}(k)\ \hat{\eta}(k)]$$
$$\boldsymbol{\phi}(k-1)^T = [-y(k-1)\ u(k-1)]$$

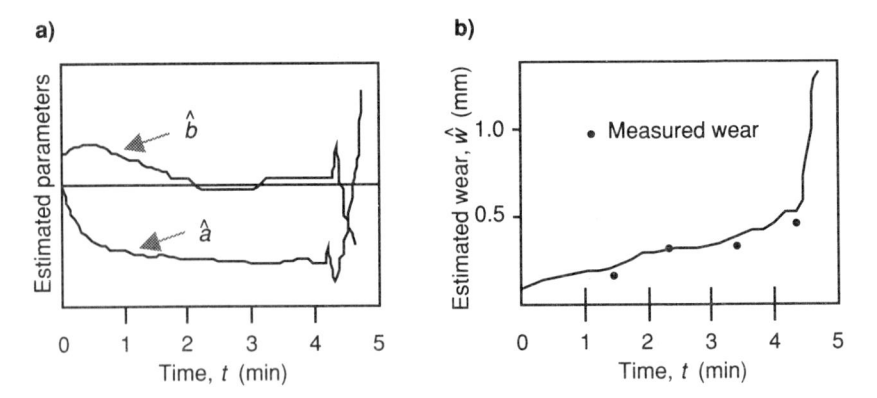

Figure 7.9 Wear estimation in turning (see Example 7.7): (a) estimated model parameters, (b) estimated and measured wear

The results from actual machining tests as reported in [DaUl87b] are shown in Figure 7.9.

<<>>

7.6 SUMMARY

Mathematical modeling of dynamic systems is useful for various monitoring, signal processing, and control functions. Mathematical models can be developed based on fundamental physical laws, on measurements of system inputs and outputs, or on some combination. In this chapter we presented the mathematical representation of dynamic systems in terms of state equations and transfer functions. We also discussed solution of the system equations to investigate stability and response for both continuous and discrete time systems.

Methods for determination of model structure and order and for estimation of model parameters were presented. In particular we introduced both the off-line and on-line versions of the least squares parameter estimation algorithm.

7.7 PROBLEMS

1. In the forest-products industry, lumber must often be kiln dried before it can be sold. You are asked to design a microprocessor-based system for kiln temperature control. Given the model of the open loop system:

$$\frac{dT}{dt} = -T(t) + 10V(t)$$

where $T(t)$ is the kiln temperature, $V(t)$ is the voltage input to the heater, and t is time:

   a. Determine, for a sampling period of $\Delta t = 0.1$, the corresponding difference equation for the system.
   b. Using the difference equation found in (a), determine $T(t = 3\Delta t)$ given $T(0) = 0$ and $V(0) = 1$, $V(1) = 2$, and $V(2) = 0$.
   c. Find the transfer function $T(s)/V(s)$ from the given differential equation.
   d. Find the pulse transfer function $T(z)/V(z)$.

2. Assume that the scalar differential equation

$$\frac{dT}{dt} = -T(t) + 20\,V(t)$$

describes the relationship between the temperature T of a mold cavity and the voltage V applied to a heater for preheating the mold. The voltage will be adjusted by a microprocessor through the DAC at intervals of $\Delta t = 0.1$ sec.

   a. Write the difference equation describing the temperature as a function of the input voltage.
   b. Assume that $T(0) = 25°C$, and calculate $T(0.2)$ given $V(0) = 4$ V and $V(0.1) = 3$ V.
   c. Given $T(0) = 25°C$, and $V(0) = 4$ V, $V(0.1) = 2$ V, and $V = 5$ V for all $t_i \geq 0.2$, determine $T(0.1)$, $T(0.2)$, and $T(0.3)$.
   d. If the input voltage from the DAC remained at 5 V for a long time, what would your estimate of $T(10)$ be?
   e. Given $T(0) = 25°C$, $V(k) = 0$, and $\Delta t = 0.1$ sec, use the difference equation to generate data for least squares estimation of the time constant and initial condition.

3. Consider a microcomputer-based digital controller for liquid level x in a tank. The system can be modeled by

$$\frac{dx}{dt} = -10x + u$$

where u is the voltage signal sent from the DAC and is proportional to the liquid flow rate into the tank.

   a. Write the transfer function $X(s)/U(s) = G(s)$ and determine the gain and time constant for this system.
   b. Assume a sampling period of $\Delta t = 10$ msec and determine the difference equation describing this system.
   c. Given $x(0) = 1.0$, $u(0) = 0$, and $u(0.01) = 5$; what is $x(0.02)$?

4. The differential equation for the speed of a particular dc motor is given as

$$\frac{d\omega}{dt} = -\omega(t) + V(t)$$

a. What is the unit step response of the motor for zero initial conditions [i.e., what is $\omega(t)$ if $\omega(0) = 0$ and $V(t) = 0$ for $t < 0$ and $V(t) = 1$ V for $t \geq 0$]?
b. Roughly sketch the motor speed versus time. Include appropriate numerical annotation of the axes.
c. What is the motor speed after 1 sec?
d. What is the motor speed after 5 sec?
e. Transform the differential motor equation into a difference equation.
f. Again if the input is a unit step, and the sampling period, Δt, is 1 sec, what is the motor speed after 2 sec $[\omega(2\Delta t)]$?
g. If $\Delta t = 0.1$, what is the approximate speed after 5 sec, [i.e., $\omega(50\Delta t)$]?
h. What is the motor speed $\omega(3\Delta t)$ if $V(0\Delta t) = 0$, $V(1\Delta t) = 0.5$ and $V(2\Delta t) = 1$?
i. Discuss briefly the advantages and disadvantages of the differential versus difference form for the motor equation.

5. For the dc motor in Examples 7.2 through 7.5:
a. What is the motor time constant? What is the steady-state gain?
b. Use the difference equation and a unit step input $V(k)$ to generate rotational speed data versus time. Use this input and output data to estimate the parameters p and q of the difference equation by both the off-line and on-line versions of the least squares algorithm.

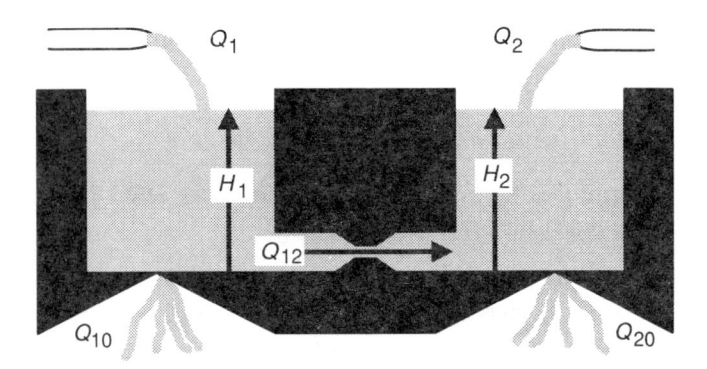

Figure 7.P7 Liquid level in two coupled tanks

6. For the wear estimation problem discussed in Example 7.7 we first estimated d, then c, then a and b of the wear model. Instead formulate first an off-line estimation problem to determine c and d simultaneously from measurements of f, w, and F. Then formulate an on-line estimation problem, given c and d, and measurements f and F, to estimate a, b, and, consequently, the wear w. Explain your formulation clearly.

7. Develop the dynamical mathematical model for the liquid level H_1 and H_2 in the two coupled tanks illustrated in Figure 7.P7. Assume that the tank cross sections are uniform with areas A_1 and A_2, respectively. Also the orifices are characterized by resistance values R_1 and R_2 for flow out of the two tanks, and by R_{12} for flow between the two tanks. Write the final model equations in MIMO state equation format. (Hint: Use the ideal linear element relationships provided in Table 7.1, and assume that the density of liquid in the tanks is constant.)

Chapter 8
DIGITAL CONTROL

This chapter describes how a computer can be used to control a dynamic system, such as a manufacturing process. We have previously discussed the computer hardware and programming issues needed for computer control. Here we introduce the basic concepts and methods from control theory. This chapter builds on the ideas of mathematical modeling and system identification introduced in the previous chapter.

First, a general description of computer control systems is given, and basic concepts from control theory are introduced. Then we describe the digital control of single-input single-output (SISO) dynamic systems using the idea of pole assignment. Finally, we conclude the chapter with a discussion of multi-input multi-output (MIMO) system control via decoupling.

8.1 COMPUTER CONTROL SYSTEMS

The basic idea in automatic control is to design a controller or compensator such that when it is combined with the controlled system or process into a single system the combined system will have some desired specified behavior. This is illustrated in Figure 8.1 for both an *open loop* configuration (i.e., there is no measurement of the controlled variable y), and for a *closed loop* (or feedback) configuration. In most applications, feedback is used to counteract the effects of process disturbances and other unknown factors not explicitly included in the process model.

A computer control system is schematically illustrated in Figure 8.2 for a typical sampled data control system. The controller is implemented on a digital computer and interacts with the controlled system through ADCs for analog sensors and DACs for analog actuators. Digital sensors and/or actuators would

156 Chapter 8 DIGITAL CONTROL

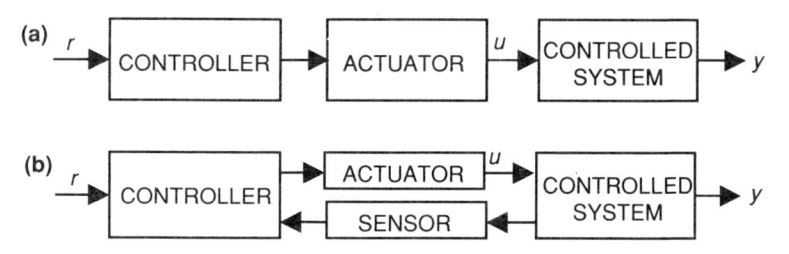

Figure 8.1 Schematic of a control system: (a) open loop, (b) feedback (closed loop)

interact directly through digital I/O ports. In either case, the computer control system is treated mathematically as a discrete time system. This is illustrated in Figure 8.3, where the hybrid digital/analog system in part (a) is represented as a completely digital system in part (b). The plant pulse-transfer function $G_p(z)$ is obtained from $G_p(s)$ using Eq. 7.3.8, as illustrated in Figure 7.6. The sensor and actuator characteristics must be included in $G_p(s)$ as needed. Recalling the discussion of sampling in Section 5.3, an antialiasing filter must be included between the analog sensor and the ADC to prevent aliasing, and, if the filter dynamics is significant, then it must also be accounted for in $G_p(s)$.

Thus, as shown in Figure 8.2, a computer control system consists of a computer that implements a digital controller. A digital control $u(k)$ is then sent to a DAC that converts it to an analog signal $u(t)$, usually by implementing a zero-order hold (ZOH). This signal drives an actuator, which is the energy-delivering component of the system, to act on the controlled system. The response $y(t)$ is measured by a sensor whose output is filtered to prevent aliasing before it is sampled and quantized by the ADC for use in the digital controller.

With this combination of hardware and software, control objectives for the closed-loop system are

1. Stability
2. Little or no steady-state error as y follows r

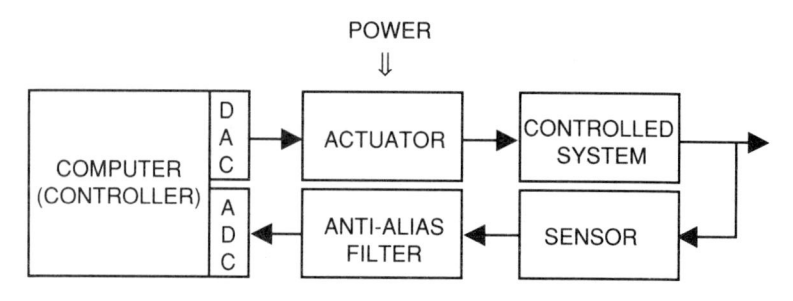

Figure 8.2 A typical sampled data computer control system

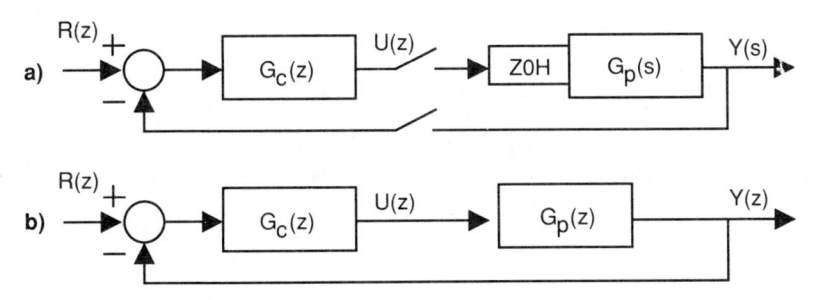

Figure 8.3 Block diagram: (a) sampled data system, (b) discrete-time representation

3. Rapid transient response (i.e., high bandwidth)
4. Insensitivity (robustness) to disturbance inputs, to variations in the process parameters, and to modeling errors.

Some of the control designs that will be discussed will be better than others at meeting these objectives; in general, the first requirement of stability must always be met.

8.2 DIGITAL CONTROL OF SISO SYSTEMS

In this section, we describe some simple digital control algorithms and controller design strategies for single-input single-output (SISO) systems. Although we are considering SISO systems, they may be multivariable in nature. More specifically, we consider systems of the form

$$\mathbf{x}(k+1) = \mathbf{P}\,\mathbf{x}(k) + \mathbf{q}u(k) \tag{8.2.1}$$

$$y(k) = \mathbf{c}^T \mathbf{x}(k) \tag{8.2.2}$$

Notice that $\mathbf{x}$ is a state vector of order n, y is a scalar output (measured) variable, and u is a scalar input (manipulated) variable. As shown in Figure 8.4, the digital computer must produce the signal u using the feedback information y. The goal is to maintain y at some desired reference r, despite disturbances $\mathbf{d}$ that may act on the system. For the linear continuous-time system described by Eqs. 7.1.2 and 7.1.3, note that $\mathbf{P}$ and $\mathbf{q}$ in Eq. 8.2.1 can be determined from Eqs. 7.3.4 and 7.3.5.

We consider here a few traditional algorithms that are commonly used and their implementation on a digital computer. There are all different ways of generating a control input based on errors, with their names indicating how the error is manipulated: Proportional (P) Control, Integral (I), Control, and Derivative (D) Control. Generally, these laws are combined to provide algorithms such

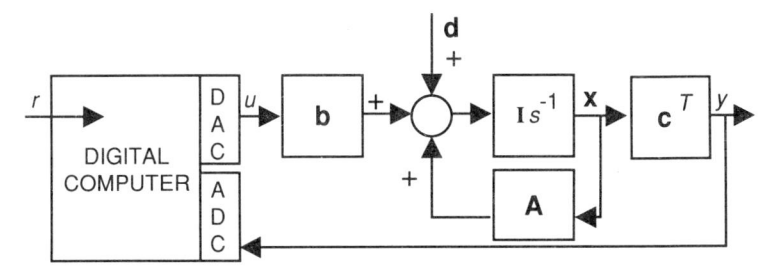

Figure 8.4 Computer control of a linear SISO system

as PI, PD, or PID Control [CaMa70; FrPE86; TaRA74]. Here we consider the discrete-time version of these algorithms for a given constant sampling interval Δt, with $e(k) = r(k) - y(k)$:

1. Proportional (P) Control:

$$u(k) = k_p e(k) \qquad (8.2.3)$$

2. Integral (I) Control:

$$u(k) = k_i \sum_{j=0}^{k} e(j)\Delta t = u(k-1) + (k_i \Delta t) e(k) \qquad (8.2.4)$$

3. Derivative (D) Control:

$$u(k) = k_d[e(k) - e(k-1)]/\Delta t = (k_d/\Delta t)[e(k) - e(k-1)] \qquad (8.2.5)$$

A discrete-time PI control algorithm is

$$u(k) = k_p e(k) + (k_i \Delta t) \sum_{j=1}^{k} e(j) \qquad (8.2.6)$$

If we note that

$$u(k-1) = k_p e(k-1) + (k_i \Delta t) \sum_{j=1}^{k-1} e(j) \qquad (8.2.7)$$

Then we can rewrite the PI control law in the more usual alternative form:

$$u(k) = u(k-1) + (k_p + k_i \Delta t) e(k) - k_p e(k-1) \qquad (8.2.8)$$

Although we may have an algorithm such as the PI control in Eq. 8.2.8, we must still determine values for the controller parameters or gains. In general, one considers the performance specifications for a particular problem and tries to select the controller gains such that the specified performance of the closed-loop system is obtained. Although we cannot discuss this topic in detail here, we briefly consider stability, steady-state error, and transient response. The issue of robustness (i.e., insensitivity to disturbances, parameter variations, and modeling errors) is a very important consideration in controller design. However, it is beyond the scope of our discussion here. We only mention that typically there is a tradeoff between rapid transient response (high bandwidth) and robustness, and refer you to [AsWi84; FrPE86; Frie86].

The *stability of linear discrete time systems* can be investigated by determining the roots of the characteristic equation. The characteristic equation is the denominator of the pulse transfer function set equal to zero, for example,

$$G(z) = N(z)/D(z) \tag{8.2.9}$$

then,

$$D(z) = 0 \tag{8.2.10}$$

is the characteristic equation. Equivalently, the characteristic equation can be found from the difference equations, Eq. 8.2.1:

$$|z\mathbf{I} - \mathbf{P}| = 0 \tag{8.2.11}$$

In either case, this is a homogeneous polynomial equation in z. The roots (poles or eigenvalues) of the equation must *all* satisfy the following criterion for the system to be stable:

$$|z_i| \leq 1; \quad i = 1, 2, \ldots n \tag{8.2.12}$$

In general, the roots z are complex, so this implies that they must lie inside the unit circle in the z-plane for stability (see Fig. 8.5). If any of the roots of the

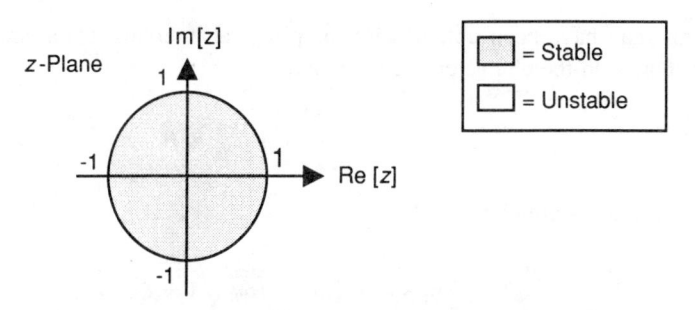

Figure 8.5 Unit circle stability criterion for linear discrete-time systems

characteristic equation lie outside the unit circle, then the system is unstable. If a root lies on the unit circle, the system is limitedly stable. So we must select the controller parameters (i.e., k_p, k_i, k_d) such that the closed-loop system poles or eigenvalues all lie inside the unit circle. This stability criterion is very closely related to the one discussed in Section 7.2 for continuous time systems. Note that a pole z_i of a discrete time system obtained by sampling a continuous time system at intervals Δt is given by $z_i = \exp(s_i \Delta t)$, where the s_i's are the poles (eigenvalues) of the continuous time system.

The steady-state response of a discrete time system can be found from Eq. 8.2.1 by setting $x(k+1) = x(k) = x(k-1), \ldots$ at the steady state. Alternatively, we can determine the steady-state value of a discrete time function $f(k)$ from its Z transform by using the final value theorem [Jury64]:

$$\lim_{k \to \infty} f(k) = \lim_{z \to 1} (z-1) F(z) \qquad (8.2.13)$$

For controller design, we are interested in the steady-state value of the error $e(k) = r(k) - y(k)$. Ideally, we would like to select the control algorithm and controller parameters such that

$$\lim_{k \to \infty} e(k) = 0 \qquad (8.2.14)$$

This can generally be achieved for constant reference inputs $r(k) = r_0$ when the controller contains I action.

For a linear SISO system the closed-loop poles not only determine the stability of the system, but their location in the z-plane determines the closed-loop system transient response characteristics. Thus, one method for selecting controller parameters is to specify closed-loop pole locations to achieve certain desired performance characteristics. This is referred to as *pole* or *eigenvalue assignment*. Consider, for example, a closed loop system with the characteristic equation:

$$z^2 + bz + c = (z - z_1)(z - z_2) = 0 \qquad (8.2.15)$$

This may have been obtained by sampling (at intervals Δt) a continuous-time system with the characteristic equation:

$$s^2 + 2\zeta \omega_n s + \omega_n^2 = 0 \qquad (8.2.16)$$

and the roots are (for $\zeta < 1$):

$$s_{1,2} = -\zeta \omega_n \pm i \omega_n \sqrt{1 - \zeta^2} \qquad (8.2.17)$$

The *maximum percent overshoot* of a second-order system to a step input is given by [AsWi84; FrPE86]:

$$p_o = 100\exp(-\zeta\pi/\sqrt{1-\zeta^2}) \approx 100(1-\zeta)^{2.603} \qquad (8.2.18)$$

and the *settling time* for such a system in response to a unit step reference input is:

$$t_s \approx 4.5/\zeta\omega_n \qquad (8.2.19)$$

Thus, the designer can specify desired values for p_o and t_s to determine corresponding values for ζ and ω_n from Eqs. 8.2.18 and 8.2.19. Then s_1 and s_2 can be calculated from Eq. 8.2.17, and z_1 and z_2 are then given by:

$$z_i = \exp[s_i \Delta t] \qquad (8.2.20)$$

These values of z_1 and z_2 are then used in Eq. 8.2.15 to determine b and c. Although the preceding discussion is limited to second-order dynamic systems, the same approach can be extended to higher-order SISO systems using the concept of dominant poles as discussed in [AsWi84; GoSi84].

Example 8.1 PI Control of the Speed of a dc Motor

Consider the PI algorithm in Eq. 8.2.8 as applied to the dc-motor speed-control problem. We will use the foregoing stability, steady-state error, and transient response concepts to select appropriate values for k_p and k_i.

Figure 8.6 shows the block diagram for the dc motor with a PI controller, where $p = 0.923$ and $q = 0.769$ for our particular problem (see Example 7.4). The closed-loop pulse-transfer function is

$$Y(z)/R(z) = \frac{q(k_p + k_i\Delta t)z - qk_p}{z^2 + (qk_p + qk_i\Delta t - 1 - p)z + (p - qk_p)}$$

and the roots of the characteristic equation are

$$z_{1,2} = -(b/2) \pm (1/2)\sqrt{b^2 - 4c}$$

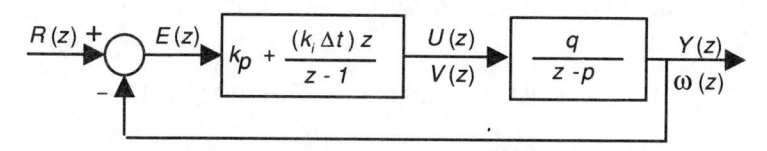

Figure 8.6 Block Diagram of the PI dc-motor speed-control system

where

$$b = qk_p + qk_i \Delta t - 1 - p = 0.769(k_p + k_i \Delta t) - 1.923$$
$$c = p - qk_p \qquad\qquad = 0.923 - 0.769 k_p$$

It is possible, then, to determine the values of z for various Δt, k_p, and k_i and to check the system stability. Some sample calculations are given in Table 8.1 for $\Delta t = 0.1$ sec. In fact it can be shown that for stability we have the following conditions on k_p and k_i [AsWi84; Jury64]:

$$k_i \geq 0$$
$$k_p \geq -0.077$$

and

$$2k_p + k_i \Delta t \leq 5$$

Let us also check the steady-state error conditions for the closed-loop system. If we consider a unit step reference input, $R(z) = z/(z - 1)$, then the steady-state value of $Y(z)$ is

$$\lim_{z \to 1} \frac{q(k_p + k_i \Delta t)z^2 - qk_p z}{z^2 + (qk_p + qk_i \Delta t - 1 - p)z + (p - qk_p)} = \frac{qk_i \Delta t}{qk_i \Delta t} = 1$$

Thus, the steady-state error will be zero for this system for all $k_i > 0$, and we see that the I action ensures desirable steady-state performance for constant reference inputs.

Finally, let us consider pole assignment as a way to determine specific values for k_p and k_i. The simplest approach is to select $z_1 = z_2 = 0$ in Eq. 8.2.15; this is referred to as a *deadbeat controller* or a finite time-settling control [AsWi84;

Table 8.1 Closed-Loop Poles for Various PI Controller Gains in Example 8.1 with $\Delta t = 0.1$ sec

k_p	k_i	z_1	z_2	Comment
0	1	0.923 + 0.266i	0.923 − 0.266i	Stable
0	10	0.577 + 0.768i	0.577 − 0.768i	Stable
1	1	0.170	0.908	Stable
1	10	0.193 + 0.312i	0.193 − 0.312i	Stable
10	1	0.990	−6.834	Unstable
10	10	0.909	−7.455	Unstable

TaRA74]. Note that we must have $s_{1,2}$ approaching negative infinity to obtain discrete-time closed-loop poles at the origin of the z plane (see Eq. 8.2.20). Thus, a deadbeat control corresponds to the fastest-possible-responding closed-loop system and often leads to excessive values of the control effort (the manipulated variable or motor voltage in this example). The deadbeat control leads to the gains:

$$k_p = p/q = 1.2$$
$$k_i \Delta t = 1/q = 1.3$$

If the deadbeat control is not satisfactory, for example, the control voltage $V(k)$ may saturate, then we can use Eqs. 8.2.18 and 8.2.19 to design the controller. Choosing, for example, $p_o \leq 15\%$ and $t_s \approx 1$ sec, for $\Delta t = 0.1$ sec, we can assign $\zeta = 0.517$ and $\omega_n = 8.704$ to give

$$s_1 = -4.5 + 7.45i$$
$$s_2 = -4.5 - 7.45i$$

and

$$z_1 = 0.469 + 0.432i$$
$$z_2 = 0.469 - 0.432i$$

which gives

$$k_p = 1.16$$
$$k_i \Delta t = 0.124$$

This controller will respond more slowly than the deadbeat design, but it will be less likely to have problems with saturation of the armature voltage.

<<>>

8.3 DIGITAL CONTROL OF MIMO SYSTEMS

In the previous section, we considered systems with a scalar input u and a scalar output y. Here we consider the general multi-input multi-output (MIMO) linear discrete time system:

$$\mathbf{x}(k+1) = \mathbf{P}\mathbf{x}(k) + \mathbf{Q}\mathbf{u}(k) \quad (8.3.1)$$
$$\mathbf{y}(k) = \mathbf{C}\mathbf{x}(k) \quad (8.3.2)$$

Since a general discussion of MIMO digital control is beyond the scope of this introduction, we simply rely on the following example to illustrate the basic concepts and methods. For a more detailed discussion, refer to [TaRA74].

Example 8.2 Control of a Paper-making Process

There are times when two outputs from a process cannot be adjusted independently because they are physically linked. One example from the paper industry is the relationship between what is called basis weight (the mass per unit area of paper) and moisture [DePW77; DeVW78]. Moisture is more or less independent of basis weight, and it can be manipulated by adjusting the steam pressure in the dryers of a paper machine. Basis weight, on the other hand, is coupled with moisture in that when measurements are made, the paper mass and water mass are measured at the same time. The main way that this mass can be affected is by changing the mass flow of fiber to the paper machine.

This is a fairly typical example where there are two outputs, coupled together, and two inputs. By defining state variables:

$$\mathbf{x}(t) = \begin{Bmatrix} x_1(t) \\ x_2(t) \end{Bmatrix} = \begin{Bmatrix} \text{Basis weight} \\ \text{moisture} \end{Bmatrix} \quad (8.3.3)$$

and an input vector:

$$\mathbf{u}(t) = \begin{Bmatrix} u_1(t) \\ u_2(t) \end{Bmatrix} = \begin{Bmatrix} \text{Stock flow} \\ \text{steam pressure} \end{Bmatrix} \quad (8.3.4)$$

a state variable model can be written.

The rate of change of $x_1(t)$, which depends on both the levels of $x_1(t)$ and $x_2(t)$ and the stock flow, is

$$\frac{dx_1(t)}{dt} = -4x_1(t) - x_2(t) + 0.5u_1(t) \quad (8.3.5)$$

and the rate of change of moisture only depends on the level of moisture and steam pressure:

$$\frac{dx_2(t)}{dt} = -3x_2(t) + u_2(t) \quad (8.3.6)$$

Writing these two equations in the standard form gives

$$\dot{\mathbf{x}}(t) = \begin{bmatrix} -4 & -1 \\ 0 & -3 \end{bmatrix} \mathbf{x}(t) + \begin{bmatrix} 0.5 & 0 \\ 0 & 1 \end{bmatrix} \mathbf{u}(t) \quad (8.3.7)$$

To develop a discrete time model, we start with finding the eigenvalues and vectors:

8.3 Digital Control of MIMO Systems

$$|\lambda \mathbf{I} - \mathbf{A}| = \begin{vmatrix} (\lambda + 4) & 1 \\ 0 & (\lambda - 3) \end{vmatrix} = \lambda^2 + 7\lambda + 12 = 0 \quad (8.3.8)$$

which gives $\lambda_1 = -4$ and $\lambda_2 = -3$. Defining

$$\Lambda = \begin{bmatrix} -4 & 0 \\ 0 & -3 \end{bmatrix} \quad (8.3.9)$$

Since Λ is diagonal, we can readily calculate

$$\exp[\Lambda t] = \begin{bmatrix} e^{-4t} & 0 \\ 0 & e^{-3t} \end{bmatrix} \quad (8.3.10)$$

and it can be shown that [TaRA74]:

$$\exp(\mathbf{A}t) = \Phi(t) = \mathbf{T} \exp(\Lambda t) \mathbf{T}^{-1} \quad (8.3.11)$$

where $\mathbf{T}$ is the modal transformation matrix whose columns are the eigenvectors of $\mathbf{A}$. These eigenvectors $\mathbf{v}^{(i)}$ corresponding to λ_i can be found from $(\lambda_i \mathbf{I} - \mathbf{A})\mathbf{v}^{(i)} = 0$. Note that the eigenvectors are only determined to within an arbitrary constant, and if we arbitrarily select $v_1^{(i)} = 1$ for both $i = 1$ and 2, we obtain the modal transformation matrix $\mathbf{T}$ and its inverse $\mathbf{T}^{-1}$:

$$\mathbf{T} = \begin{bmatrix} 1 & 1 \\ 0 & 1 \end{bmatrix}, \mathbf{T}^{-1} = \begin{bmatrix} 1 & 1 \\ 0 & -1 \end{bmatrix}$$

We now assume that the processor clock will be set so that $\Delta t = 0.05$ s, then using Eqs. 7.3.4 and 7.3.5 gives

$$\mathbf{P} = \Phi(\Delta t = 0.05) = \mathbf{T} \exp[\Lambda 0.05] \mathbf{T}^{-1} = \begin{bmatrix} 0.819 & -0.042 \\ 0 & 0.861 \end{bmatrix} \quad (8.3.12)$$

$$\mathbf{Q} = (\mathbf{P} - \mathbf{I})\mathbf{A}^{-1}\mathbf{B} = \begin{bmatrix} 0.23 & -0.001 \\ 0 & 0.046 \end{bmatrix} \quad (8.3.13)$$

It will be assumed that the inputs to the system are step changes in both flow rate and steam pressure at time intervals Δt. Thus, for modeling the behavior of this system with two input variables $u_1(t)$ and $u_2(t)$, that send out step changes via DACs, the difference equation for modeling the coupled behavior of this system is

$$\mathbf{x}(k + 1) = \begin{bmatrix} 0.819 & -0.042 \\ 0 & 0.861 \end{bmatrix} \mathbf{x}(k) + \begin{bmatrix} 0.23 & -0.001 \\ 0 & 0.046 \end{bmatrix} \mathbf{u}(k) \quad (8.3.14)$$

166 Chapter 8 DIGITAL CONTROL

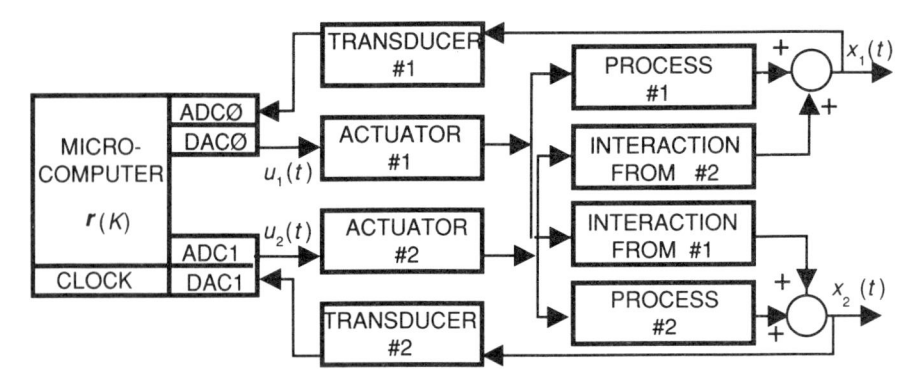

Figure 8.7 Computer control of the MIMO process in Example 8.2

Suppose the complete system and control can be represented as shown in Figure 8.7. The two outputs, $x_1(t)$ and $x_2(t)$, will be sampled at the same time with an ADC. Then the errors will be computed and used to calculate the manipulations $u_1(k)$ and $u_2(k)$ that are sent to the actuating device through DACs. The calculations will be done at clock intervals $\Delta t = 0.05$ sec.

The synthesis problem is to compute the gains so that when references $r_1(k)$ and $r_2(k)$ are changed independently the controller implemented by the processor decouples the outputs so that they react independently. Although this is the major part of the decoupling algorithm, there is another aspect that has to be specified, which is the speed of response to these changes, that is, the time constants for the decoupled processes.

The solution can be developed for the general case first, considering that the system can be written as a difference equation:

$$\mathbf{x}(k+1) = \mathbf{P}\mathbf{x}(k) + \mathbf{Q}\mathbf{u}(k) \qquad (8.3.15)$$

The decoupling algorithm will be based on the error vector

$$\mathbf{e}(k) = \mathbf{r}(k) - \mathbf{x}(k) \qquad (8.3.16)$$

So that the manipulations are a function of the errors, that is, a special case of proportional feedback, so that

$$\mathbf{u}(k) = \mathbf{D}\mathbf{e}(k) = \mathbf{D}[\mathbf{r}(k) - \mathbf{x}(k)] \qquad (8.3.17)$$

This expression can be substituted into Eq. 8.3.15 to get

$$\mathbf{x}(k+1) = (\mathbf{P} - \mathbf{Q}\mathbf{D})\mathbf{x}(k) + \mathbf{Q}\mathbf{D}\mathbf{r}(k) = \mathbf{P}^0\mathbf{x}(k) + \mathbf{Q}^0\mathbf{r}(k) \qquad (8.3.18)$$

which is the same form as the original model; however, by selecting $\mathbf{P}^0$ to be a diagonal matrix with specified elements on the diagonal, the gain matrix can be computed, that is,

$$\mathbf{D} = \mathbf{Q}^{-1}(\mathbf{P} - \mathbf{P}^0) \qquad (8.3.19)$$

which is the required result. For this particular example, select

$$\mathbf{P}^0 = \begin{bmatrix} 0.8 & 0 \\ 0 & 0.8 \end{bmatrix} \qquad (8.3.20)$$

Then, finding the gain matrix amounts to solving

$$\mathbf{D} = \begin{bmatrix} q_{11} & q_{12} \\ q_{21} & q_{22} \end{bmatrix}^{-1} \begin{bmatrix} (p_{11} - p_{11}^0) & p_{12} \\ p_{21} & (p_{22} - p_{22}^0) \end{bmatrix} \qquad (8.3.21)$$

With the given numerical values we obtain

$$\mathbf{D} = \begin{bmatrix} 0.826 & -1.768 \\ 0 & 1.326 \end{bmatrix} \qquad (8.3.22)$$

From this, the manipulated variables are

$$\mathbf{u}(k) = \begin{Bmatrix} u_1(k) \\ u_2(k) \end{Bmatrix} = \begin{bmatrix} 0.826 & -1.768 \\ 0 & 1.326 \end{bmatrix} \begin{Bmatrix} e_1(k) \\ e_2(k) \end{Bmatrix} \qquad (8.3.23)$$

or in scalar form, the manipulations are

$$u_1(k) = 0.826 e_1(k) - 1.768 e_2(k) \qquad (8.3.24)$$

$$u_2(k) = 1.326 e_2(k) \qquad (8.3.25)$$

As a result of using this algorithm, set points for moisture can be changed without changing the basis weight. This problem is discussed in more detail in [DePW77; DeVW78].

8.4 SUMMARY

Computer control of manufacturing processes requires some knowledge of control theory as well as computer hardware and software. This chapter has introduced computer control systems by first describing typical control system configurations and their implementation. Next, an introduction to the digital control of SISO systems was given, and concepts of stability, steady-state response, and transient response were introduced as they relate to digital controller design. A dc-motor speed-control example was described in detail. Finally, the chapter concluded with an example to illustrate digital control of a MIMO system and to introduce modal transformation methods and decoupling.

8.5 PROBLEMS

1. Refer to Problem 7.1, and consider the control of the kiln temperature.

    a. For proportional control, $V(k) = k_p e(k) = k_p[R(k) - T(k)]$ and $R(k)$ is the reference temperature at time $t = k\Delta t$. Select a value of k_p such that for a step-reference input $R(k)$, the steady-state value of $T(k)$ is within 10% of $R(k)$.
    b. Repeat part (a) using a PI algorithm with controller gains selected to ensure stability and zero steady-state error for step-reference inputs $R(k)$. Can this PI controller also have a faster transient response than the P controller?

2. Refer to Problem 7.2; temperature control of a mold cavity.

    a. Write a FORTRAN routine to implement PI control of this process. Assume that your FORTRAN routine will be called by a Z80 assembler service routine:

    ```
    SERVICE:    DI              ; disable interrupts
                CALL PICONT     ; call PI controller subroutine
                CLKEOI          ; resume clock interrupts
    ```

    Further assume that FORTRAN callable routines ADCONV(AD-VAL) and DACONV(DAVAL) are available and return 16-bit INTEGER arguments.
    b. Also develop P, PD, and PID controller subroutines similar to the PI routine in part (a).
    c. Simulate and compare the closed-loop system performance with P, PD, PI, and PID controllers.

3. Refer to Problem 7.3 for tank liquid level control. Design a PI controller for this system to be implemented on a microcomputer. Select controller gains such that the closed-loop characteristic equation has poles at $z_1 = z_2 = 0$ (deadbeat control).

4. In Example 8.1, obtain values of k_p and k_i when the maximum percent overshoot $p_0 = 0$ and the settling time $t_s = 1$ sec.

5. Describe how you might extend the dc-motor speed-control described in Example 8.1 to both speed and position control of the motor. Draw a block diagram to illustrate the control system structure.

6. Consider the liquid level control of two coupled tanks using decoupling and pole assignment. Refer to Problem 7.7 and Figure 7.P7. Assume the following values are given: $A_1 = A_2 = 1\text{m}^2, R_1 = R_2 = 1 \text{ sec/m}^2$, and $R_{12} = 0.5 \text{ sec/m}^2$. The flows into the two tanks, Q_1 and Q_2, are the manipulated inputs with units of m^3/sec, and the goal is to maintain the liquid heights H_1 and H_2 at independent reference values r_1 and r_2. If each decoupled loop is to have a time constant of 1 sec, what should the controller gains be? What will the steady-state performance of the system be like for step changes in the reference inputs?

Part IV
CASE STUDIES

Chapter 9
SETTING CLEARANCES ON A ROLLING MILL: MODULAR PROGRAMMING

In the first of four case studies we consider the problem of elastic deformation in a rolling mill, which leads to a product thickness greater than the one desired. This is an interesting engineering problem, and as described in the first section below, we calculate the roll separating force for the operation, and the mill structural stiffness to account for the elastic deformation in the mill. From an implementation point of view, our objectives are rather modest, but very important. We consider the preparation of a high level language program to calculate the expected mill deformation so that we can adjust the mill setting appropriately. In terms of programming experience, this example is aimed at examining the engineering problem and writing code that is clear and breaks down the programming task, in the same way that a design problem is broken down before analysis.

9.1 DESCRIPTION OF THE ENGINEERING PROBLEM

A rolling mill is schematically illustrated in Figure 9.1. A strip of width w enters the rolls with an initial thickness h_0. Depending on the amount of thickness reduction, geometry of the rolls, and material properties, a roll separating force is produced which tends to push the rolls apart. This force can cause elastic deformation of the roll stand such that the actual stand clearance is greater than the desired thickness at the exit h_f. The engineering objective of the case study is to compute a clearance setting that takes into account the elastic deformation

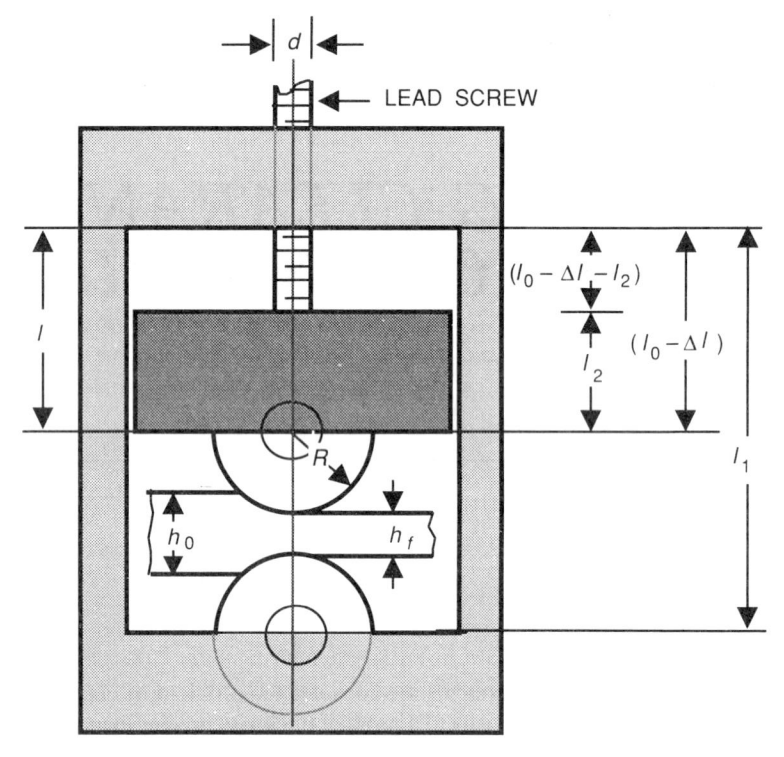

Figure 9.1 *Side view of a rolling mill with the rolls loaded*

caused by the roll separating force. Referring to Figure 9.1 we define the parameters,

h_0 = entry thickness
h_f = desired exit thickness
w = width of the sheet (assumed constant)
R = roll radius

We can then compute the following from the geometry:

$$\Delta h = (h_0 - h_f) \qquad = \text{thickness reduction} \qquad (9.1.1)$$

$$\bar{h} = \frac{h_0 + h_f}{2} \qquad = \text{average thickness} \qquad (9.1.2)$$

$$L = \sqrt{R^2 - (R - (\Delta h/2))^2} = \text{contact length} \qquad (9.1.3)$$

To estimate the roll separating force per unit width, we will use some results from plasticity theory [Mill79; HoCa82]. We will need to know the following material properties:

K = strain hardening coefficient
n = strain hardening exponent
cw = amount of prior cold work
μ = coefficient of friction between roll and work

The flow strains and stresses (see Fig. 9.2) can be computed as [HoCa82]:

$$\epsilon_1 = \ln(\tfrac{1}{1-cw}) \qquad \text{= flow strain at entry} \tag{9.1.4}$$

$$\sigma_1 = K\,\epsilon_1^n \qquad \text{= flow stress at entry} \tag{9.1.5}$$

$$\epsilon_2 = \epsilon_1 + \ln\tfrac{h_0}{h_f} \qquad \text{= flow strain at exit} \tag{9.1.6}$$

$$\sigma_2 = K\,\epsilon_2^n \qquad \text{= flow stress at exit} \tag{9.1.7}$$

$$\sigma_0 = \tfrac{\sigma_1+\sigma_2}{2} \qquad \text{= average flow stress} \tag{9.1.8}$$

From these expressions, assuming no back tension or roll deformation, the roll separating force per unit width can be calculated:

$$F_{s/w} = \bar{h}\sigma_0 \frac{(\exp[\mu L/\bar{h}] - 1)}{\mu} \tag{9.1.9}$$

Except for the roll radius, the previous relations are independent of the rolling mill configuration. To calculate the elastic deformation in the mill, consider the particular mill geometry in Figure 9.1. We want to determine l_0, the free length required to produce h_f. We make the following assumptions: (1) the frame, in which the bottom roll and lead screws for adjusting stand clearance

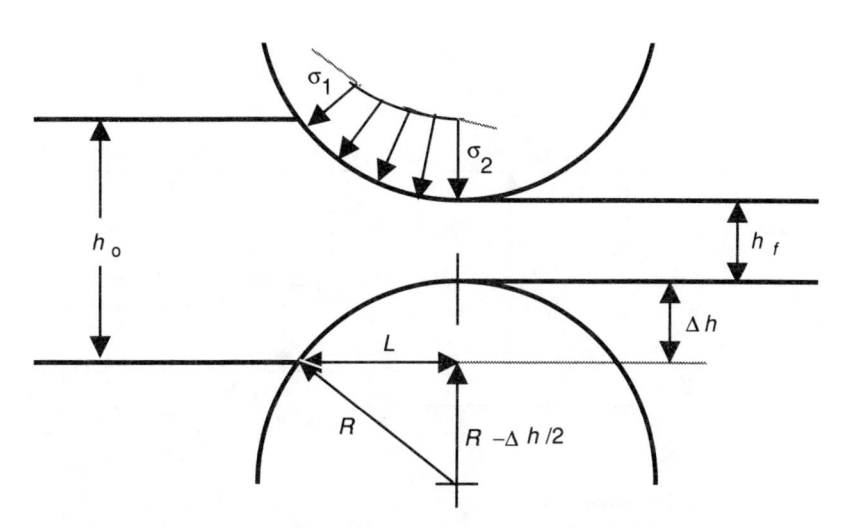

Figure 9.2 *Parameters to compute the roll separating force*

are mounted, is perfectly rigid, (2) the guide for the top rolls prevents any lateral motion, (3) all deflection is confined to vertical motion in the lead screws. The geometry of the rolling mill is described by:

l_1 = distance from top cross member to centerline of the bottom roll
d = pitch diameter of the lead screws
E = elastic modulus of lead screw material
l_2 = distance from center of top roll to lead screw

An idealized sketch of the rolling mill is shown in Figure 9.3, where the answer sought is the free length to be set on the lead screws without any loading (l_0). From geometry (see Figs. 9.1–9.3), the loaded length (l) is

$$l = (l_0 - \Delta l) = (l_1 - 2R - h_f) \qquad (9.1.10)$$

And the mill modulus (K_{mill}) is given by,

$$K_{mill} = \frac{2\pi E d^2}{4(l_0 - \Delta l - l_2)} \qquad (9.1.11)$$

Thus,

$$\Delta l = \frac{w F_{s/w}}{K_{mill}} \qquad (9.1.12)$$

is the mill deflection, and the required free length is,

$$l_0 = l + \Delta l \qquad (9.1.13)$$

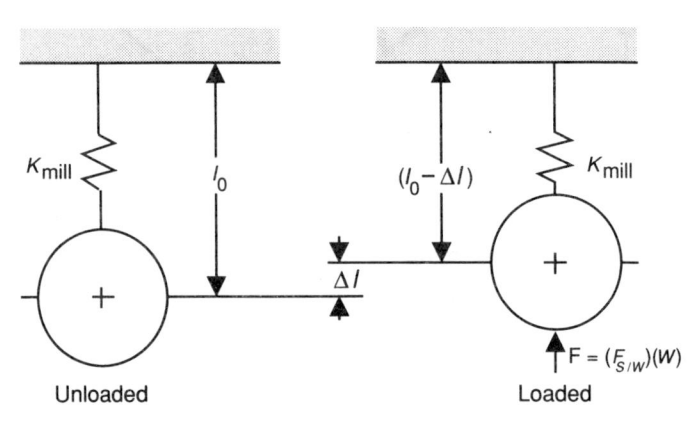

Figure 9.3 *Ideal model of the mill deflection showing the unloaded (or free length) position and loaded position*

We can use Eqs. 9.1.9, 9.1.11 and 9.1.12 to estimate the required free length l_0 for a particular mill given the geometric and material properties of the mill and the workpiece. A computer program that would prompt for the required input, perform the calculations, and display the results would be a very useful aid to the rolling mill operator. Such a program is described in the next section.

9.2 PROGRAMMING OBJECTIVES

Our goal is to develop a computer program that can aid the rolling mill operator to determine the appropriate free length setting on the mill. In the last section we developed the required engineering background and appropriate equations. Most often it is the engineer's responsibility not only to identify what has to be done, what engineering expertise is needed, what the reasonable simplifying assumptions are, but also to then direct or actually write and implement the required software. Many engineers are not trained in programming style, however it is a skill like writing or design synthesis that can be learned. Keep in mind that software design takes practice, judgment, and a realization that there is no right or wrong way, but that there are poor and good ways of writing software. The case study in this chapter is aimed at trying to make engineers aware of programming style. Refer to the programming guidelines presented in Table 2.2 as you prepare your program. The program should run on a microcomputer so that it can be readily available to the operator at the mill, it should be

- Interactive with checks of the input data to minimize mistakes
- Modular so that if other mills of similar design are encountered some (and preferably most) of the software can be reused
- Capable of repeating calculations with new sets of data

Example 9.1 Rolling Mill Simulation

Using the sample parameter values given in Table 9.1, we first determine the free length l_0 by hand calculation:

Δh = 0.1 $\bar{h}$ = 0.15
L = 0.7053368 σ_0 = 56190.995
$F_{s/w}$ = 44680.402 K_{mill} = 70108807.
Δl = 0.019119 l = 13.90

l_0 = 13.919119 is the desired free length to be set on the mill

Now we write a FORTRAN program to compute l_0 given material properties and work geometry. We assume that the rolling mill geometry is fixed, although

Table 9.1 Sample Values for Rolling Mill Setup Program

h_0	= 0.2 in.	h_f	= 0.1 in.
w	= 30 in.	R	= 5 in.
K	= 80,000 lbf/in^2	n	= 0.25
cw	= 0.05	μ	= 0.05 (steel on steel)
l_1	= 24 in.	l_2	= 9 in.
d	= 2.7 in.	E	= 30 × 10^6 lbf/in.^{2} (steel)

the rolls may be changed from time to time. Program listings are included in Appendixes A and B for 8088- and Z80-based systems, respectively. Both programs, although different, are reasonably well organized and documented. They are modular, so that tasks such as the input and modification of data are separated from the calculations. They are designed to be easily modified if our engineering assumptions are changed. The programs include modules to perform the following tasks:

- Get, check, and display the required input data
- Perform preliminary geometric calculations
- Calculate the strains, stresses, and roll separating force
- Calculate the mill modulus, and mill deflection
- Determine and display the required free-length setting for the mill

Describe how the programs given in the Appendixes could be improved, and develop a flowchart from which such programs could be written. Extend the program to include its own internal material and geometry data base so that user input can be reduced, and suggest other extensions to make the program more useful.

9.3 SUMMARY

This chapter, the first case study, considered the problem of deformation in a rolling mill due to the roll separating force. Such a deformation leads to a product thickness greater than the desired value. We developed equations to determine the roll-separating force and the mill modulus (or stiffness) so that the mill deformation could be estimated and compensated for by appropriately setting the mill free length before rolling. This case study also gave us the opportunity to familiarize ourselves with a particular microcomputer system by developing a modular and interactive program using a high level language. The program was designed to be used interactively by the rolling mill operator to aid in estimating the appropriate mill free length setting for a particular rolling mill.

9.4 PROBLEMS

1. The case study described the development of a computer code to set up a rolling mill. Now suppose that you wanted to make the processor actually run the mill with these functions to be performed:

    - The exit sheet thickness must be maintained at the value used in your set up routine
    - If no material is entering the rolls, a warning is printed and one second later the mill is shut down
    - The rotational speed of the rolls must be controlled, with the set points provided by another processor that is controlling a mill "down the line" (this would be for slowing down or speeding up a stand of rolls so they will not buckle or break the sheet)

    Approach this problem as if you were a consultant that would handle the whole job, and this is your first attempt at looking at the problem. With this in mind:

    a. Indicate to the best of your ability the hardware needed to do the job, and make a rough sketch of how it would be used.
    b. What type of computer interfaces (e.g., DACs, parallel or serial I/O) would have to be purchased for the processor?
    c. What software would have to be developed to do the job?

2. A new material is to be selected for manufacturing shafts, and machinability on a tool life basis is being considered. It is known that material properties affect machinability in the following manner:

$$B = \frac{k}{H_B} \sqrt{1 - \frac{A_r}{100}}$$

    where

    B = Work material constant directly proportional to the cutting velocity for a given tool life
    k = Thermal conductivity of the material at 600° F
    H_B = Brinell hardness at 600° F
    A_r = Area reduction at 600° F

    Flowchart, write, and test a high-level language (e.g., FORTRAN) program that can interactively help in material selection for machinability.

Chapter 10
MONITORING A TURNING PROCESS: REAL-TIME DATA ACQUISITION

The on-line monitoring of manufacturing processes is receiving increasing attention owing to the industrial goal of untended manufacturing. The data collected on-line by a computer can lead to a better understanding of the particular process, and can provide useful management and diagnostic information. Also, the first step in most control applications is instrumenting the system to collect data, since this step is necessary and is unobtrusive. In this case study we consider the problem of using a microcomputer system for monitoring a turning process.

First we outline the relationships in turning between measurable variables, such as force and spindle speed, and quantities of engineering interest such as metal removal rate and horsepower consumption. In this case study, we also introduce real-time data acquisition using a microcomputer. In particular, we use digital input of the spindle speed and analog-to-digital conversion to sample the cutting force. We also apply assembly language as well as high-level language programming.

10.1 DESCRIPTION OF THE ENGINEERING PROBLEM

The engineering objective of the case study is to compute, from on-line measurements of spindle speed and cutting force, the metal removal rate (Q), the horsepower at the cutting edge (*HP*), and the unit horsepower (hp_u) during turning.

The unit horsepower (hp_u), or specific cutting energy, can be used to characterize how well a particular material machines [AmOB84; Kalp84]. This property is usually determined experimentally, and is defined as

$$hp_u = \frac{HP}{Q} \qquad (10.1.1)$$

where *HP* is the total horsepower at the cutting edge and *Q* is the instantaneous metal removal rate in (in.^{3}/min). For a turning operation:

$$Q = fN \pi \frac{(D^2 - (D - 2d)^2)}{4}$$

$$= fN \pi Dd\left(1 - \frac{d}{D}\right)$$

and for $d \ll D$:

$$Q \approx 12fvd \qquad (10.1.2)$$

where

$v = N\pi D/12$ is average cutting speed (ft/min)
D = outside diameter of the workpiece (in.)
N = average rotational speed (rev/min) of the workpiece
f = feed (in./rev)
d = depth of cut (in.)

The horsepower at the cutting edge is approximately equal to the product of the cutting speed and the cutting force measured in the same direction (it is approximate in that the power in the feeding direction is neglected because the velocity in that direction is very small):

$$HP = \frac{F_c v}{33000} \qquad (10.1.3)$$

where

F_c = average cutting force (lbf) at the outside diameter of the rotating workpiece

All the quantities required (hp_u, *HP*, and *Q*) can be calculated given *D*, *f*, *d*, *N*, and F_c. (Note that 33000 in Eq. 10.1.3 is the unit conversion factor from ft-lbf/min to *HP*, and the 12 in Eq. 10.1.2 is the unit conversion factor from feet to inches.)

The experimental system, as shown in Figure 10.1, uses a lathe that is instrumented with a pulse-generating device for rotational speed measurement and a force dynamometer for cutting-force measurements. The relationships that have been developed are simply definitions of quantities that can be measured with this experimental system using data-acquisition techniques. The next section is aimed at developing the software to handle this hardware so that these machining properties can be easily calculated. However, this setup is only one possible way to handle the problem. There are, in fact, some design decisions that would have to be made when setting up such a system that can be outlined briefly before going to the software considerations.

As was pointed out in Chapter 5, a tachometer that puts out an analog voltage proportional to rotational speed is an alternative to the magnetic pickup used in this case study. The tachometer might be a better alternative if timing in a control loop is critical because, as will be seen later, the counting and timing of pulses will be an added computational task for the microcomputer. The counting task could be reduced or eliminated if a special counter timer device were added to the data acquisition system and the microcomputer only had to read this device when it needed the pulse rate; the number of gear teeth generating the pulse train would still need to be known. The magnetic pickup was actually chosen because of its availability, simplicity, and relatively low cost. Also, it is a fairly common way to measure speed on rotating machines and may already be installed on a machine tool used for data acquisition.

Measuring the force, both how and why only one component, is another design and engineering question. The second part, why only one component, is the easiest to start with and may be obvious to those familiar with machining. Power is a scalar quantity, so it is possible to add the product of force and velocity in the tangential (cutting) and axial (feeding) directions to come up with the horsepower HP in Eq. 10.1.3. But even though the two force components

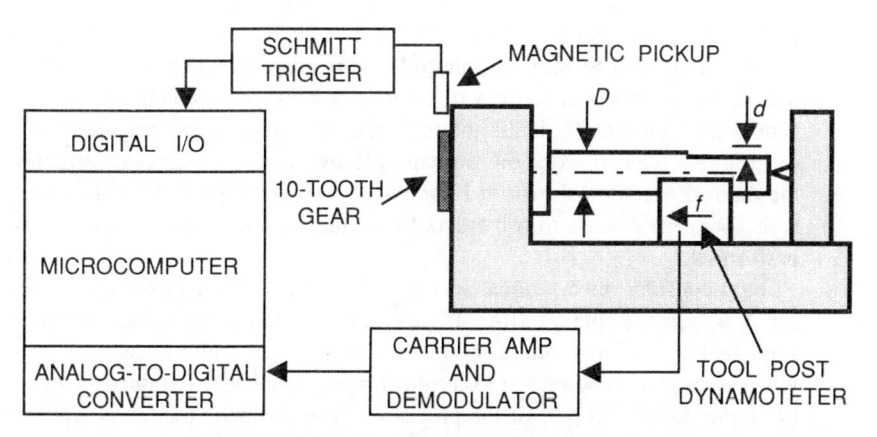

Figure 10.1 Schematic of data acquisition setup for turning

are on the same order of magnitude, the velocity in the feeding direction is about two orders of magnitude less than in the cutting direction. This means that the feeding force contribution to the total power is of the order of 1%; about the same order of magnitude as errors in the measurement. So common practice often uses only the tangential force and velocity when estimating hp_u.

What principle to use to measure force will depend on both cost and the frequency response needed to calculate a value for F_c in Eq. 10.1.3. For this case study, the force measurements will be averaged over several revolutions, so a high bandwidth of the signal is not a high priority. If this were not the case and techniques like spectrum analysis, discussed in Chapter 6, were to be used, the frequency response of the dynamometer would be important. Piezoelectric devices are usually preferred in such situations because of their stiffness and frequency response, however, at the expense of reliable dc measurements. In the setup used in this case study, a tool post dynamometer with strain gages as sensing elements was used to measure two force components, even though only the cutting force was used in the calculations.

10.2 PROGRAMMING OBJECTIVES

We will write a computer program that, given D, f, and d, measures N and F_c and calculates Q, HP, and hp_u. In writing the code for this experiment you will program at two levels. FORTRAN conveniently handles the entry and display of information and the floating-point computations. Assembly language routines, which can be called by the FORTRAN program, are faster and more efficient for the data acquisition functions. The program may be developed and tested on a development system, then down-loaded to a smaller and more rugged target system that can be located near the lathe on the shop floor. The details of this process are described in Appendixes A and B for two particular microcomputer systems.

With your experience from the previous case study, the preparation of the FORTRAN code should not be difficult, but this is the first time that real-time programming will be part of your code. This makes modular code important so that you can pin down sources of errors when you are debugging; in fact, you should plan on how to debug the real-time aspects. For the calculations, Table 2.2 reminds you to work out an example by hand. A suggested structure for the main program is shown in Figure 10.2, which helps with modularizing the code. We will discuss in some detail how the data-acquisition functions can be performed.

First consider the rotational speed measurement. The hardware for determining N is fairly simple. A 10-tooth gear is connected to the spindle of the lathe, a magnetic proximity probe produces a pulse every time a gear tooth passes by, and an electronic circuit (Schmitt trigger) conditions the pulses before they enter the digital I/O port of the processor. The magnetic pickup is a permanent

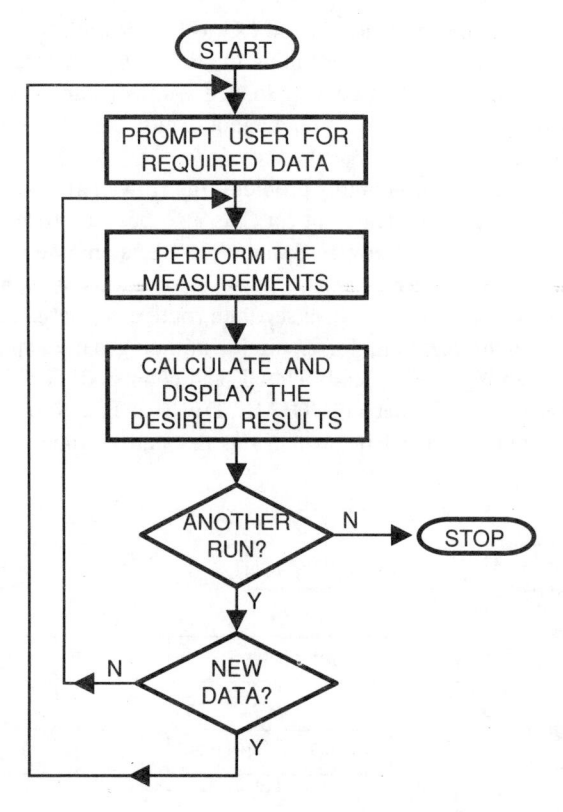

Figure 10.2 Structure of the main program for data acquisition on a lathe

magnet that has a small coil of wire wrapped around one of its poles. When the gear tooth passes near the magnet, the flux density changes, which produces a current in the wire. This current, when passed through a resistor produces a voltage. The Schmitt trigger is designed so that if this voltage exceeds a certain threshold, a 5-V signal is produced; otherwise, the signal from the Schmitt trigger is 0 V. Thus, a pulse train is produced whose frequency is proportional to the rotational speed of the spindle,

$$N = \frac{60 N_p}{N_t \Delta t} \qquad (10.2.1)$$

where N is the average rotational speed (rev/min), N_p is the number of pulses in an interval of time Δt (sec), and N_t is the number of teeth on the gear. This is essentially a digital tachometer.

One way to determine N, given N_t, is to determine the amount of time it takes for a fixed number of pulses to occur. This can be done by synchronizing on either a leading (transition from LO to HI) or trailing (transition from HI to LO) edge of a pulse and determining the time when this transition occurs. After reading the time, a down counter loaded with N_p can be started, and, when it reaches zero, the time can be read again. Thus, for a constant specified N_p we determine Δt.

After "talking" through this problem, the flowcharts for doing this, shown in Figure 10.3, become clear, but this has also been a process of breaking down and modularizing the software design. TRAIL is an assembly-language routine that detects the trailing edge of a pulse; it could just as well have detected the leading edge and is written as a separate routine so that it can be used for other types of synchronization, for example, initiating data acquisition. DWNCNT, also an assembly routine, calls TRAIL and counts NPULSE pulses. PULRAT is a FORTRAN routine that calls TRAIL, DWNCNT, and two additional assembly language routines: TIMEIN, which sets the starting time, and TIMOUT, which

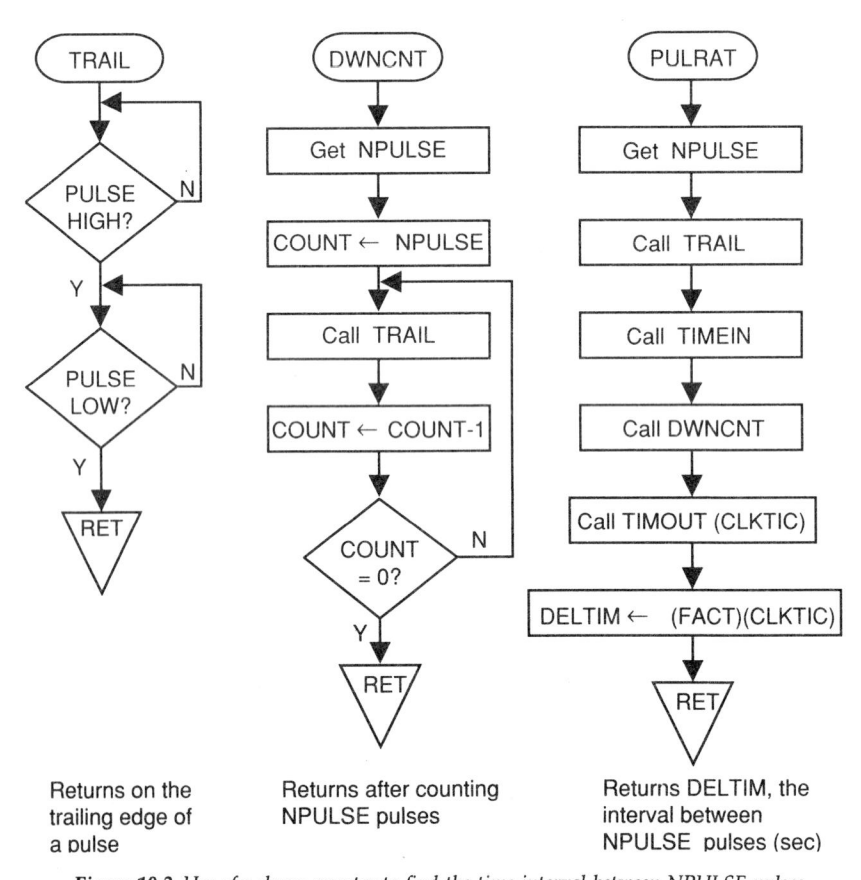

Figure 10.3 *Use of a down counter to find the time interval between NPULSE pulses*

returns the elapsed time so that the time interval Δt can be calculated. Recall and review the related program listings discussed in Chapter 5.

Consider the precision of this setup for measuring the rotational speed. To analyze this problem, Eq. 10.2.1 points out that the accuracy of this method of measurement depends on the number of teeth, the number of pulses counted, and the least count or resolution of the timer. Figure 10.1 shows that for this system $N_t = 10$ and, as this software is written N_p, is a parameter decided on initially. Synchronizing on the trailing edge means there should be no uncertainty in the value of N_p. The time interval, Δt, is really the variable quantity with its precision determined by the clock resolution; assume it is fixed and is 0.010 sec. Then, if $N_p = 60$ pulses (i.e., the time for 6 complete revolutions), the system will return a Δt accurate to the lowest one-hundredth of a second. Now suppose that the value of Δt returned was 0.10, which could correspond to the interval [0.10, 0.11], taking into account the least count of the clock. Using the values for N_p and N_t already given and this time interval, the range of values for N in Eq. 10.2.1 is [3600, 3273] rev/min. If, on the other hand, the value of Δt returned was 1.00 so that the appropriate interval is [1.00, 1.01], then using the same approach, the range of values for N is [360, 356]. For this approach, the accuracy improves at lower speeds or if the value of N_p increases; in either case, the time to make a measurement will be a variable and take longer as the accuracy increases.

Looking at this algorithm critically, there are two undesirable aspects of this particular approach for finding the rotational speed. If the speed is so high that it takes less than the least count of the timer, in this setup 0.01 sec, the value computed will be $\Delta t = 0.00$ sec, leading to a division by zero in Eq. 10.2.1. If, on the other hand, the speed is very low, that is, the lathe spindle stops rotating after the routine DWNCNT has been entered, then there will essentially never be a return from DWNCNT. This discussion corresponds to the comments on programming style in Table 2.2 regarding checking the program by hand at its boundaries; in this case at the high and low speed limits.

What are the possible solutions to these potential problems? We could, of course, use software checks with the existing algorithm. If the counting is to be done in software, interrupts generated by the spindle rotation pulse are also a solution. But up to this point interrupts have not been used. This would correspond to making the time interval a specified parameter and N_p a variable. A key point to remember is that in real-time programming, errors can cause serious problems; program "crashes" may be physical crashes.

A tool post dynamometer is used to measure the cutting force (F_c). This dynamometer uses strain gages, excited by a carrier signal and then demodulated, to produce an analog dc voltage proportional to the cutting force. This analog signal will be connected to an analog-to-digital converter (ADC). Because there is some fluctuation in force level while cutting, it is advisable to take an average of the measured values during the period of the cut. Then the average force would be

$$\bar{F}_c = \frac{K}{M} \sum_{i=1}^{M} x_i \qquad (10.2.2)$$

where

K = gain to convert from ADC values to lbf
x_i = the 12-bit ADC values
M = number of samples used for the average

The gain, K, should be entered from the console before starting the test and includes the force dynamometer and the ADC gains. The force dynamometer must first be calibrated.

A FORTRAN routine, GETDAT, is flowcharted in Figure 10.4, which combines the function of the PULRAT routine in Figure 10.3 with the cutting-

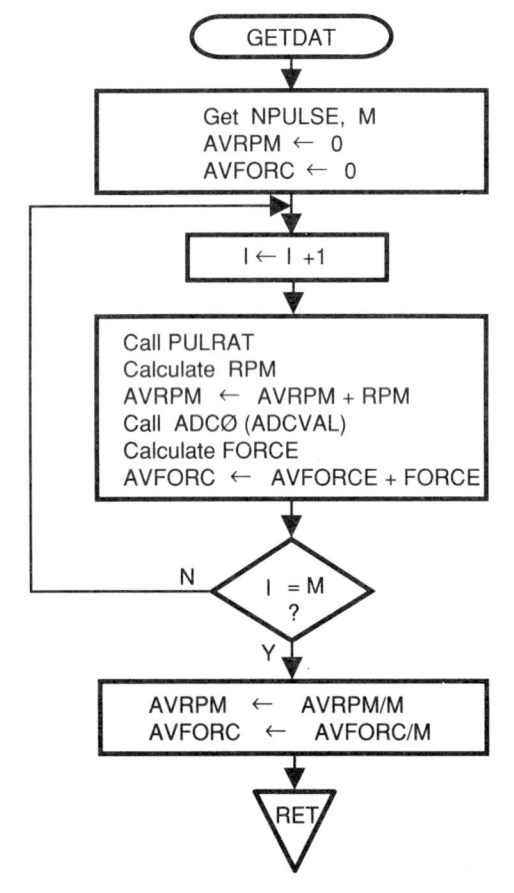

Figure 10.4 *Average rotational speed and force calculations*

force measurement. Note that the DWNCNT routine is also used here to initiate an ADC. Thus, a cutting-force measurement is made every NPULSE pulses and M cutting-force and rotational-speed measurements are averaged. In your program, NPULSE can be fixed and M should be specified by the user at the start of a run.

Example 10.1 Data Acquisition on a Lathe

The program listing in Appendixes A and B demonstrate how the foregoing procedure can be implemented on a particular microcomputer system. In studying these listings also refer to the discussions in the Appendixes on programming the ADCs, on linking FORTRAN and assembly-language routines, and on other system-specific features. Note that these programs include modules to perform the following tasks:

- Initialize variables; get, check, and display the required data
- Count NPULSE pulses using the parallel I/O port
- Calculate the time elapsed in counting NPULSE pulses using the clock-interrupt routines
- Use the ADC to sample the cutting force
- Repeat the foregoing calculations M times to get average spindle speed and force values
- Calculate Q, HP, and hp_u; display the results

The following parameter values were used in the implementation:

$$D = 1.0 \text{ in.}$$
$$f = 0.1 \text{ in./rev}$$
$$d = 0.01 \text{ in}$$
$$N_t = 10$$
$$N_p = 20$$
$$M = 10$$

For a $\pm$ 10 V, 12-bit ADC, with a dynamometer gain = 1 lbf/V:

$$K = (1.0)\frac{10 - (-10)}{2^{12}} = \frac{1}{204.8}$$

or for a $\pm$ 5 V, 8-bit ADC, with a dynamometer gain = 1 lbf/V:

$$K = (1.0)\frac{5 - (-5)}{2^8} = \frac{1}{25.6}$$

Discuss how this data-acquisition system for turning might be improved and extended, and how it might be used for machine and process diagnostics.

10.3 SUMMARY

This case study considers data acquisition on a lathe and builds on the programming experience from the previous case study by introducing assembly-language programming, digital input, and analog-to-digital conversion.

The engineering problem, monitoring of a turning process, is an important one owing to rapidly increasing automation in manufacturing. Here we consider force measurement and signal processing (averaging) together with pulse counting to obtain the spindle speed. These values are then used to compute metal removal rate and the specific cutting energy.

10.4 PROBLEMS

1. Suppose that you wanted to take on the job of "retrofitting" the lathe in this case study to make it a computer numerically controlled (CNC) lathe that could be controlled by a small computer. To do this, you must be able to control:

    - The lateral motion of the cutting edge (feed)
    - The in and out motion of the cutting edge (depth)
    - The rotational speed of the workpiece (spindle speed)

    In addition, there must be provisions for avoiding "crashes" that could occur at random times—when the cutting edge comes too close to either end of the machine. Given this problem statement:

    a. What type of hardware would be required for the lathe, either for measurement or control? (If a sketch helps to describe this, make one.)
    b. What type of hardware would be required for the microcomputer to interface with the lathe and the rest of the outside "world"? (This might include some simple logic gates to take some of the burden off the microcomputer.)
    c. What type of software would have to be written to meet the control objective?

2. Debugging a real-time programming application many times is more difficult than a simulation or a computational algorithm, in part because

physical devices and interfaces are involved. With that in mind, make a list of the equipment you would need to test the spindle-speed measurement algorithm flowcharted in Figure 10.3 and make a schematic of how you would set it up and what you would test.

3. Assuming that you copied the FORTRAN code in either Appendix A or B to do you own setup, what if the plant or lab where you worked:

   a. Only used SI units, rather than English, what changes would you have to make in this code?
   b. Might have to provide results in either SI or English units to make customers happy, then how would you change the code?
   c. Required a "fancy" graphical display of the results on a printer or graphics terminal rather than a display of numerical values. What change in the program would be needed?

Chapter 11
CONTROL OF A STEPPING-MOTOR DRIVEN x-y TABLE: INTERRUPTS AND DIGITAL DIFFERENTIAL ANALYZERS

Among the sensing and actuation devices used in many computer-based systems are digital devices such as optical encoders and stepping motors. Stepping motors are the most widely used of the digital actuating devices. The input to a stepping motor is a train of digital pulses, as can be directly provided by a digital input/output port on a computer. In this chapter we discuss the use of digital computers and stepping motors to produce a desired mechanical motion [Boll72; SeBo71]. This material is fundamental to the use of digital computers in many manufacturing operations (e.g., CNC machine tools, parts handling, inspection). In fact, commercially available stepping-motor driver chips can and would normally be purchased to perform the motion-generation tasks described in this case study.

Figure 11.1a shows how stepping motors are used in an *open-loop digital-control* system. DC motors are used in *closed-loop digital control*, as shown in Figure 11.1b and discussed in Chapter 12. Stepping motors are incremental actuation devices used in many open-loop digital position control systems. Figure 5.3 is a more detailed schematic of the system shown in Figure 11.1a. The digital differential analyzer (DDA) can be used to control linear motion of an axis driven by a stepping motor through a lead screw [SeBo71]. This case study will cover how to implement these algorithms on a microprocessor.

194 Chapter 11 CONTROL OF A STEPPING-MOTOR DRIVEN x-y TABLE

The engineering objective of this case study is to develop the computer code necessary to move a stepping-motor driven x-y table such that:

- A single axis can be moved with a velocity profile that will accelerate to a maximum velocity and then decelerate
- A two-axis constant velocity move can be made along a line and optionally along one quadrant of a circular arc

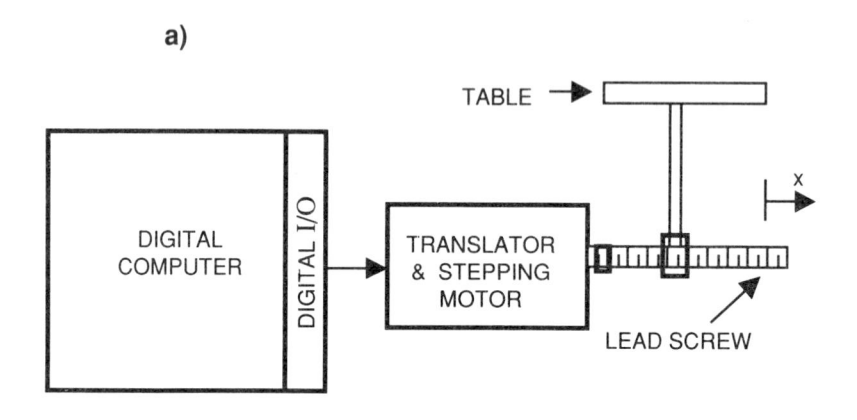

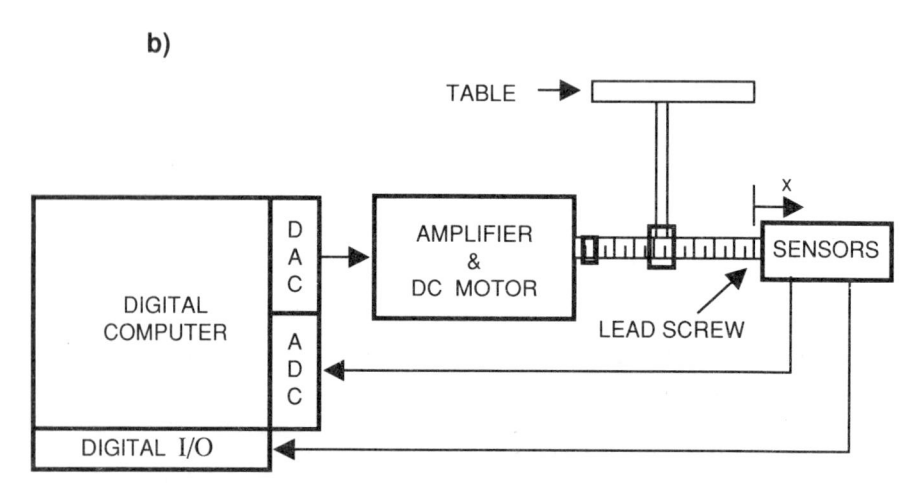

Figure 11.1 Digital control of motion: (a) Open-loop control system with stepping motor, (b) Closed-loop control system with dc motor

11.1 DESCRIPTION OF THE ENGINEERING PROBLEM   195

The programming objectives are to

- Demonstrate the use of macros in assembly language programming
- Implement a reasonably complex assembly language program that uses interrupts

11.1 DESCRIPTION OF THE ENGINEERING PROBLEM

In Chapter 5, the operation of stepping motors was discussed. We will use the microcomputer to generate a pulse train that will be sent to the stepping-motor driven axis of an x-y table. The number of pulses sent to each axis determines the displacement of that axis, and its velocity depends on the pulse frequency. The generation of motion along a path in the x-y plane requires the simultaneous coordinated motion of both axes. To accomplish this task, we first introduce the digital differential analyzer (DDA) algorithm for generation of pulses, then we consider a single axis move (first at constant velocity, then with nonconstant velocity profiles), and finally we discuss multiaxis continuous-path control (with linear and circular interpolation).

The Digital Differential Analyzer Algorithm

We now consider how to generate in the computer the pulse trains required to produce the desired motion. One obvious approach is to perform the calculations indicated by Eqs. 5.1.2 through 5.1.4 in Chapter 5, considering the limitations of the motor drive such as the slew rate, then using a timing device (such as a clock or delay loops) to determine when a pulse should be sent. The digital differential analyzer (DDA) concept provides a method for performing these calculations efficiently on a small digital processor [Size68]. A hardware DDA is a digital computer very similar in capability to an analog computer. The integration function (implemented with op-amp circuits in an analog computer) is approximately performed by summations using binary words. The hardware implementation uses flip-flops, as described in Chapter 3, for counters and registers. Such hardware DDA devices are commonly used for the interpolators found in NC machine tools. We will discuss here how to implement the DDA algorithm in software on a microcomputer.

The basic idea behind the DDA algorithm is to approximate integrations by summations. For example, displacement, denoted here by $d(t)$, is the time integral of the velocity, $v(t)$:

$$d(t) = \int_0^t v(\tau)d\tau \qquad (11.1.1)$$

so that integrating a specified time history of velocities allows the displacement to be generated. As shown in Figure 11.2, Eq. 11.1.1 can be approximated by:

$$d_k = \sum_{i=1}^{k} v_i \Delta t = \sum_{i=1}^{k-1} v_i \Delta t + v_k \Delta t \qquad (11.1.2)$$

Since the second summation in Eq. 11.1.2 is the same form as the first, except for its upper limit, this expression can be written as a recursive algorithm of the form:

$$d_k = d_{k-1} + v_k \Delta t \qquad (11.1.3)$$

The binary words d_k and v_k can be displacement and velocity, respectively, for a motion control system such as we are considering here. From basic kinematics, it is clear that velocity could be specified as the integral of acceleration, so a recursion could be developed there too. This is also true for jerk, the third derivative of displacement. The basis for using digital differential analyzers is that differential equations that govern the kinematics of the motion can be solved by numerical integration. Numerical integration is basically addition, something microcomputers do extremely well. This idea holds for dc drives as well as for the stepping motors that will be the focus of this case study.

As was discussed in Chapter 5, the main ways to control stepping motors are by the number and rate at which pulses are sent to the motor. Counting is easy. Generating and coordinating variable-frequency pulse trains to a single stepping motor, or to several stepping motors so that a desired displacement versus time profile can be obtained, is a little more complicated. However, since microcomputers use clocks to coordinate their execution, all that has to

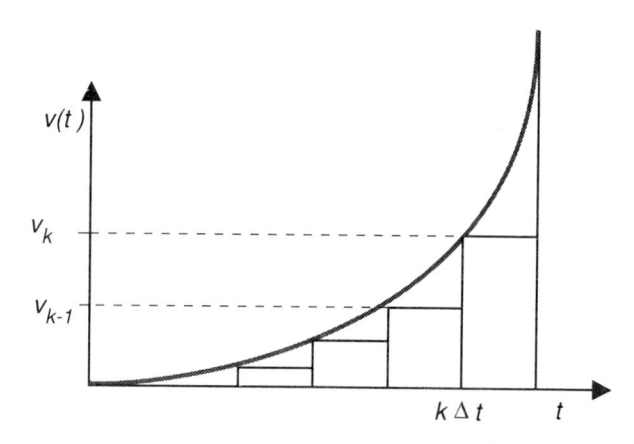

Figure 11.2 Approximating integration by summation in the digital differential analyzer (DDA)

11.1 DESCRIPTION OF THE ENGINEERING PROBLEM

be done is to develop an algorithm to generate variable frequencies. In doing this, two characteristics of the processor are used to advantage:

1. Fixed word length
2. Clock interrupts

If integer arithmetic is used for the additions in Eq. 11.1.3, the fixed word size means that a word can only represent 2^N values, where N is 8, 16, or 32. When an addition is performed that results in a number greater than 2^N, an overflow flag is set and the resulting sum is truncated. This overflow condition can be used to signal when a pulse should be sent out from the processor. If the simple arithmetic in Eq 11.1.3 is done as part of a clock interrupt service routine at fixed steps Δt, the value added to d_{k-1} is chosen to cause overflows at the rate needed to generate pulses for the stepping motors. This is the basic DDA algorithm; it will be further developed for different velocity profiles and for two axes.

Controlling One Axis

We now discuss how the DDA algorithm can be used on a microprocessor to control the motion of a single axis and for the simultaneous control of two axes of motion. In the following examples, it is assumed that the number of steps to be sent to the motor, N, and the frequencies of the pulse train, f, are computed from Eqs. 5.1.3 and 5.1.4.

Constant Velocity Profile

The simplest case of motion generation is illustrated by the profile shown in Figure 11.3. The motor has to move at constant velocity in segments for a distance equivalent to the area under the profile. To generate the desired velocity and displacement profile, the frequency of the pulse train must be proportional to the velocity and the number of pulses has to be equal to the distance moved.

If the DDA algorithm is to work, the highest pulse rate will occur when an overflow occurs on every call to the service routine. This means the clock frequency, f_c, must be greater than the frequency to be generated (usually four to five times the largest pulse-train frequency). The number stored in the velocity register VR is added to the displacement register DR every time there is a clock pulse, and if there is an overflow from DR, a pulse is sent to the motor. The rate at which there are overflows from DR depends on the capacity of DR, the value of VR, and the clock frequency, f_c. For a fixed clock frequency, the output rate may be changed by altering the value of VR by simple ratio and proportion. The following expression governs the behavior of the DDA shown in Figure 11.3b:

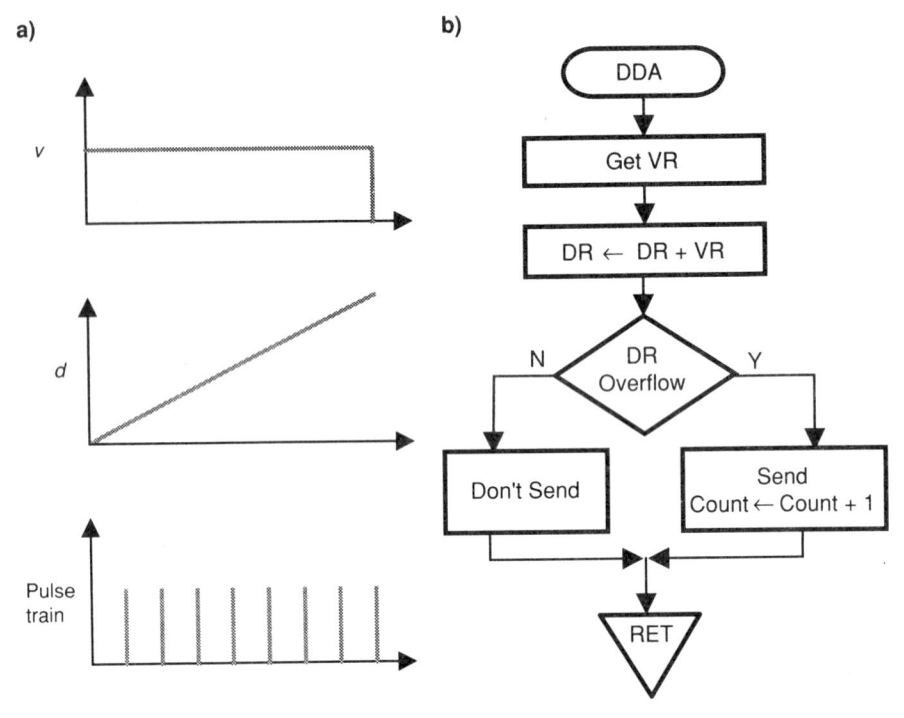

Figure 11.3 Constant velocity profile and the DDA subroutine to generate it: a) pulse train, b) DDA flowchart

$$VR = \frac{\text{(Capacity of DR)(Desired Output Rate)}}{\text{(Clock frequency)}} \qquad (11.1.4)$$

Flowcharts using this algorithm are shown in Figure 11.4. When there is an overflow, the program sends a pulse to the motor. Then the total number of pulses sent to the motor is checked to see if the move is completed. Because DR and VR are integer values, there will usually be an error in generating the desired pulse rate, but the error can be reduced with a higher clock frequency or a larger word size for the DR register (e.g., using 32, rather than 16, bit words). This is illustrated in Example 11.1.

Example 11.1 DDA Algorithm for Constant Velocity

Total Distance $(N) = 1000$ steps
Clock frequency $(f_c) = 200$ Hz
Word size $= 16$ bits
Velocity $= 51$ steps/sec
$DR = 2^{16} = 32,768$

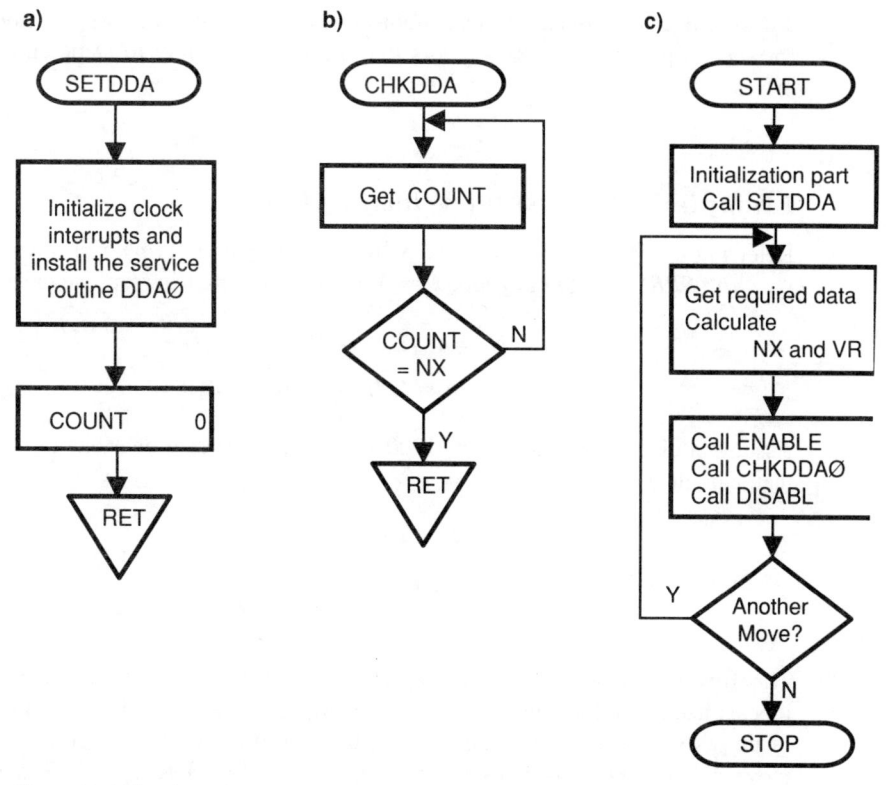

Figure 11.4 Flowcharts for constant velocity DDA (a) Setup routine, (b) check for end of move, (c) driver program

$$VR = Round\frac{(32,768)(51)}{(200)} = 8356$$

The value of VR is rounded to the nearest integer, $VR = 8356$. Therefore, every 1/200 sec, 8356 will be added to the DR register. When it overflows, a pulse will be sent to the motor. Because of the integer arithmetic and the rounding in calculation of VR, there is a velocity error. The specified velocity was 51 steps/sec. By back substitution, the actual velocity is

$$\text{Actual velocity} = \frac{(200)(8356)}{(32,768)} = 51.001 \text{ steps/sec}$$

The velocity is in error by less than 0.002%, which is quite small. The accuracy in velocity depends on the desired velocity and the clock frequency. For any noninteger determined for VR, the accuracy increases for greater clock frequencies and word size; however, the computer time required increases. Thus, there is often a trade-off between accuracy and computer time, which is really a design decision. Because these algorithms are very simple, they are suitable for special-purpose microprocessors that are dedicated only to stepping-motor drives.

<<>>

Example 11.2 Constant Velocity Move

Referring to Figure 11.3, and given SPR = 200 steps/rev, $(1/p)$ = 1 thread/in., $\Delta x = p/(SPR)$ = 1/200 in./step, x = 0.05 in., and v = 1 in./sec, we calculate:

$$N = (x/\Delta x) = 10 \text{ steps}$$
$$f = (v/\Delta x) = 200v = 200 \text{ steps/sec}$$

Assume a clock frequency f_c = 1000 Hz and a word size of 3 bits ($2^3 = 8$), then calculate

$$DR = (2^3 - 1) = 7$$
$$VR = \frac{(DR)(f)}{f_c} = Round(7/5) = 1$$

The first nine clock cycles are shown in Table 11.1. Notice that the cycle repeats itself and that a pulse is sent every eighth clock pulse. The total number of pulses sent is 10 in 73 clock pulses; thus the time required to complete the move is (73/1000) = 0.073 sec. The initial value of the DR register (7 in Table 11.1) determines when the pulses start being sent; in this example, after the first clock pulse. The actual velocity is

$$f = \frac{(1000)(1)}{7} = 136.99 \text{ steps/sec}$$

Table 11.1 Nine Clock Cycles for the Constant Velocity DDA

Clock Pulse	DR	Output Pulse
0	7	
1	0	1
2	1	
3	2	
4	3	
5	4	
6	5	
7	6	
8	7	
9	0	1

and the velocity error is

$$\left(\frac{136.99 - 200}{200}\right)(100) = -31.5\%$$

so in this case, the small word size, 3 bits, leads to very large velocity errors.

<<>>

Constant Acceleration–Constant Velocity Profile

When the load inertia is high or the pulse rate approaches the slewing rate of the motor, the magnitude of the step change in velocity is limited. In order to handle large loads or to generate smooth motion, the motor may be made to accelerate up to speed and decelerate to a halt as shown in the velocity profile of Figure 11.5a.

The desired velocity increases uniformly from zero to the maximum, remains constant, and then decreases uniformly. The DDA setup for achieving such a

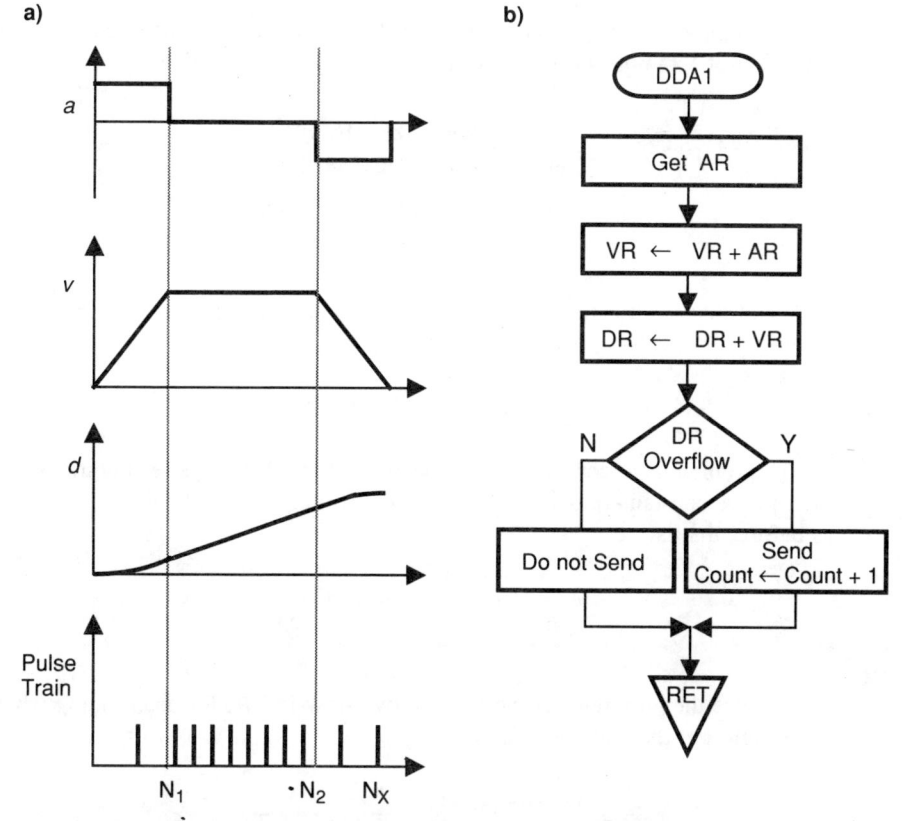

Figure 11.5 Constant acceleration profile and DDA, subroutine to generate it: (a) pulse train, (b) DDA flowchart

profile is given in Figure 11.5b. For every clock pulse, AR is added to VR and VR is added to DR. The number of additions to cause an overflow from the DR register is the number of clock pulses before a pulse is sent to the motor. VR starts with value zero and increases to a final value dependent on $N1$; the rate at which DR overflows increase at a uniform rate. At the end of the acceleration segment (detected by the number of overflows corresponding to the distance $N1$), the acceleration register AR is made zero and then the situation is exactly the same as for the constant velocity move. At the end of the constant velocity segment (again detected by the number of overflows $N2$), AR is loaded with the negative of the value used in the acceleration segment and addition to the VR register continues. The move is completed when the number of overflows of DR is equal to NX. The capacity of a binary register is fixed by the number of bits it contains. However, the actual capacity can be varied by partly filling it before any additions take place. The algebra for finding the capacity of DR will be outlined in Example 11.3.

Example 11.3 Constant-Acceleration Profile

As an example of the generation of a constant-acceleration profile, consider an extension of Example 11.1. There, the following data was given:

$$\text{Word size} = 16 \text{ bits}$$
$$\text{Clock frequency } (f_c) = 200 \text{ Hz}$$
$$\text{Total distance } (N) = 1000 \text{ steps}$$
$$\text{Velocity} = 51 \text{ steps/sec}$$

Now add the specification that the acceleration and deceleration rates be given by:

$$\text{Acceleration} = 20 \text{ steps/sec}^2$$

The problem is to compute the values for AR and VR so these conditions are met. In the constant-velocity portion of the profile, the VR should be the same as before, that is,

$$VR = \frac{(32,768)(51)}{200} = 8356.$$

Now, since the only register we want to overflow is DR, for determining what AR should be, the computation is

$$AR = \frac{(8356)(20/51)}{200} = Round(16.38) = 16$$

The actual acceleration will be 19.53 steps/sec^2 owing to rounding. The procedure for establishing the DDA values is as follows:

1. During the acceleration segment, $AR = 16$ and initially $VR = 0$
2. During the constant velocity segment $AR = 0$ and $VR =$ final value obtained at the end of the acceleration segment
3. During the deceleration segment $AR = -16$ and $VR = 0$ at the end of the profile

The software for generating the profile will be slightly more complicated than for the constant-velocity case. One pass through the DDA now requires two additions and an overflow check and hence takes longer.

<<>>

Constant-Jerk–Constant-Velocity Profile

With heavier loads on the motor, or more delicate control of the motor desired, it may be necessary to bring the motor up to speed using a constant-jerk profile as illustrated in Figure 11.6a. For the case of constant velocity, the DDA solved the differential equation $d\theta/dt =$ constant; $d^2\theta/dt^2 =$ constant was used for constant acceleration; and so $d^3\theta/dt^3 =$ constant is used for constant jerk.

The number of registers in the DDA will now increase to four for the constant-jerk portion of the profile. The DDA set up is shown in Figure 11.6b. The DDA works the same way as before, starting with JR as positive for the segment from 0 to N1, JR is then changed to negative for the segment from N1 to N2; both AR and JR are zero from N2 to N3, and JR is again negative for the portion from N3 to N4 and switches sign from N4 to NX.

The procedure for computing JR is similar to that used for finding AR. The problem is worked backward, starting with a VR that will give overflows in DR at the proper rate for constant velocity. Then AR is computed, as in the previous subsection, and finally the AR result is used to compute JR.

Once these values are computed, the program works as before. The total number of steps and register values for each segment are first calculated and stored in memory. The number of additions to be performed during each pass of the DDA increases, but the flowchart remains basically the same as those shown previously.

Multiaxis Continuous-Path Control

In the previous subsections, the methodology of software DDAs has been explained. The procedure can now be extended to two or more axes to achieve continuous-path control. Every contour can be said to be made up of a set of velocity profiles for each axis; such that when every axis follows a certain

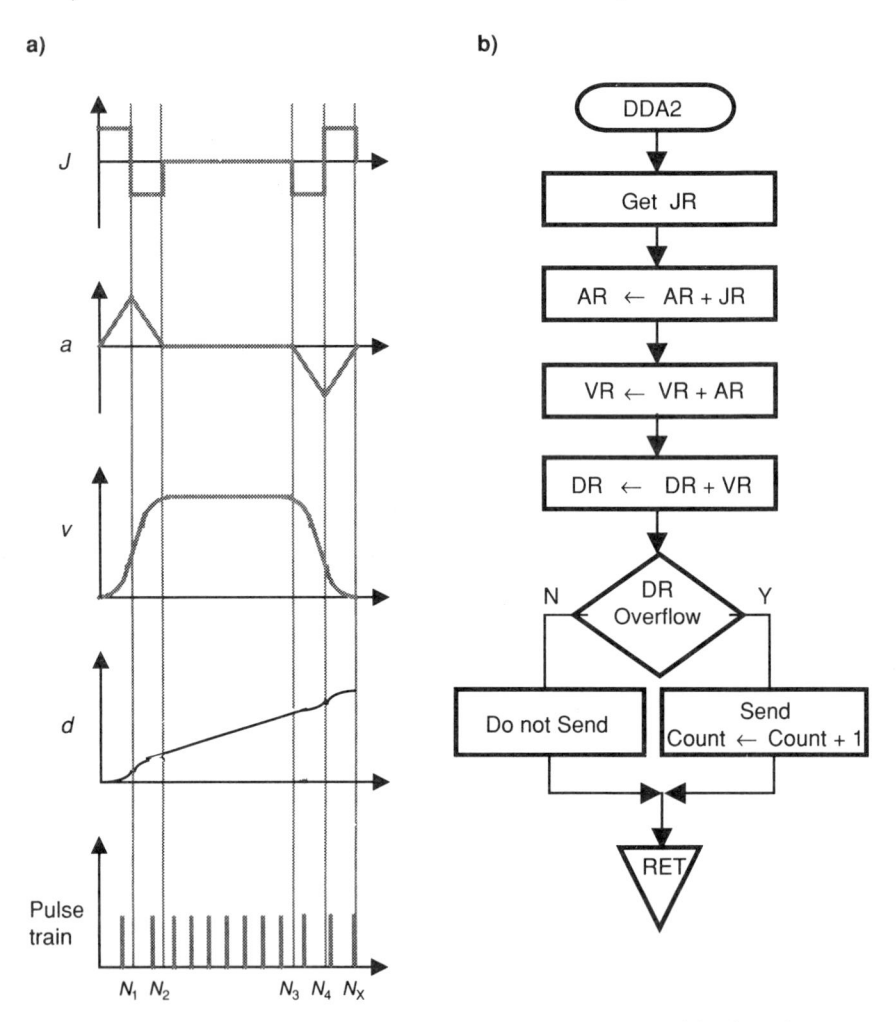

Figure 11.6 Constant jerk profile and the DDA subroutine to generate it: (a) pulse train, (b) DDA flowchart

profile, the contour is generated. It is therefore possible to build separate DDAs for each axis and use the overflows from the displacement registers for controlling each motor.

In the two cases considered next, generation of linear and circular paths in the x-y plane will be considered. Most NC and CNC interpolators generate cutter paths using linear and circular segments[Kore76;KoBe78;Kore83]. Both the path and the velocity along the path are characteristics that have to be controlled, but for these examples the higher-order derivatives, that is, acceleration and jerk, are assumed to be zero.

Linear Interpolation at Constant Velocity

For the simplest case of a straight line in the x-y plane, the slope of the line gives the desired velocity ratio between the two axes. The axes could be at constant velocity or constant acceleration, but as long as the instantaneous velocity ratio is maintained, the desired straight line will be generated.

Figure 11.7 shows such a straight line move, going from the origin to the coordinates x' and y', where the velocity along the path, V, is constant. The equations for this motion can be written in either rectangular or polar coordinates. In rectangular form, the trajectory is

$$y = mx$$

where $m = y'/x'$, the velocity is

$$\frac{dy}{dt} = V(y) = m\frac{dx}{dt} = mV(x)$$

which means that the respective velocity components are

$$V(x) = \frac{V}{\sqrt{m^2 + 1}}$$

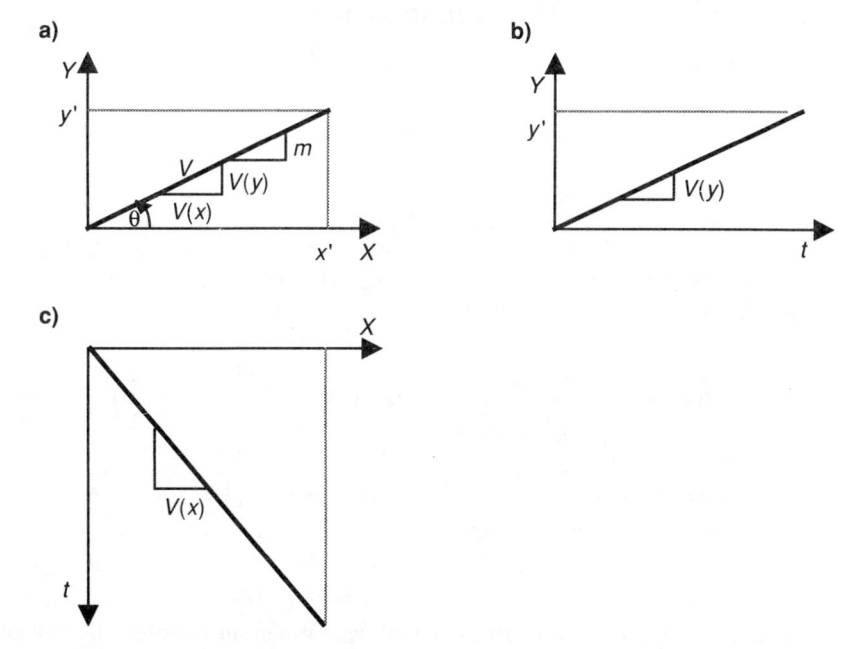

Figure 11.7 *Straight-line motion at constant velocity: (a) Path on xy axis, (b) projected motion on yt axis, (c) projected motion on xt axis*

$$V(y) = \frac{Vm}{\sqrt{m^2 + 1}}$$

Similarly, if polar coordinates are used, the independent trajectories are

$$y = \sin\theta \sqrt{x'^2 + y'^2} = R \sin\theta$$
$$x = \cos\theta \sqrt{x'^2 + y'^2} = R \cos\theta$$

where $\theta = \arctan(y'/x')$, and $R = \sqrt{x'^2 + y'^2}$. The velocity components, from the geometry, are

$$V(x) = V \cos\theta$$
$$V(y) = V \sin\theta$$

When both axes are moving at constant velocity, a DDA like those described here will be maintained for each axis; but in general, the velocity registers will cause overflows at different rates; these rates being the ratio of the slope desired.

Example 11.4 Linear Interpolation

Let the number of steps to the x and y axes be

$$NX = 400$$
$$NY = 300$$

and the velocity along the trajectory correspond to 51 steps/sec. As before, the word size and clock rate are 16 bits and 1000 Hz, respectively. In setting up the DDAs, the velocity registers are found as follows:

$$VRX = \frac{(32,768)(400)(51)}{\sqrt{400^2 + 300^2}}(1000) = Round(1336.9344) = 1337$$

$$VRY = \frac{(32,768)(300)(51)}{\sqrt{400^2 + 300^2}}(1000) = Round(1002.7008) = 1003$$

These values would be loaded in the respective DDAs, and every clock pulse would cause a pass through both routines. When an overflow is detected in either displacement register (DRX or DRY), a pulse is sent to the corresponding stepping motor.

The errors, due to integer rounding affect both the slope of the interpolated line and the velocity along the trajectory. The slope error in this case would be

$$\text{Slope error} = \frac{0.75 - (1003/1337)}{0.75} \cdot 100 = -0.025\%$$

and the error in velocity would be

$$\text{Velocity error} = \frac{51 - 51.00708}{51} \cdot 100 = -0.014\%$$

<<>>

Circular Interpolation

For the more complicated case of a circle, the velocity along the two axes continuously changes and is proportional to the radius vector component along the other axis, as shown in Figure 11.8, which is for the case of a counterclockwise path.

$$V(y) = V\cos\theta = (V)(x/R)$$

$$V(x) = V\sin\theta = (V)(y/R)$$

The value of the velocity registers VR in the two DDAs will have to be changed at every overflow to conform to this changing velocity. The computer also has to keep track of the direction of motion along the circle and the quadrant of the circle in which the motion is desired, as shown in Table 11.2. These decide the

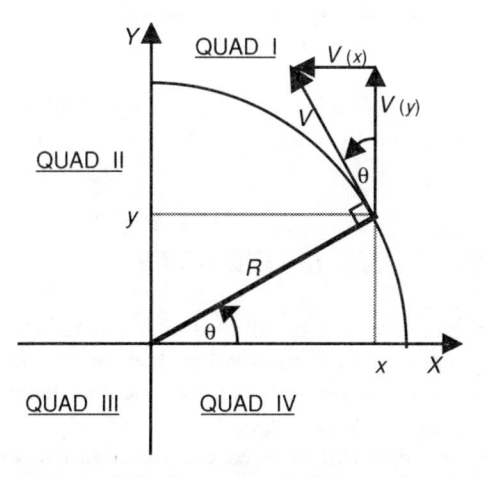

Figure 11.8 Circular interpolation for a counterclockwise arc through the first quadrant

Table 11.2 Initial and Final Points for Circular Interpolation

Motion Through	Direction CCW	Direction CW
Quad I	Start: $x = R$ & $V(x) = 0$ $y = 0$ & $V(y) = V$ Stop: $x = 0$ & $V(x) = -V$ $y = R$ & $V(y) = 0$	Start: $x = 0$ & $V(x) = V$ $y = R$ & $V(y) = 0$ Stop: $x = R$ & $V(x) = 0$ $y = 0$ & $V(y) = -V$
Quad II	Start: $x = 0$ & $V(x) = -V$ $y = R$ & $V(y) = 0$ Stop: $x = -R$ & $V(x) = 0$ $y = 0$ & $V(y) = -V$	Start: $x = -R$ & $V(x) = 0$ $y = 0$ & $V(y) = V$ Stop: $x = 0$ & $V(x) = V$ $y = R$ & $V(y) = 0$
Quad III	Start: $x = -R$ & $V(x) = 0$ $y = 0$ & $V(y) = -V$ Stop: $x = 0$ & $V(x) = V$ $y = -R$ & $V(y) = 0$	Start: $x = 0$ & $V(x) = V$ $y = -R$ & $V(y) = 0$ Stop: $x = -R$ & $V(x) = 0$ $y = 0$ & $V(y) = V$
Quad IV	Start: $x = 0$ & $V(x) = V$ $y = -R$ & $V(y) = 0$ Stop: $x = R$ & $V(x) = 0$ $y = 0$ & $V(y) = V$	Start: $x = R$ & $V(x) = 0$ $y = 0$ & $V(y) = -V$ Stop: $x = 0$ & $V(x) = -V$ $y = 0$ & $V(y) = 0$

direction in which the motor is to move. The final radius vector components help determine when the motion is to be terminated.

11.2 PROGRAMMING OBJECTIVES

In this case study we will develop assembly language software to implement the motion control algorithms developed in the previous section. Because of computation speed requirements we use assembly language programming with digital output and clock interrupts.

The hardware that will be used consists of a two-axis x-y table, stepping motors to control each axis, translator cards that will convert a pulse sent from the processor to a fraction of a motor revolution, and a connection from the parallel I/O board on the microprocessor. In addition, it is assumed that the processor has a clock that can initiate interrupts at a fixed rate (e.g., 1 kHz). A sketch showing the details of the hardware and interface is given in Figure 11.9 (also see Fig. 5.3).

For the single-axis move, the desired acceleration, velocity, and displacement profiles are shown in Figure 11.10. Thus, given the following nominal values:

1. Length of travel = 1.5 in.
2. Maximum velocity = 0.1 in./sec
3. Acceleration = ± 0.05 in./sec^2

The program must calculate the number and frequency of pulses to be sent to the stepping motor using a software DDA implemented on the microprocessor. The basic DDA algorithm should be implemented as a macro, so that it can be called as needed from the routine that services clock interrupts (see Fig. 11.11).

For the two-axis move with constant velocity, you should be able to use the same DDA macro with the acceleration parameter (AR) set to zero. For the linear two-axis move, assume that the motion will start at (0.0, 0.0) and terminate at (0.9, 1.2) and that the tangential velocity is 0.1 in./sec. Optionally, you may want to generate a circular arc, as follows:

1. Assume a center at (0.0, 0.0) and start at (1.0, 0.0)

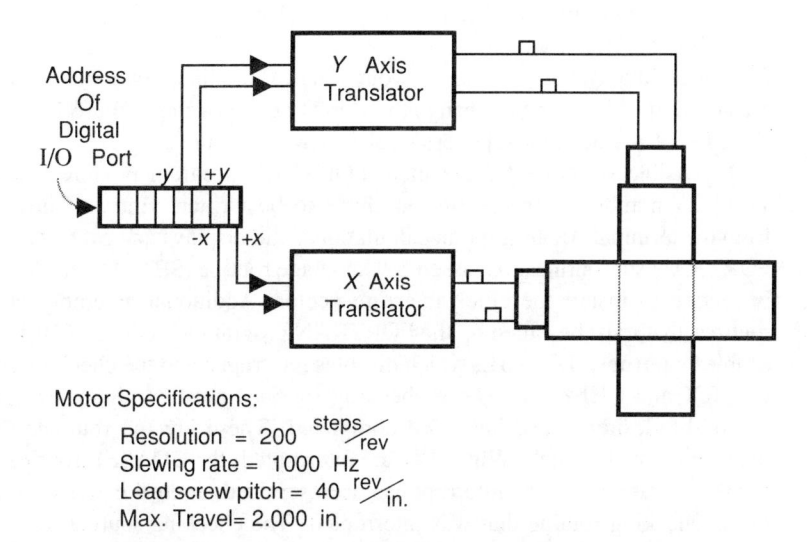

Motor Specifications:
Resolution = 200 steps/rev
Slewing rate = 1000 Hz
Lead screw pitch = 40 rev/in.
Max. Travel = 2.000 in.

Figure 11.9 Stepping-motor hardware

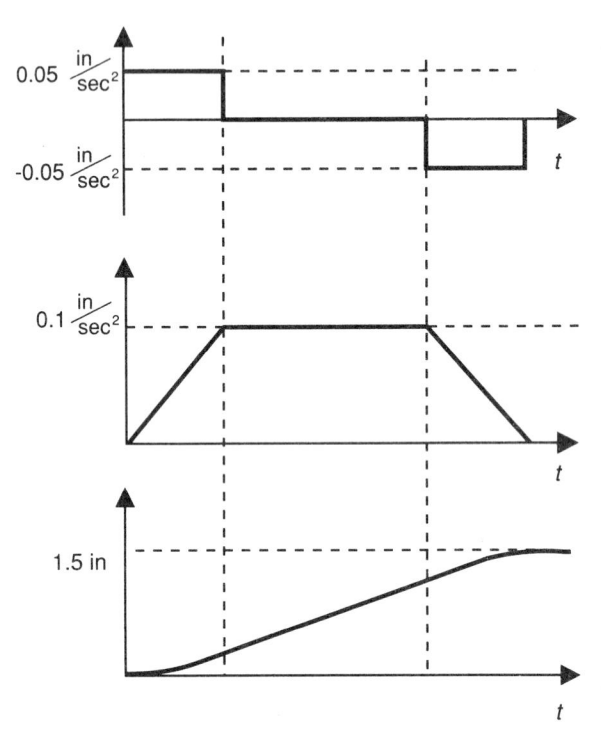

Figure 11.10 Desired motion for the single-axis move: (a) Acceleration, (b) velocity, (c) displacement

2. Make a counter clockwise move through the first quadrant
3. Use a constant tangential velocity of 0.1 in./sec

For the circular move you can use your code for the linear two-axis move, with the addition of a routine to change the velocity components. Figure 11.12 shows sketches of the desired trajectories for the two-axis moves.

A possible structure for the main FORTRAN program is shown in Figure 11.13. A number of pieces of code have to be written. Entry of information from the terminal, floating-point calculations, and display can best be handled by FORTRAN subroutines. An assembly language routine (SETDDA in Fig. 11.13) is needed to install the clock interrupt vector. Additional assembly language routines that must be called by the FORTRAN program include ENABLE, which enables interrupts; DISABL, which disables interrupts; and the checking routines CHEKX and CHEKXY. These checking routines must continuously check to see if the desired move has been completed. These are the routines that are interrupted by the clock. When they are interrupted, the DDASER routine begins to execute; and when the interrupt has been serviced, control is transferred back to the checking routine that was interrupted. The checking routine for the two-axis constant velocity case (CHEKXY) is flowcharted in Figure 11.11. The

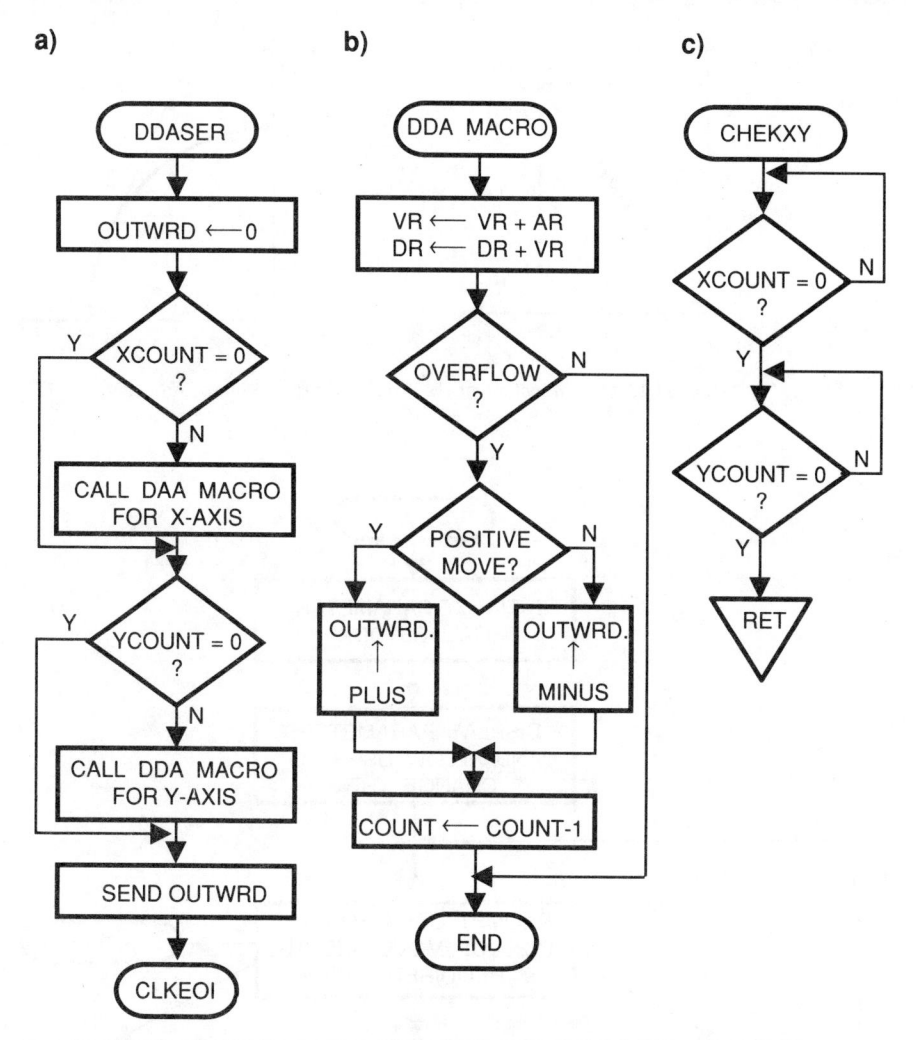

Figure 11.11 Flowcharts for some assembly language routines: (a) DDA service routine, (b) DDA macro, (c) checking routine for constant velocity move

checking routine for the single-axis constant acceleration case (CHEKX) will be somewhat more complicated (see Fig. 11.5).

Example 11.5 Programs for x-y Table Motion Control

Included in the listings in Appendixes A and B are programs implementing the single-axis constant velocity move, the single-axis constant acceleration move, and the two-axis constant velocity move with linear interpolation. These programs use assembly language routines for the time-critical task of sending

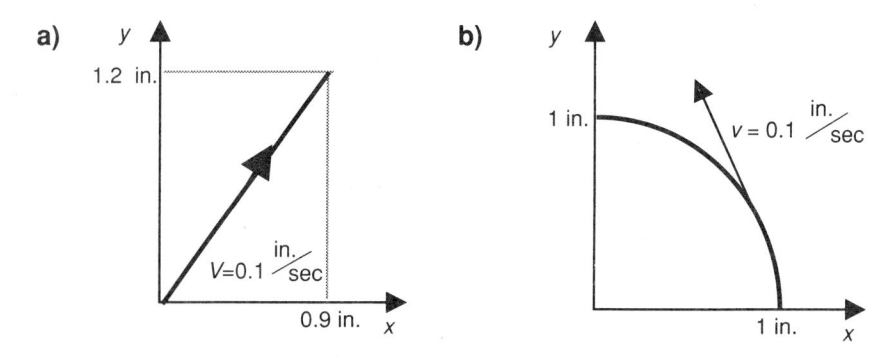

Figure 11.12 Desired motion for the two-axis moves: (a) linear, (b) circular

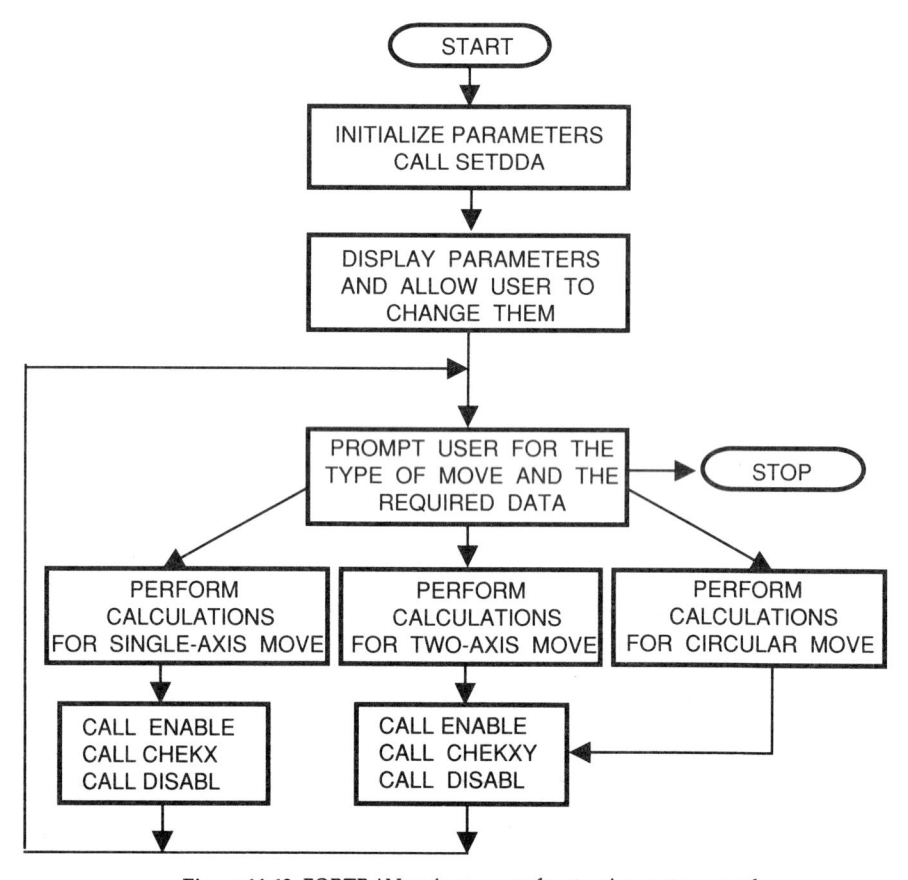

Figure 11.13 FORTRAN main program for stepping-motor control

pulses and FORTRAN for other aspects of the program. The program listings contain system-specific aspects, such as the parallel I/O port address and configuration; these are discussed in Appendixes A and B. Also note that although the programs given in the two Appendixes are different, they both contain modules to perform the following tasks:

- Get, check, and display the required input data
- Install the clock-interrupt service routine
- Select the type of move and perform required calculations
- Check to see if the move is completed
- Enable/disable clock interrupts
- DDA macro routine
- Clock-interrupt service routine

Extend these programs by including nonconstant acceleration and/or circular interpolation or other extensions that might be useful. Check your program to make sure that it is operating correctly (i.e., generating the correct number and frequency of pulses).

<<>>

11.3 PROBLEMS

1. Write the Z80 Assembler code to implement a 16-bit DDA macro. Assume that:

    a. We are interested in constant velocity motion.
    b. All arithmetic is done using 16-bit words and the parity flag is used to detect an overflow.
    c. The macro will assemble the appropriate bit pattern to be sent in location OUTWRD.

    Explain the operation of your macro, define all labels, and make liberal use of comments.

2. Repeat Problem 11.1 using 8088 Assembly language.

3. Write the Z80 or 8088 assembler code necessary to implement the first-order digital differential analyzer (DDA) in Figure 11.5b. To do this, assume:

    a. A positive value of AR has been calculated and is transferred as a single subroutine argument.

b. VR and DR are local variables.

c. All arithmetic is done using 8-bit words, and the carry flag is set when an overflow occurs.

d. A subroutine without arguments called SENDIT is external to your subroutine and is called when a pulse has to be sent.

4. Consider an x-y table with $(1/p)$ = 20 threads/in. and SPR = 200 steps/revolution for each axis. Using 16-bit words and a clock frequency of f_c = 10,000 Hz, what should the velocity registers contain for a constant velocity (v = 1 in./sec.) move from (0.0, 0.0) to (1.0, 2.5)?

5. How is AR computed in Example 11.3? Can you give a general expression similar to Eq. 11.1.4, for computing AR?

6. How are JR, AR, and VR to be calculated for the constant jerk profile in Fig. 11.6?

7. How do we arrive at the values given in Table 11.2 for circular interpolation?

Chapter 12
SPEED CONTROL OF A dc MOTOR

As discussed previously in Chapters 5 through 8 and 11, the dc motor is a common actuating device in many motion-control systems. For example, many machine tools and robots employ dc-motor-driven axes. This type of closed-loop digital control system is more suitable for variable load applications, where stepping motors may not be adequate because they may miss pulses or because very smooth motion is desired. Here we consider only the armature-controlled dc motor commonly used in servo control systems and generally considered to be the most satisfactory adjustable speed motors. The field current is kept constant, and the motor speed is controlled by manipulating, the armature voltage V (see Fig. 12.1).

As discussed in Chapter 8, microprocessors are commonly used as the controller element in many closed-loop (feedback) control applications, and dc motors are widely used as actuators in such systems. In this case study we will use a microcomputer for closed-loop digital control of the rotational speed of a dc motor. This might, for example, apply to the rotational speed control of a workpiece in turning or the spindle-speed control in drilling or milling [Boll72; Kore77; KoBe78; KoBo78].

12.1 DESCRIPTION OF THE ENGINEERING PROBLEM

The engineering objective is to implement a discrete-time (digital) version of the proportional plus integral (PI) regulator to maintain a desired rotational speed of a dc motor. We have selected a PI controller because it is often used

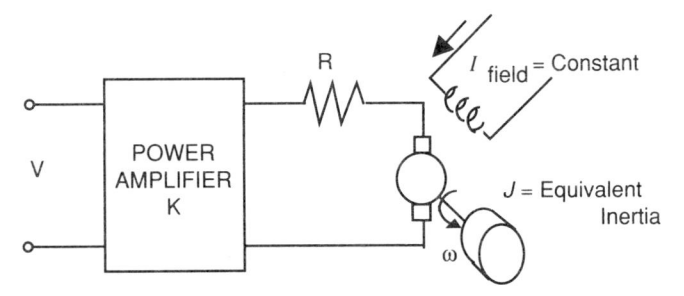

Figure 12.1 dc motor schematic

for control of first-order processes such as the dc motor. It will ensure zero steady-state error for constant reference and disturbance inputs. It will also provide for complete assignment of the closed-loop poles and thus enable us to select the desired transient response characteristics. It is a simple and robust control strategy for first-order processes. There are, of course, other control strategies that could be used to provide satisfactory results. The hardware setup is schematically illustrated in Figures 12.1 and 12.2. A power amplifier takes the low power signal from the DAC and provides enough current to drive the dc motor. The dc motor drives an inertial load and it in turn is connected to a matched dc generator (tachometer) producing a voltage proportional to the speed of the shaft. The shaft velocity could also have been measured using an incremental optical encoder, which would have eliminated the need for the ADC. However, an optical encoder would have required counting of pulses in a given time interval to obtain rotation speed (e.g., see [ChUl87]). A hardware counter would have been required, since counting pulses in software as well as calculating the control signal would have lead to large sampling intervals. Thus, the tachometer appears to be a simpler solution in this application.

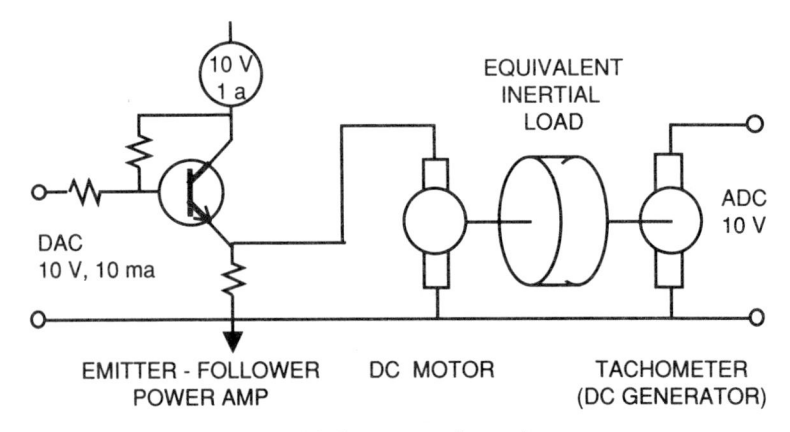

Figure 12.2 dc motor speed-control setup

12.1 Description of the Engineering Problem

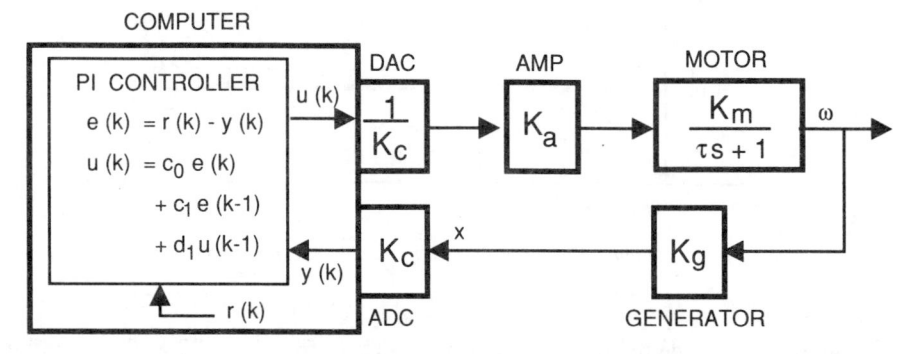

Figure 12.3 Digital PI motor speed control

There is a gear train (not shown in Figs. 12.1 and 12.2) with a gear ratio of one connecting the dc motor shaft to the tachometer shaft; the load inertia is mounted directly to the dc motor shaft. Because the motor and generator are matched, there is no need to consider the speed and torque constants, except to know that zero volts corresponds to no rotation and full-scale voltage is the maximum rotational speed. In most servo applications, this is not the case. The tachometer is a small dc device for providing a feedback signal, whereas the dc motor is a large device that provides the power to drive the mechanical system.

A block diagram representation of the system is shown in Figure 12.3, where we have represented the analog (or continuous-time) portion of the system as,

$$\frac{X(s)}{U(s)} = \frac{K}{\tau s + 1} \quad (12.1.1)$$

and $K = K_a K_m K_g$. Rather than modeling the components of the system from first principles, we will assume the first-order open loop structure in Eq. 12.1.1 and apply system identification methods. This approach is simple and has the advantage of including any effects of the amplifier and gear train without separate and detailed analysis of these components. For a step input $u(t) = u_0$ and the initial condition $x(0) = 0$, the response of the system is given by,

$$x(t) = K u_0 (1 - e^{-t/\tau}) \quad (12.1.2)$$

and illustrated in Figure 12.4. If u_0 is known, then K and τ can be determined from the response curve. This can be done graphically, as indicated in Figure 12.4, or using a least squares estimation procedure similar to the one in Example 7.6.

The PI algorithm, one of the algorithms described in Section 8.2, is based on manipulation of the error, defined as:

$$e(k) = r(k) - y(k) \quad (12.1.3)$$

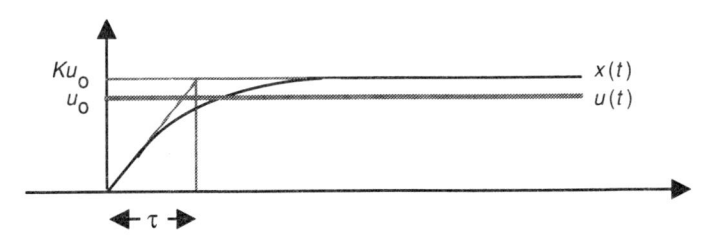

Figure 12.4 Response of a first-order system to a step

where $r(k)$ is the reference value of $y(k)$. Then, from Eq. 8.2.8, the algorithm can be written as:

$$u(k) = u(k-1) + (k_p + k_i \Delta t)e(k) - k_p e(k-1) \qquad (12.1.4)$$

where the sampling period is Δt. We can also rewrite the PI algorithm as:

$$u(k) = (k_p + k_i \Delta t)e(k) - k_p e(k-1) + u(k-1)$$
$$= c_0 e(k) + c_1 e(k-1) + d_1 u(k-1) \qquad (12.1.5)$$

Comparing this to Figure 12.3, we note that

$$c_o = (k_p + k_i \Delta t) \qquad (12.1.6)$$
$$c_1 = -k_p \qquad (12.1.7)$$
$$d_1 = 1 \qquad (12.1.8)$$

In the next section the implementation of this digital PI controller for dc-motor speed control is discussed.

12.2 PROGRAMMING OBJECTIVES

The programming objectives for this case study are to develop a real-time, clock-interrupt-driven computer code for digital control of the dc motor rotational speed. To accomplish this, we will need to program the ADCs, the DACs, and the clock, and we will need to consider implementation issues such as word size, sampling frequency, saturation, and computational speed [AsWi84].

With real-time control applications it is good practice not to assume anything about the hardware. So the first thing we will want to do is to develop the code to be sure the motor is off when the program begins and ends executing. This can be accomplished by sending zero volts out through the DAC channel

12.2 Programming Objectives 219

connected to the motor. The computer program can be written primarily in FORTRAN, except for a short assembly language interrupt service routine. (See Fig. 12.5). The interrupt service routine will simply call a FORTRAN subroutine to implement the PI algorithm whenever a clock interrupt occurs. This latter routine should employ integer arithmetic to reduce computation time. In using integer arithmetic, one should check to make sure that the resulting gains are reasonable; for example, they should be such that if they are multiplied by an

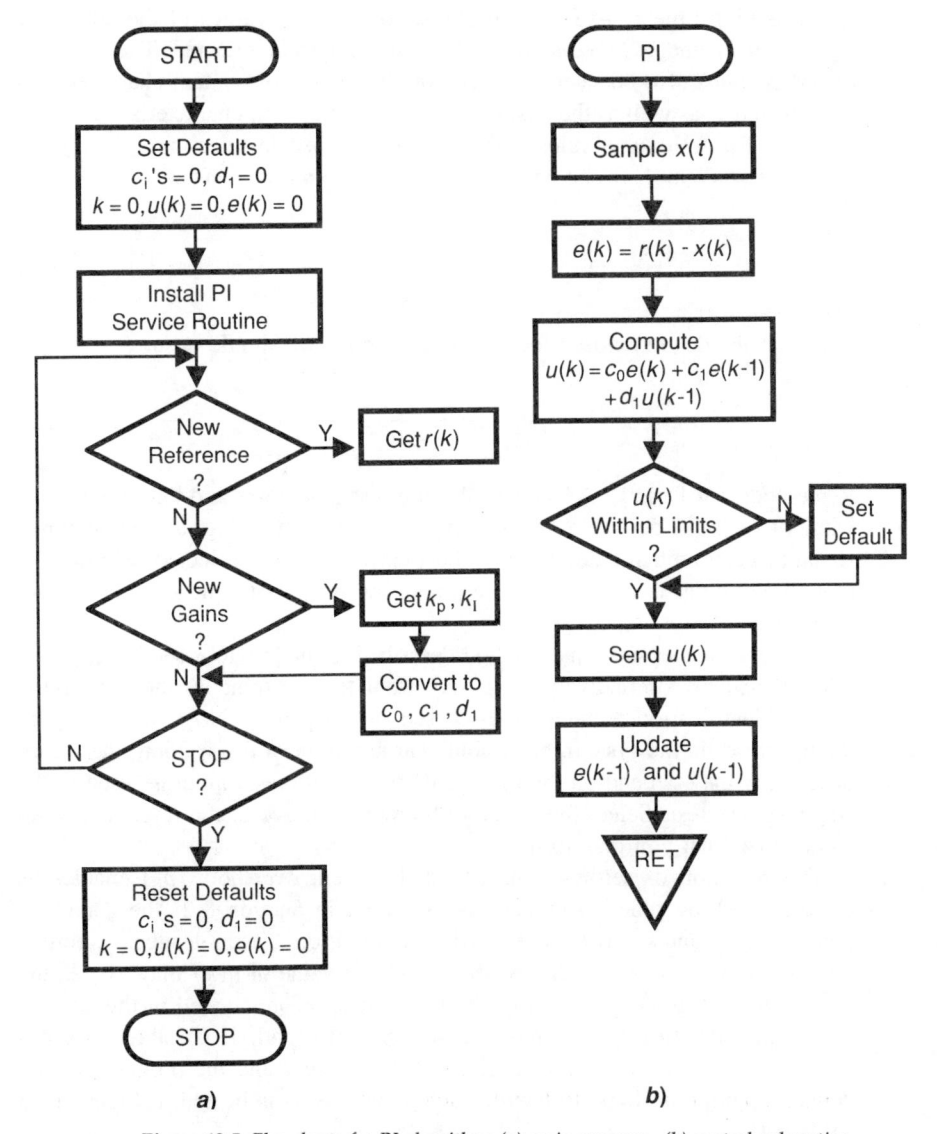

Figure 12.5 Flowcharts for PI algorithm: (a) main program, (b) control subroutine

integer in the range -128 to $+127$ (from an 8-bit ADC) the result will be in the range -32768 to $+32767$ for 16-bit integer arithmetic. Also remember that gains less than one will be truncated to zero, so you may need to use scaling. Another practical consideration is that the value sent to the DAC should be in the proper range (i. e., -128 to $+127$ for an 8-bit converter). Finally, to eliminate aliasing due to noise, it may be necessary to filter the signal $x(t)$ from the tachometer before analog-to-digital conversion. An appropriate filter frequency must be determined, as discussed in Chapter 5.

Based on the desired performance characteristics of the closed-loop system (e.g., settling time t_s and percent overshoot p_0), the values of the controller parameters k_p and k_i (or equivalently c_0 and c_1) must be selected. The sampling period Δt must also be selected, based on the open-loop system characteristics (i.e., K and τ) as well as the desired closed-loop response characteristics (i.e, t_s and p_0). The following "rules of thumb" may be used in selecting the sampling period Δt [AsWi84]. Based on the open-loop system time constant τ, we should ensure:

$$\Delta t \leq \frac{\tau}{5} \qquad (12.2.1)$$

Based on the desired closed loop settling time t_s, we should ensure:

$$\Delta t \leq \frac{t_s}{20} \qquad (12.2.2)$$

Remember that Eqs. 12.2.1 and 12.2.2 are just guidelines and that too small a value of Δt is not desirable owing to limitations of computing time and control-signal (voltage) magnitude. The antialiasing filter must be selected such that the filter break point frequency f_f is less than the Nyquist frequency $f_s/2$, where $f_s = 1/\Delta t$.

First use the step-response method described in the previous section to determine K and τ. Alternatively, you may want to determine K and τ by other system-identification techniques discussed in Chapter 7. Note that owing to nonlinearities in the system (e.g., nonlinear amplifier or dry friction) the values of K and τ may depend on the magnitude of the reference input u_0 used in the step-response test. Select this value to be representative of the reference speed values $r(k)$ that you plan to use.

The selection of settling time (t_s) and percent overshoot (p_0) enables us to determine, by pole assignment as discussed in Section 8.2, the controller parameters k_p and k_i. If t_s is selected very small (e.g., as in deadbeat control), the control may saturate. That is, the calculated value of $u(k)$ may exceed the maximum voltage output level of the DAC. It is recommended in this case to reduce the specified t_s to a more reasonable level and, if possible, to use a larger Δt value. However, saturation may still occur and the program should contain appropriate logic to handle such situations as indicated in Figure 12.5 [AsWi84].

Example 12.1 Speed Control of a dc Motor

In Appendixes A and B the program listings for this case study are included for an 8088-based and a Z80-based system, respectively. Both programs include the following features and capabilities:

- Identification of K and τ from the open-loop response
- Turn the motor off to start and end gracefully.
- Performance specifications, selection of Δt, given desired closed-loop poles calculation of controller gains.
- Closed loop PI control, including clock interrupt service routine; sampling with the ADC; calculation of control signal with scaling, finite word size, and saturation; and use of DAC to send out control.

Examine these program listings and suggest improvements to the programs.

<<>>

12.3 PROBLEMS

1. In addition to the PI controller for the dc motor speed-control problem, implement and compare the performance of P, PD, and PID controllers.

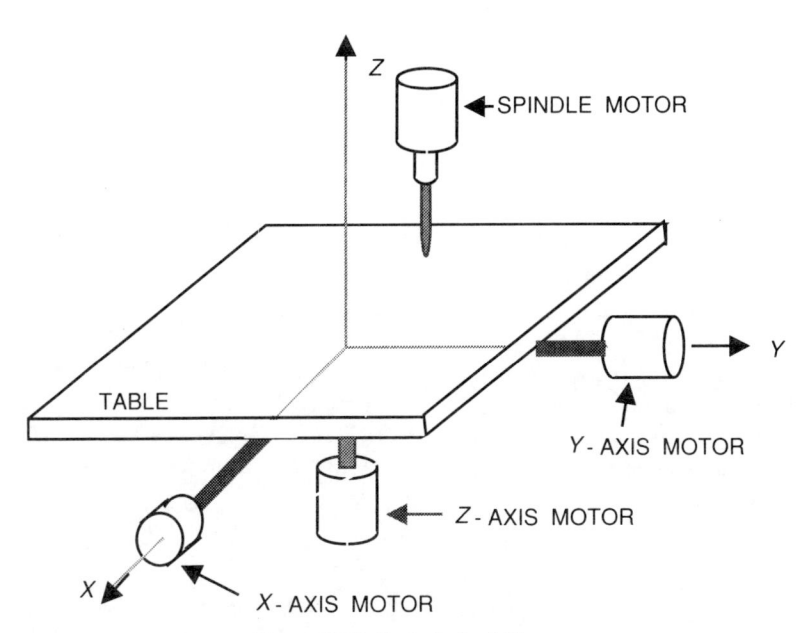

Figure 12.P2 *Control of a drill press*

2. You are asked to develop a microcomputer-based control system for a small drill press. As sketched in Figure 12.P2, there are three dc motor-driven axes (x, y, and z) and a dc motor is also used to provide the spindle rotation. Once a workpiece is mounted on the table, the table must be positioned properly under the drill and the drill fed in (along the z-axis) and the spindle rotational speed controlled. Describe what hardware and software you will need, and the overall function and logic of the software that will have to be developed.

3. Describe how you would perform the coordinated control of an x-y table driven by dc servomotors.

4. Write a PI dc motor speed-control program, and include a subroutine to calculate the control gains k_p and k_i from given specifications for settling time and percent overshoot. Make the PI controller adaptive by including recursive least squares parameter estimation of the process parameters and updating the calculation of the control parameters at each time step based on these parameter estimates.

BIBLIOGRAPHY

[AmOB84] Amstead, B. H., Ostwald, P. F., and Begeman, M. L., 1984, *Manufacturing Processes,* Wiley, New York.

[Andr82] Andrews, M., 1982, *Programming Microprocessor Interfaces for Control and Instrumentation,* Prentice–Hall, Englewood Cliffs, N.J.

[Anon74] Anonymous, 1974, *The Value of Power,* Prepared for General Automation by A. Osborne, Berkeley.

[AsWi84] Astrom, K. J., and Wittenmark, B., 1984, *Computer Controlled Systems: Theory and Design,* Prentice–Hall, Englewood Cliffs, N.J.

[AuSa81] Auslander, D. M., and Sagues, P., 1981, *Microprocessors for Measurement and Control,* Osborne/McGraw–Hill, Berkeley.

[Barn84] Barnes, J. G. P., 1984, *Programming in Ada,* Addison–Wesley, Reading, Mass.

[Bart85] Bartee, T. C., 1985, *Digital Computer Fundamentals,* McGraw–Hill, New York.

[Batc79] Batchelor, B. G., 1979, *Pattern Recognition Ideas in Practice,* Plenum, New York.

[BePi66] Bendat, J. L., and Piersol, A. G., 1966, *Measurement and Analysis of Random Data,* Wiley, New York.

[Berg69] Bergland, G. D., 1969, "A Guided Tour of the Fast Fourier Transform," *IEEE Spectrum,* Vol. 6, July, pp. 41–52.

[Boll72] Bollinger, J., 1972, "Computer Control of Machine Tools," *CIRP Annals,* Vol. 21, No. 2.

[Booc83] Booch, G., 1983, *Software Engineering with Ada,* Benjamin/Cummings, Menlo Park, Calif.

[Bowl77] Bowles, K. L., 1977, *Microcomputer Problem Solving Using Pascal,* Springer-Verlag, New York.

[Brig74] Brigham, E. D., 1974, *The Fast Fourier Transform,* Prentice–Hall, Englewood Cliffs, N.J.

[CaMa70] Cadzow, J. A., and H. R. Martens, 1970, *Discrete-Time and Computer Control Systems,* Prentice–Hall, Englewood Cliffs, N.J.

[Cand88] Candy, J. V., 1988, *Signal Processing: The Modern Approach,* McGraw–Hill, New York.

[Cann67] Cannon, R. H., 1967, *Dynamics of Physical Systems,* McGraw–Hill, New York.

[Cent64] Centner, R., 1964, *Final Report on the Development of the Adaptive Control Technique for a Numerically Controlled Milling Machine,* USAF Technical Documentary Report ML-TDR-64-279:

[ChUl86] Chalhoub, N. G., and Ulsoy, A. G., 1986, "Dynamic Simulation of a Flexible Robot Arm and Controller," *ASME Journal of Dynamic Systems, Measurement and Control,* Vol. 108, No. 2, pp. 119–126.

[ChUl87] Chalhoub, N. G., and Ulsoy, A. G., 1987, "Control of a Flexible Robot Arm: Experimental and Theoretical Results," *ASME Journal of Dynamic Systems, Measurement and Control,* Vol. 109, No. 4, pp. 299–309.

[Ciar86] Ciarcia, S., 1986, "Build an Analog to Digital Converter", *Byte*, Vol. 11, No. 1, pp. 104–116.

[CoMD78] Colwell, L. V., Mazur, J. C., DeVries, W. R., 1978, "Analytical Strategies for Automatic Tracking of Tool Wear," *Proceedings of the 6th North American Manufacturing Research Conference*, pp. 276–282.

[Cons74] Considine, D. M., 1974, *Process Instruments and Controls Handbook*, McGraw–Hill, New York.

[Cran84] Crandall, R. E., 1984, *Pascal Applications for the Sciences*, Wiley, New York.

[DaOr83] Dale, N., and Orshalick, D., 1983, *Introduction to Pascal and Structured Design*, Heath, Lexington, Mass.

[DaUl87a] Danai, K., and Ulsoy, A. G., 1987a, "An Adaptive Observer for On-Line Tool Wear Estimation in Turning, Part I: Theory and Part II: Results," *Mechanical Systems and Signal Processing*, Vol. 1, No. 2, pp. 211–240.

[DaUl87b] Danai, K., and Ulsoy, A. G., 1987b, "A Dynamic State Model for On-Line Tool Wear Estimation in Turning," *ASME Journal of Engineering for Industry*, Vol. 109, No. 4, pp. 396–399.

[DeAs81] Deshpande, P. B., and Ash, R. H., 1981, *Elements of Computer Process Control*, Instrument Society of America, Research Triangle Park, N.C.

[deSi85] deSilva, C. W., 1985, "Motion Sensors in Industrial Robots," *Mechanical Engineering*, June, pp. 40–51.

[DeVE84] DeVries, W. R., and Evans, M. S., 1984, "Computer Graphics Simulation of Metal Cutting," *CIRP Annals*, Vol. 33, No. 1, pp. 15–18.

[DeVL85] DeVries, W. R., and Li, C. J., 1985, "Algorithms to Deconvolve Stylus Geometry from Surface Profile Measurements," *ASME Journal of Engineering for Industry*, Vol. 107, pp. 167–184.

[DeVr79] DeVries, W. R., 1979, "Chatter Vibration Monitoring Using On-Line Identification of Discrete Second Order Autoregressive Time Series Models," *Proceedings of the 7th North American Manufacturing Research Conference*, pp. 275–278.

[DePW77] DeVries, W. R., Pandit, S. M., and Wu, S. M., 1977, "Evaluation of the Stability of Paper Basis Weight Using Multivariate Time Series," *IEEE Transactions on Automatic Control*, Vol. AC-22, No. 4, pp. 590–594.

[DeRM81] DeVries, W. R., Raski, J. Z., and Mazur, J. C., 1981, "Investigation of Adaptive Exponential Smoothing Algorithms in Monitoring Tool Wear," *Proceedings of the 9th North American Manufacturing Research Conference*, pp. 523–527.

[DeRU83] DeVries, W. R., Raski, J. Z., and Ulsoy, A. G., 1983, "Microcomputer Applications in Manufacturing: A Senior Laboratory Course," *Proceedings of the International Computers in Engineering Conference*, Chicago.

[DeVW78] DeVries, W. R., and Wu, S. M., 1978, "Evaluation of Process Control Effectiveness and Diagnosis of Variation in Paper Basis Weight via Multivariate Time Series Analysis," *IEEE Transactions on Automatic Control*, Vol. AC-23, No. 4, pp. 702–708.

[Doeb83] Doeblin, E. O., 1983, *Measurement Systems: Application and Design*, McGraw–Hill, New York.

[Doyl85] Doyle, L. E., 1985, *Manufacturing Processes and Materials for Engineers*, Prentice–Hall, Englewood Cliffs., N. J.

[Ette83] Etter, D. M., 1983, *Structured FORTRAN 77 for Engineers and Scientists*, Benjamin/Cummings, Menlo Park, Calif.

[FlKL84] Flom, D. G., Komanduri, R., and Lee, M., 1985, "High Speed Machining of Metals," *Annual Review of Material Science*, Vol. 14, pp. 231–278.

[Fost81] Foster, C. C., 1981, *Real Time Programming—Neglected Topics*, Addison-Wesley, Reading, Mass.

[FrPE86] Franklin, G. F., Powell, D. J., and Emami-Naeini, A., 1986, *Feedback Control of Dynamic Systems*, Addison–Wesley, Reading, Mass.

[Frie86] Friedland, B., 1986, *Control System Design*, McGraw–Hill, New York.

[Gilb50] Gilbert, W. W., 1950, "Economics of Machining," *Machining Theory and Practice*, American Society of Metals, Cleveland.

[Gold65] Golden, J. T., 1965, *FORTRAN IV Programming and Computing*, Prentice–Hall, Englewood Cliffs, N.J.

[GoSi84] Goodwin, G. C., and Sin, K. S., 1984, *Adaptive Filtering Prediction and Control*, Prentice–Hall, Englewood Cliffs, N.J.

[Groo80] Groover, M. P., 1980, *Automation, Production Systems, and Computer-Aided Manufacturing*, Prentice–Hall, Englewood Cliffs, N.J.

[HeDL86] Hens, K. F., DeVries, W. R., and D. Lee, "Integration of Die Design and Manufacture for Plastic and Metal Parts Processing," *Proceedings of the Japan/USA Symposium on Flexible Automation*, Osaka, Japan, pp. 625–628.

[HoCa82] Hosford, W. F., and Caddell, R. M., 1982, *Metal Forming Mechanics and Metallurgy*, Prentice–Hall, Englewood Cliffs, N.J.

[Huds82] Hudson, C. A., 1982, "Computers in Manufacturing," *Science*, Vol. 215, February 12.

[Hugh69] Hughes, J. L., 1969, *Digital Computer Lab Workbook*, Digital Equipment Corporation, Maynard, Mass.

[Inte81] Intel, 1981, *iAPX 86, 88 User's Manual*, Santa Clara, Calif.

[Jone79] Jones, B. E., 1979, *Instrumentation, Measurement and Feedback*, McGraw–Hill, New York.

[Jury58] Jury, E. I., 1958, *Sampled Data Control Systems*, Wiley, New York.

[Jury64] Jury E. I., 1964, *Theory and Application of the Z-Transform Method*, Wiley, New York.

[Kalp84] Kalpakjian, S., 1984, *Manufacturing Processes for Engineering Materials*, Addison-Wesley, Reading, Mass.

[KePl74] Kernighan, B. W., and P. J. Plauger, 1974, *The Elements of Programming Style*, McGraw–Hill, New York.

[KoBe78] Koren, Y., and Ben-Uri, J., 1978, *Numerical Control of Machine Tools*, Khanna, Delhi.

[KoBo78] Koren, Y., and Bollinger, J., 1978, "Design Parameters for Sampled-Data-Drives for CNC Machine Tools," *IEEE Transactions on Industrial Applications*, Vol. IA-14, No. 3, pp. 255–264.

[KoBu83] Kochar, A. K., and Burns, N. D., 1983, *Microprocessors and Their Manufacturing Applications*, Arnold, London.

[KolD87] Kolarits, F. M., and DeVries, W. R., 1987, "An Improved Model for the

Cost of Product Assembly Using a Flexible Robotic Workstation," *Proceedings of the 15th North American Manufacturing Research Conference,* pp. 656–662.

[Koma85] Komanduri, R., 1985, "High Speed Machining," *Mechanical Engineering,* Vol. 107, No. 12, pp. 64–76.

[Kore76] Koren, Y., 1976, "Interpolator for a Computer Numerical Control System," *IEEE Transactions on Computers,* Vol. C-25, No. 1, pp. 32–37.

[Kore77] Koren, Y., 1977, "Computer Based Machine Tool Control," *IEEE Spectrum,* Vol. 14, March, pp. 80–84.

[Kore83] Koren, Y., 1983, *Computer Control of Manufacturing Systems,* McGraw–Hill, New York.

[Kore85] Koren, Y., 1985, *Robotics for Engineers,* McGraw–Hill, New York.

[LaUl85] Lauderbaugh, L. K., and Ulsoy, A. G., 1985, "Dynamic Modeling for Control of the Milling Process," in Kannatey-Asibu, Jr., E., Ulsoy, A. G., and Komanduri, R. (eds.), *Sensors and Controls for Manufacturing,* ASME, New York, pp. 149–158.

[Leve78] Leventhal, L. A., 1978, *Introduction to Microprocesors: Software, Hardware, Programming,* Prentice–Hall, Englewood Cliffs, N.J.

[Leve79] Leventhal, L. A., 1979, *6502 Assembly Language Programming,* Osborne/McGraw–Hill, Berkeley.

[Leve80] Leventhal, L. A., 1980, *6800 Assembly Language Programming,* Osborne/McGraw–Hill, Berkeley.

[Ljun87] Ljung, L., 1987, *System Identification,* Prentice–Hall, Englewood Cliffs, N.J.

[LjSo83] Ljung, L., and Soderstrom, T., 1983, *Theory and Practice of Recursive Identification,* MIT Press, Cambridge, Mass.

[LuCA87] Ludema, K. C., Caddell, R. M., and Atkins, A. G., 1987, *Manufacturing Engineering: Economics and Processes,* Prentice–Hall, Englewoods Cliffs, N.J.

[Micr81] *IBM Macro Assembler by Microsoft,* 1981, IBM(P/N 6172234), P.O. Box 1328-W, Boca Raton, FL 33432.

[Micr85a] *Microsoft FORTRAN Compiler User's Guide,* 1985, MS(P/N 005-014-029, D/N 8206L-330-05), 10700 Northup Way, Bellevue, WA 98004.

[Micr85b] *Microsoft FORTRAN Compiler Reference Manual,* 1985, MS(P/N 005-014-029, D/N 8205-330-09), 10700 Northup Way, Bellevue, WA 98004.

[Mill79] Miller, W. E., 1979, *Automation of Metallurgical Processes—An Overview,* General Electric Research Report No. 3059A.

[Mone87] Money, S. A., 1987, *Practical Microprocessor Interfacing,* Wiley–Interscience, New York.

[Mors82] Morse, S. P., 1982, *The 8086/8088 Primer,* Hayden, New York.

[Moto84] Motorola, 1984, *M68000 16/32 Bit Microprocessor: Programmer's Reference Manual,* Prentice–Hall, Englewood Cliffs, N.J.

[MuSh81] Murray, S. and R. L. Shoemaker, 1981, *Interfacing Microcomputers to the Real World,* Microcomputer Books/ Addison-Wesley, Reading, Mass.

[Oles70] Oleston, N. O.,1970, *Numerical Control,* Wiley, New York.

[Osbo78] Osborne, A., 1978, *Introduction to Microcomputers,* Osborne/McGraw–Hill, Berkeley.

[Palm86] Palm, W. J., 1986, *Control Systems Engineering,* Wiley, New York.

[PaWu83] Pandit, S. M., and Wu, S. M., 1983, *Time Series and System Analysis with Applications,* Wiley, New York.

[Potv85] Potvin, J., 1985, *Applied Process Control Instrumentation,* Reston, Reston, Va.

[PrWi77] Pressman, R. S., and Williams, J. E., 1977, *Numerical Control and Computer-Aided Manufacturing,* Wiley, New York.

[ReAl80] Rector, R., and Alexy, G., 1980, *The 8086 Book,* Osborne/McGraw–Hill, Berkeley.

[Ruoc87] Ruocco, S. R., 1987, *Robot Sensors and Transducers,* Halstead Press, New York.

[Scie85] *Scientific Solution LabMaster Installation Manual User's Guide,* 1985, No. 931654 Rev. A, Scientific Solution Inc., 6225 Cochran Road, Solon, OH 44139-3377.

[Sedg83] Sedgewick, R., 1983, *Algorithms,* Addison-Wesley, Reading, Mass.

[SeBo71] Seth, M. K., and Bollinger, J. G., 1971, *Incremental Control in Computer-Aided Manufacturing,* SME Paper MM71-285.

[Size68] Sizer, T. R., 1968, *The Digital Differential Analyzer,* Chapman & Hall, London.

[TaRA74] Takahashi, Y., Rabins, M. J., and Auslander, D. M., 1974, *Control,* Addison–Wesley, Reading, Mass.

[Tecm83] *PC-Mate LabTender Installation Manual User's Guide,* 1985, No. 20028-5-6-83, Tecmar Inc., 6225 Cochran Road, Cleveland, OH 44139.

[Thom71] Thomas, L. J., 1971, *N/C Handbook,* Bendix, Detroit.

[Ulso85] Ulsoy, A. G., 1985, "Applications of Adaptive Control Theory to Metal Cutting," in Donath, M. (ed), *Dynamic Systems: Modeling and Control,* ASME, New York.

[UlHa85] Ulsoy, A. G., and Han, E., 1985, "Tool Breakage Detection Using a Multi-Sensor Strategy," *Proceedings of the IFAC Conference on Control Science and Technology for Development,* Beijing, China, pp. 181–186.

[UlKR83] Ulsoy, A. G., Koren, Y., and Rasmussen, F., 1983, "Principal Developments in the Adaptive Control of Machine Tools," *ASME Journal of Dynamic Systems Measurement and Control,* Vol. 105, No. 2, pp. 107–112.

[WaDD87] Wallace, D. A., Darlow, M. S., and DeVries, W. R., 1987, "Parameter Estimation by Transform Equation Error Identification," *Proceedings of the 5th International Modal Analysis Conference,* London, pp. 1639–1644.

[WiKr83] Willen, D. C., and Krantz, J. I., 1983, *8088 Assembler Language Programming: The IBM PC,* Sams, Indianapolis.

[Wins84] Winston, P. H., 1984, *Artificial Intelligence,* Addison–Wesley, Reading, Mass.

[Wirt82] Wirth, N., 1982, *Programming in Modula-2,* Springer-Verlag, New York.

[Zaks80] Zaks, R., 1980, *Programming the Z-80,* Sybex, Berkeley, Calif.

[ZaLe79] Zaks, R., and Lesea, A., 1979, *Microprocessor Interfacing Techniques,* Sybex, Berkeley, Calif.

APPENDIX A

DESCRIPTION OF AN 8088-BASED SYSTEM

These notes provide an overview of the hardware and software aspects of a laboratory based on the IBM PC/XT microcomputers. These IBM PC/XT systems consist of

1. An 8088-based microcomputer
2. A CRT terminal
3. 5-and 1/4-in. floppy disk drive
4. 10 M byte hard disk drive
5. ADC, DAC, digital I/O port, and clock on Lab Tender or Lab Master boards from the Tecmar Co. and Scientific Solutions Inc. [Scie85; Tecm83].

Some of the systems also have printers. A complete set of manuals should be available in the laboratory. A book, such as [Mors82; ReAl80; WiKr83], describing the 8088-based IBM PC hardware and assembly language programming should also be available as a reference. Here, only a brief overview is provided.

Section A.1 provides an overview of the development system software and describes methods for passing parameters between FORTRAN and assembly language routines. Section A.2 provides necessary information for programming the ADCs and DACs. Section A.3 contains program listings for the case studies in Chapters 9 through 12.

A.1 DEVELOPMENT SOFTWARE FOR THE IBM SYSTEM

In order to develop and run programs on the IBM PC microcomputer, several pieces of software common to most small processors must be used: the operating system (DOS), editor, FORTRAN complier, assembler, and linker. The purpose of this section is to (1) briefly describe each of these various types of development software in terms of its purpose and relation to other software; (2) describe the type of files that serve as input and output; and (3) briefly describe several of the more useful commands. Further information concerning the development software can be found in the DOS, FORTRAN, and assembler manuals [Micr81; Micr85a; Micr85b].

Disk Operating System (DOS)

The DOS contains the I/O handling software (i.e., assigning a logical to a physical device) as well as providing features that allow for debugging assembly language programs. The debugger allows for the possibility of suspending a program; resuming execution; and examining, displaying, and modifying the contents of registers or memory.

The DOS is the program loaded by the bootstrap (a "read-only memory" hardware program that enables the processor to load disk files into memory) so that when the processor is turned on, the DOS is the first development program encountered. The system first runs a self-check, which ends with an audible beep and a message telling the user how much memory is available. It also asks the user for the date and time.

The DOS provides for the control and management of the hard and floppy disk drives. Once in DOS, the user is able to call other development software, such as the editor, FORTRAN compiler, assembler, and linker, necessary to create and run programs.

To interact with physical devices through FORTRAN programs, logical unit numbers must be assigned to physical devices. Console input and console output are automatically assigned by DOS to the default logical unit number '*' (or '0') for use in Microsoft FORTRAN [Micr85a; Micr85b] 'READ' and 'WRITE' statements. For example,

```
READ(*;10)x
```

will read the value of 'x' from the console input device (keyboard) using the FORMAT specified in statement number 10. Similarly a

```
WRITE(*,10)x
```

displays the value of 'x' according to the FORMAT specification in statement 10 on the console output device (CRT). When interacting with devices other than the console, such as files on the disk, the OPEN statement must be used to associate a physical device (or file) with a logical unit number. For example,

```
OPEN(7,FILE='LPT1:')
```

will assign the printer ('LPT1:') to the logical unit number 7, and output can be sent to the printer by using a statement like:

```
WRITE(7,10)x
```

As another example, consider reassigning the console to the logical unit number 5:

```
OPEN(5, FILE='CON:')
```

Now read and write statements directed to unit 5 will use the console device (keyboard and CRT). Files on disk drives can be accessed in a similar manner. To create a file 'file1' on disk drive 'A' and to assign it to logical unit number '1', use:

```
OPEN(1, FILE='A:FILE1',STATUS='NEW')
```

For more information on files and the use of the OPEN statement please consult [Micr85a; Micr85b].

DOS Commands

```
COPY          The format of the COPY command is:
              C > copy a:FILE1 b:FILE2
```

The file FILE1 on drive a will be copied to drive b and named FILE2.

```
DIR       The format of this command is:
          C > DIR d:
```

A list of all the files and directories on drive d: will be displayed along with their size and the date they were created. Adding the option /w will cause the list to be formatted differently — size and date information is not displayed so that the entire list will fit on the screen at one time.

```
DISKCOPY  The format of this command is
          C > DISKCOPY a: b:
```

The contents of the *entire* disk in drive a will be copied to drive b.

```
DISKCOMP  The format of this command is
          C > DISKCOMP a: b:
```

The disks in drives a and b will be compared, and any mismatches will be pointed out. This command is used after a DISKCOPY to ensure that a perfect copy was made.

```
ERASE      The format of this command is
               C > ERASE d: FILE
```

The file FILE on the diskette in drive d will be eliminated.

```
FORMAT     The Format of this command is
               C > FORMAT d:
```

The disk in drive d will be initialized to a recording format acceptable to DOS, and a directory, file allocation table, and system loader will be placed on the diskette.

```
PRINT      The format of this command is
               C > PRINT d:FILE
```

A prompt will appear, which should be answered with a carriage return. The file FILE on the diskette in drive d: will then be printed on the printer.

```
RENAME     The format of this command is
               C > TYPE d:FILE
```

The file FILE on drive d will be listed on the console. If the ‹CTRL›‹PRTSC› keys are first pressed, a hard copy can also be obtained. The file is printed without line numbers.

Editor

The editor is the program designed to facilitate the entry and the modification of FORTRAN or assembly language files. It allows the user to enter characters, append them, insert them, add lines, remove lines, search for characters or strings. Refer to the manual for the particular editor on your system for a list of editor commands. We have used VEDIT, a product of CompuView Products, Inc., Ann Arbor, Michigan.

FORTRAN Compiler

The Microsoft FORTRAN Compiler [Micr85a; Micr85b] is a development program that translates a FORTRAN code file into machine language (i.e., instructions in their binary form) and places the binary instructions in a new file (object file). The compiler is used after the editor and prior to the linker when writing software.

Batch files (file.bat) can be used for compiling and/or linking, with the option of linking with specific assembler object files for different case studies. Some potentially useful batch files for the case studies in Chapters 9 through 12 are indicated in tabular form here:

	Compile FORTRAN	Link FORTRAN Object Code	Link Assembler Object Code	
FTN	X			
FTNL	X	X		
LINK		X		
FTNL2	X	X	X	CHP10
LINK2		X	X	
FTNL3	X	X	X	CHP11
LINK3		X	X	
FTNL4	X	X	X	CHP12
LINK4	X	X	X	

For example the command:

`FTNL CHP9`

will take the FORTRAN source code in file CHP9.FOR, compile it to generate the object file CHP9.OBJ, and then link CHP9.OBJ with the necessary Microsoft FORTRAN libraries to produce the execution file CHP9.EXE. A sample batch file listing is provided in Figure A.1 for the batch file FTNL above.

MACRO Assembler

The MACRO assembler is the program that translates the mnemonic representation of instructions into their binary equivalent [Micr81]. It normally translates one symbolic instruction into one binary instruction (which may occupy 1, 2, 3, or 4 bytes). The resulting binary code is called object code. It is directly executable by the microcomputer. As a side effect, the assembler will also produce a complete symbolic listing of the program, as well as the equivalence tables to be used by the programmer and the symbol occurrence list in the program. In addition, the assembler will list syntax error such as misspelled or illegal instructions, branching errors, duplicate labels, or missing labels. It will not delete logical errors (this is your problem). The IBM assembler is called, while in DOS, by the command:

```
REM  Batch file to compile FORTRAN programs on the IBM PC under DOS.
REM  by Tsu Ren Ko, 1987
REM  NOTE: Do not type the .FOR extention when compiling. It
REM        is implied. If you do type .FOR, the following
REM        test will branch to FAIL.
EXT_EQS %1
IF NOT ERRORLEVEL == 1 GOTO OK
GOTO FAIL
:OK

REM  Now start compiling. This is a two pass compiler.
FOR1 %1,%1,%1,NUL;
IF ERRORLEVEL 1 GOTO FAIL
PAS2
IF ERRORLEVEL 1 GOTO FAIL

REM  In order to include more object files required for the
REM  .EXE program, change the first %1 to %1+obj1+...+objn.
REM  The line right below is a very long line ending with ';'.
LINKER %1,%1,NUL,C:\FORT77\CEXEC.LIB+C:\FORT77\8087.LIB+C:\FORT77\-FORTRAN.LIB;
REM  DOS can take no more than 128 characters in one line!
REM  To avoid exceeding this number, make file names short.
:FAIL
```

Figure A.1 A batch-file listing for FTNL

```
> MASM asmfile
```

The assembler will respond with the following prompts, where the file in brackets is the default response:

Prompt	Response
Object file name [asmfile.OBJ]:	⟨CR⟩
Source listing [NUL.LST]:	1. asmfile (will generate asmfile.LST listing)
	2. ⟨CR⟩ for no source listing.
Cross reference [NUL.CRF]:	⟨CR⟩

The source listing will contain the assembler source code and the corresponding assembler hexadecimal instructions. Any errors found will be flagged with a message at the appropriate line in the code. The error messages are also shown on the console whether or not a source listing is produced.

Example A.1 Creating and Running a Program

Turn on the computer, and wait for the message:

CURRENT DATE?

Respond by entering the date or pressing RETURN. Repeat when prompted for time. Place your disk in drive A (the floppy drive), and in response to the prompt "C:\USER > ", type

C:\USER >FORMAT A: /V

to initialize your disk (Warning: Do this the first time only!). Now use the editor Vedit to create a file by typing

C:\USER > Vedit A:FCHP9.FOR

This will create the file named FCHP9.FOR on your disk. You can now use editor commands to type a FORTRAN program. Then when finished, save the file and return to DOS. You can now compile your FORTRAN program by typing:

C:\USER > FTN A:FCHP9

Now invoke the editor again, as needed, to correct your FORTRAN program:

C:\USER > Vedit A:FCHP9.FOR

After your editing session, FCHP9.FOR will contain the corrected FORTRAN program and FCHP9.BAK will contain the original version. You can recompile it as follows:

C:\USER > FTN A:FCHP9

You can now invoke the linker to link your program with any required system routines:

C:USER > LINK A:FCHP9

and to run the program type:

C:\USER > A:FCHP9

<<>>

Sharing Data between FORTRAN and Assembly Routines

When calling an assembly subroutine from a FORTRAN program, there are two ways to transfer data. One is by parameter transfer, and the other is by using named common blocks. Regardless of which method is used, before the FORTRAN program transfers control to the assembly subroutine, it saves all the general registers' contents and the returning address into the stack, so that by referring to the correct stack position, the assembly subroutine knows where to return the control. After returning from the assembly subroutine, the FORTRAN program can restore the values in the registers, and the operations in the assembly subroutine will not affect the execution of the FORTRAN program after the assembly language subroutine call.

For the parameter transfer method, the FORTRAN program also uses the stack to transfer the addresses of the parameters. It pushes the parameters' addresses, in sequence, right before the return address. The structure of the stack before transferring to an assembly subroutine is shown in Figure A.2a. The assembly subroutine can, then, access these addresses by locating the correct positions in the stack. A formal method using the BP register in locating the positions is shown in the program listings that follow for DWNCNT. The stack structure after the command 'MOV BP SP' is shown in Fig. A.2b. One important detail to note is that, depending on how many parameters are transferred, a number equal to four times the number of parameters should be appended to the end of the RET command in the assembly subroutine.

For using named common blocks, the only thing required is to define the segments of the named common blocks in the assembly subroutine according to FORTRAN conventions so that after linking the named common blocks in the

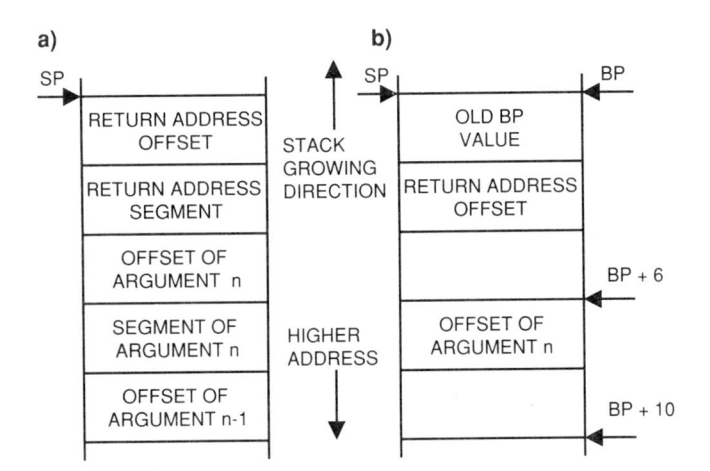

Figure A.2 Passing parameters between FORTRAN and Assembly routines: (a) Stack before transfering to Assembly subroutine, (b) Stack after "MOV BP SP" command

two different languages will match. The conventions include appending the sign '$A' at the end of the segment names, appending the sign '$' at the beginning of the segment classes, defining the variables inside the segments with the correct sequence and memory spaces, and using the GROUP assembler directive to form DGROUP, as shown in the program listings that follow for CHECKX.

After these procedures, the assembly subroutine can refer to the transferred variables in the same way as it does to the ordinary variables.

A.2 PROGRAMMING THE DACs, ADCs, AND DIGITAL I/O PORTS

The IBM PC/XTs each contain either a Lab Tender or Lab Master board from Tecmar or Scientific Solutions for ADC, DAC, Parallel I/O, and clock interrupts. For detailed information on these boards, their manuals should be consulted [Scie85; Tecm83]. Here we provide a brief overview of their use and programming.

The Lab Tender features 8-bit DACs and ADCs, digital I/O, and a timer with counters and is used with the IBM PC/XT to perform the data acquisition in the programs listed in Section A.3. There are 16 differential ADC channels in the Lab Tender with a 50-kHz conversion rate. They take input voltages in the range of $\pm$ 5 V, and convert them to an unsigned integer number from 0 to 255. In order to have two's complement integers, it is necessary to perform a conversion by subtracting the 0 volt value (i.e., 128) from the unsigned integers. There are also 16 DAC channels available in the Lab Tender, with a 3-μsec conversion rate. They take unsigned integer numbers from 0 to 255, and output voltages from -5 to $+4.96$ V. Conversion is again needed for two's complement integers. The different channels in DACs and ADCs are controlled separately by two multiplexers that can be programmed to select one channel at a time for each function. The only timer in the Lab Tender is a programmable timer. It contains five 16-bit counters and a pulse generator (clock) with programmable rates of up to 1 million pulses per second. There is a hardware setup within the IBM PC/XT so that when the counter overflows, a pulse can be sent through the IRQ4 line to the 8259 interrupt controller chip. This is used to control the stepping motors in the program listings for Chapter 11 that follow.

The parallel I/O is used to communicate with the IBM PC/XT. With this connection, the IBM PC/XT can receive data from the Lab Tender (e.g., results from ADCs, counter contents) and send data to the Lab Tender (e.g., setups of the Lab Tender control register, data for DACs). This requires 16 consecutive I/O locations on the IBM PC/XT, and the default starting address is set to be H0330. For detailed descriptions of the Lab Tender and the procedures to program it, refer to its installation manual and user guide [Tecm83]. A similar but more powerful product, the Lab Master, can also be used and is described in [Scie85].

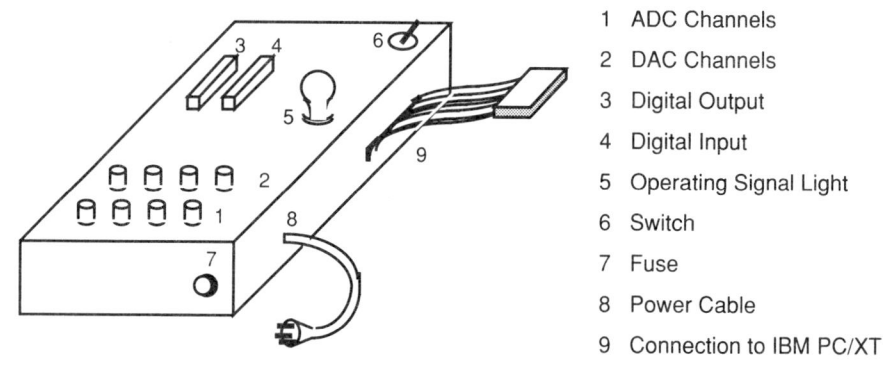

238 APPENDIX A DESCRIPTION OF AN 8088-BASED SYSTEM

1 ADC Channels
2 DAC Channels
3 Digital Output
4 Digital Input
5 Operating Signal Light
6 Switch
7 Fuse
8 Power Cable
9 Connection to IBM PC/XT

Figure A.3 Layout of the buffer box.

To interface signals to the Lab Tender, it is recommended that a buffer circuit box be constructed so that when the outside signals exceed the acceptable voltage range of the Lab Tender, the buffer box can disconnect the circuit to prevent damage. A suggested layout of the box is shown in Figure A.3 and was used in testing the program listings given in section A.3.

A.3 PROGRAM LISTINGS

In this section are complete program listing for each of the case studies described in Chapters 9 through 12 as implemented on the IBM PC/XT-based systems described in Sections A.1 and A.2. The programs are written in Microsoft FORTRAN and 8088 Assembly language. The versions given here are for the Tecmar Lab Tender board, which has 8-bit DACs and ADCs. They have all been successfully tested on actual laboratory systems. Real-time microcomputer systems, as we have seen in this textbook, are complex, so it is unlikely that these programs would run without modification on a similar system. These listings are intended to provide detailed examples that can help you to develop your own programs on similar systems or in similar applications. We encourage you to gain experience by actually applying the material covered in the book, and close with some words of advice:

- Back up all your programs on a diskette; it is only a matter of time before you will need these backups.
- Develop methodical procedures for troubleshooting your system to find malfunctioning (hardware or software) components.
- Just as in software development, develop and test all modules (hardware and software) of your system independently (e.g., use a function generator to test parallel I/O and ADC systems).

```
C      LAB1
C      ****************************************************************************
C      *    COMPUTATION OF A SETTING CLEARANCE FOR A ROLLING MILL.                *
C      ****************************************************************************
C      ****************************************************************************
C      *                                                                          *


C      *                        MAIN PROGRAM                                      *
C      *                                                                          *
C      ****************************************************************************
       CHARACTER ANS
       COMMON/PARA/PHO,PHF,PWD,PRD,PKC,PNC,PCW,PMU, PL1,PL2,PDI,PEM
       COMMON /RESULT/RFSW,RKML,PLO
C      ---------------------- SAMPLE DATA BLOCK --------------------------------
       SHO=0.2
       SHF=0.1
       SWD=30.
       SRD=5.
       SKC=80000.
       SAN=0.25
       SCW=0.05
       SMU=0.05
       SL1=24.
       SL2=9.
       SDI=2.7
       SEM=3.0E7
C      -------------------------------------------------------------------------
       CALL DTIN
       CALL DTDP
       CALL DTVLD
       CALL CALC
       CALL RTDP
810    WRITE (*,900)
       READ (*,809) ANS
       IF (ANS .EQ. 'Y') GO TO 800
       IF (ANS .EQ. 'y') GO TO 800
       GO TO 999
800    CALL ANTCAL
       CALL DTDP
       CALL DTVLD
       CALL CALC
       CALL DTDP
       CALL RTDP
       GO TO 810
999    WRITE (*,901)
```

```
C
809     FORMAT (A1)
900     FORMAT (' DO YOU NEED ANOTHER CALCULATION?(Y/N) '\)
901     FORMAT (' THIS IS THE END OF THE CALCULATION. GOOD BY!')
        STOP
        END
C
C       ***********************************************************************
C       *                          DATA INPUT
C       ***********************************************************************
        SUBROUTINE DTIN
C
        COMMON/PARA/PHO,PHF,PWD,PRD,PKC,PNC,PCW,PMU, PL1,PL2,PDI,PEM
        WRITE (*,101)
        READ (*,*) PHO
        WRITE (*,102)
        READ (*,*) PHF
        WRITE (*,103)
        READ (*,*) PWD
        WRITE (*,104)
        READ (*,*) PRD
        WRITE (*,105)
        READ (*,*) PKC
        WRITE (*,106)
        READ (*,*) PNC
        WRITE (*,107)
        READ (*,*) PCW
        WRITE (*,108)
        READ (*,*) PMU
        WRITE (*,109)
        READ (*,*) PL1
        WRITE (*,110)
        READ (*,*) PL2
        WRITE (*,111)
        READ (*,*) PDI
        WRITE (*,112)
        READ (*,*) PEM
        WRITE (*,113)
C
101     FORMAT ('1THE FOLLOWING DATA ARE NEEDED TO CALCULATE THE'/
       1        ' REQUIRED FREE LENGTH. PLEASE INPUT THE DESIRED'/
       2        ' VALUES ACCORDING TO THE INSTRUCTION.'/
       3        '0******************* DATA INPUT ********************'/
       4        '  1.ENTRY THICKNESS.(IN.)--------------------PHO= '\)
102     FORMAT ('  2.EXIT THICKNESS.(IN.)---------------------PHF= '\)
```

```
103     FORMAT (' 3.WIDTH OF THE SHEET.(IN.)------------------PWD= '\)
104     FORMAT (' 4.ROLL RADIUS.(IN.)-------------------------PRD= '\)
105     FORMAT (' 5.STRAIN HARDENING COEFFICIENT.(lbf/in2)----PKC= '\)
106     FORMAT (' 6.STRAIN HARDENING EXPONENT.----------------PNC= '\)
107     FORMAT (' 7.AMOUNT OF PRIOR COLD WORK.----------------PCW= '\)
108     FORMAT (' 8.FRICTION COEFFICIENT.---------------------PMU= '\)
109     FORMAT (' 9.DISTANCE FROM TOP CROSS MEMBER TO'/
       1        '    THE CENTERLINE OF THE BOTTOM ROLL.(IN.)---PL1= '\)
110     FORMAT (' 10.DISTANCE FROM THE CENTER OF THE'/
       1        '    TOP ROLL TO THE LEAD SCREW.(IN.)----------PL2= '\)
111     FORMAT (' 11.PITCH DIAMETER OF THE LEAD SCREW.(IN.)----PDI= '\)
112     FORMAT (' 12.ELASTIC MODULUS '/
       1        '    OF THE LEAD SCREW MATERIAL.(lbf/in2)------PEM= '\)
113     FORMAT ('0***************END OF THE DATA INPUT****************')
C
        RETURN
        END
C
C       ************************************************************************
C       *                         DATA DISPLAY                                 *
C       ************************************************************************
C
        SUBROUTINE DTDP
C
        COMMON/PARA/PHO,PHF,PWD,PRD,PKC,PNC,PCW,PMU,PL1,PL2,PDI,PEM
        WRITE (*,201) PHO,PHF,PWD,PRD,PKC,PNC,PCW,PMU,PL1,PL2,PDI,PEM
C
201     FORMAT ('1*********************************************'/
       1        ' *                 INPUT DATA           *'/
       2        ' *********************************************'/
       3        ' *                                      *'/
       4        ' *       1. PHO= ',F10.4,' (in.)        *'/
       5        ' *       2. PHF= ',F10.4,' (in.)        *'/
       6        ' *       3. PWD= ',F10.4,' (in.)        *'/
       7        ' *       4. PRD= ',F10.4,' (in.)        *'/
       8        ' *       5. PKC= ',F10.4,' (lbf/in2)    *'/
       9        ' *       6. PNC= ',F10.4,'              *'/
       A        ' *       7. PCW= ',F10.4,'              *'/
       B        ' *       8. PMU= ',F10.4,'              *'/
       C        ' *       9. PL1= ',F10.4,' (in.)        *'/
       D        ' *      10. PL2= ',F10.4,' (in.)        *'/
       E        ' *      11. PDI= ',F10.4,' (in.)        *'/
       F        ' *      12. PEM= ',E10.2,' (lbf/in2)    *'/
       G        ' *                                      *'/
       H        ' *********************************************')
```

```
C
      RETURN
      END
C
C     ****************************************************************
C     *                   INPUT DATA VALIDATION                      *
C     ****************************************************************
      SUBROUTINE DTVLD
C
      COMMON/PARA/PHO,PHF,PWD,PRD,PKC,PNC,PCW,PMU,PL1,PL2,PDI,PEM
C
      IF (PHO.GE.PHF) GO TO 350
      WRITE (*,301) PHO
      READ (*,*) PHO
      WRITE (*,302) PHF
      READ (*,*) PHF
350   IF (PCW .LT. 1.) GOTO 399
      WRITE (*,303) PCW
      READ (*,*) PCW
C
301   FORMAT ('    ATTENTION!! CALCULATION INTERRUPTED!!!'/
     1 ' THE EXIT THICKNESS IS GREATER THAN  THE ENTRY'/
     2 ' THICKNESS!   PLEASE CORRECT YOUR INPUT DATA!'/
     3 '     PHO= ',F10.4,'---------->PHO= '\)
302   FORMAT ('    PHF= ',F10.4,'---------->PHF= '\)
303   FORMAT ('    ATTENTION!! CALCULATION INTERRUPTED!!!'/
     1 ' THE AMOUNT OF PRIOR COLD WORK SHOULD BE LESS'/
     2 ' THAN 1.!    PLEASE CORRECT YOUR INPUT DATA!'/
     3 '     PCW= ',F10.4,'---------->PCW= '\)
399   RETURN
      END
C
C     ****************************************************************
C     *           CALCULATION FOR THE REQUIRED FREE LENGTH           *
C     ****************************************************************
      SUBROUTINE CALC
C
      COMMON/PARA/PHO,PHF,PWD,PRD,PKC,PNC,PCW,PMU,PL1,PL2,PDI,PEM
      COMMON /RESULT/RFSW,RKML,PLO
C     ---- CALCULATION FOR THE FLOW STRAINS AND STRESSES----------
C     -----flow strain at entry-----
      EPS1=ALOG(1./(1.-PCW))
C     -----flow stress at entry-----
      SIG1=PKC*EPS1**PNC
```

```
C      -----flow strain at exit------
       EPS2=EPS1+ALOG(PHO/PHF)
C      -----flow stress at exit------
       SIG2=PKC*EPS2**PNC
C      -----average flow stress------
       SIG0=(SIG1+SIG2)/2.
C      -----average thickness--------
       PH=(PHO+PHF)/2.
C      -----thickness reduction------
       DH=PHO-PHF
C      -----contact length-----------
       CL=SQRT(PRD**2-(PRD-DH/2.)**2)
C      ----roll separating force-----
       RFSW=PH*SIG0*(EXP(PMU*CL/PH)-1.)/PMU
C      ------------- CALCULATION FOR THE MILL MODULUS -------------------------
       PAI=3.14156
       RKML=(PEM*PAI*PDI**2)/(2.*(PL1-PHF-2.*PRD-PL2))
C      ------------- CALCULATION FOR THE MILL DEFLECTION ---------------------
       DL=RFSW*PWD/RKML
C      --------------- CALCULATION FOR THE FREE LENGTH ----------------------
       PLO=PL1+DL-(2.*PRD+PHF)
C      -----------END OF THE CALCULATION ------------------------------------
C
       RETURN
       END
C
C      ***********************************************************************
C      *              DISPLAY OF THE CALCULATED RESULT                       *
C      ***********************************************************************
       SUBROUTINE RTDP
C
       COMMON /RESULT/RSFW,RKML,PLO
       WRITE (*,700) PLO
700    FORMAT (' *                                           */
      1        ' *        FREE LENGTH:PLO= ',F10.4,' (in.) */
      2        ' *                                           */
      3        ' ****************************************************')
       RETURN
       END
```

```
C
C       *****************************************************************
C       *           DATA ALTERATION FOR ANOTHER CALCULATION             *
C       *****************************************************************
        SUBROUTINE ANTCAL
C
        CHARACTER AWR
        COMMON/PARA/PHO,PHF,PWD,PRD,PKC,PNC,PCW,PMU,PL1,PL2,PDI,PEM
C       -----------------------------------------------------------------
C       |  selection of the parameter                                   |
C       -----------------------------------------------------------------
90      WRITE (*,400)
        READ (*,*) N
        GO TO (1,2,3,4,5,6,7,8,9,10,11,12) N
C       -----------------------------------------------------------------
C       |  If the user select the number other than 1 through 12,       |
C       |  the computer requests the user try again.                    '
C       -----------------------------------------------------------------
13      WRITE (*,413)
        GO TO 90
1       WRITE (*,415)
        WRITE (*,401) PHO
        READ (*,*) PHO
        GO TO 99
2       WRITE (*,415)
        WRITE (*,402) PHF
        READ (*,*) PHF
        GO TO 99
3       WRITE (*,415)
        WRITE (*,403) PWD
        READ (*,*) PWD
        GO TO 99
4       WRITE (*,415)
        WRITE (*,404) PRD
        READ (*,*) PRD
        GO TO 99
5       WRITE (*,415)
        WRITE (*,405) PKC
        READ (*,*) PKC
        GO TO 99
6       WRITE (*,415)
        WRITE (*,406) PNC
        READ (*,*) PNC
        GO TO 99
```

```
7       WRITE (*,415)
        WRITE (*,407) PCW
        READ (*,*) PCW
        GO TO 99
8       WRITE (*,415)
        WRITE (*,408) PMU
        READ (*,*) PMU
        GO TO 99
9       WRITE (*,415)
        WRITE (*,409) PL1
        READ (*,*) PL1
        GO TO 99
10      WRITE (*,415)
        WRITE (*,410) PL2
        READ (*,*) PL2
        GO TO 99
11      WRITE (*,415)
        WRITE (*,411) PDI
        READ (*,*) PDI
        GO TO 99
12      WRITE (*,415)
        WRITE (*,412) PEM
        READ (*,*) PEM
9       WRITE (*,414)
        READ (*,416) AWR
        IF (AWR .EQ. 'Y') GO TO 90
        IF (AWR .EQ. 'y') GO TO 90
        GO TO 80

400     FORMAT (' INPUT THE PARAMETER NUMBER YOU WANT TO CHANGE'/
       1' N= '\)
401 FORMAT (' PHO= ',F10.4,'----------->PHO= '\)
402 FORMAT (' PHF= ',F10.4,'----------->PHF= '\)
403 FORMAT (' PWD= ',F10.4,'----------->PWD= '\)
404 FORMAT (' PRD= ',F10.4,'----------->PRD= '\)
405 FORMAT (' PKC= ',F10.4,'----------->PKC= '\)
406 FORMAT (' PNC= ',F10.4,'----------->PNC= '\)
407 FORMAT (' PCW= ',F10.4,'----------->PCW= '\)
408 FORMAT (' PMU= ',F10.4,'----------->PMU= '\)
409 FORMAT (' PL1= ',F10.4,'----------->PL1= '\)
410 FORMAT (' PL2= ',F10.4,'----------->PL2= '\)
411 FORMAT (' PDI= ',F10.4,'----------->PDI= '\)
412 FORMAT (' PEM= ',E10.2,'----------->PEM= '\)
413 FORMAT (' YOU HAVE SELECTED A WRONG NUMBER!'/
       1   ' TRY AGAIN!')
```

```
      414 FORMAT (' DO YOU WANT TO CHANGE ANOTHER PARAMETER?(Y/N)'\)
      415 FORMAT (' PLEASE INPUT A NEW VALUE.')
      416 FORMAT (A1)
C
       80 RETURN
          END
```

```
C     LAB2
C     ******************************************************************
C     *                                                                *
C     *                  DATA ACQUISITION ON A LATHE                   *
C     ******************************************************************
C     ******************************************************************
C     *                                                                *
C     *                        MAIN PROGRAM                            *
C     *                                                                *
C     ******************************************************************
C     ----------------------- NOMENCLATURE -----------------------------
C
C     Q       =METAL REMOVAL RATE --------------------------------(in.3/min)
C     HP      =TOTAL CUTTING HORSE POWER -------------------------(hp)
C     HPU     =UNIT HORSE POWER ----------------------------------(hp min/in.3)
C     FC      =AVERAGE CUTTING FORCE ----------------------------(lbf)
C     V       =AVERAGE CUTTING SPEED ----------------------------(ft/min)
C     F       =FEED --------------------------------------------(in./rev.)
C     D       =DIAMETER OF WORKPIECE ----------------------------(in.)
C     DC      =DEPTH OF CUT ------------------------------------(in.)
C     AVRPM   =AVERAGE SPINDLE SPEED ----------------------------(rev./min)
C     NP      =NUMBER OF PULSES IN AN INTERVAL
C     NT      =NUMBER OF TEETH ON A GEAR
C     DELTIM  =TIME INTERVAL -------------------------------------(sec)
C     M       =NUMBER OF SAMPLES
C     KT      =GAIN TO CONVERT VOLTS TO lbf ---------------------(lbf/VOLT)
C     K       =GAIN TO CONVERT ADC VALUE TO lbf -----------------(lbf/bit)
C     KD      =GAIN TO CONVERT bits TO VOLTS -------------------(VOLT/bit)
C     K       =KT*KD
```

```
C     ------------------------------------------------------------
      CHARACTER ANS
      INTEGER*2 TERMCT,SOURCF,ADCHN,ADCVAL,CLKTIC,NP
      REAL    KT
      COMMON /COUNT/TERMCT,SOURCF
      COMMON /PARA/D,F,DC,NT,NP,M,KT
      COMMON /RESULT/AVRPM,AVFORC,Q,HP,HPU
      COMMON /ADCBL/ADCHN,ADCVAL
      COMMON /FRQ/FREQCK
C     ------------------- SAMPLE DATA BLOCK -----------------------
      SD=1.0
      SF=0.1
      SDC=0.01
      SNT=10.
      SNP=20.
      SM=1.
      SKD=5.0/127
      SKT=40.
      SK=SKD*SKT
C     ------------------------------------------------------------
      CALL DTIN
  790 CALL DTDP
      CALL CLKFRQ
  800 CALL GETDAT
      CALL CALC
      CALL DTDP
      CALL RTDP
      WRITE (*,900)
      READ (*,901) ANS
      IF (ANS .EQ. 'Y') GO TO 810
      IF (ANS .EQ. 'y') GO TO 810
      GO TO 999
  810 WRITE (*,902)
      READ (*,901) ANS
      IF (ANS .EQ. 'Y') GO TO 820
      IF (ANS .EQ. 'y') GO TO 820
      GO TO 800
  820 CALL ANTCAL
      GO TO 790
  999 WRITE (*,903)
C
  900 FORMAT (' DO YOU NEED ANOTHER MEASUREMENT?(Y/N) '\)
  901 FORMAT (A1)
```

```
      902 FORMAT (' DO YOU WANT TO CHANGE THE PARAMETER?(Y/N) '\)
      903 FORMAT (' THIS IS THE END OF THE MEASUREMENT. GOOD BY!')
          STOP
          END
C
C     ************************************************************************
C     *                            DATA INPUT                                *
C     ************************************************************************
          SUBROUTINE DTIN
C
          INTEGER*2 ADCHN,ADCVAL,NP
          REAL    KT
          COMMON /PARA/D,F,DC,NT,NP,M,KT
          COMMON /ADCBL/ADCHN,ADCVAL
          COMMON /FRQ/FREQCK
C
          WRITE (*,101)
          READ (*,*) D
          WRITE (*,102)
          READ (*,*) F
          WRITE (*,103)
          READ (*,*) DC
          WRITE (*,104)
          READ (*,*) NT
          WRITE (*,105)
          READ (*,*) NP
          WRITE (*,106)
          READ (*,*) M
          WRITE (*,107)
          READ (*,*) KT
          WRITE (*,108)
          READ (*,*) ADCHN
          WRITE (*,109)
          READ (*,*) FREQCK
          WRITE (*,110)
C
      101 FORMAT ('1THE FOLLOWING DATA ARE REQUIRED TO MEASURE SPINDLE'/
         1        ' SPEED AND THE CUTTING FORCE. PLEASE INPUT THE DESIRED'/
         2        ' VALUES ACCORDING TO THE INSTRUCTION.'/
         3        '0******************* DATA INPUT ********************'/
         4        '0 1.OUTSIDE DIAMETER OF THE WORK.(IN.)----------D = '\)
      102 FORMAT ('   2.FEED.(IN.)--------------------------------F = '\)
      103 FORMAT ('   3.DEPTH OF CUT.(IN.)------------------------DC = '\)
      104 FORMAT ('   4.NUMBER OF THE TEETH OF THE GEAR.----------NT = '\)
      105 FORMAT ('   5.NUMBER OF PULSES.--------------------------NP = '\)
```

```
106 FORMAT (' 6.NUMBER OF SAMPLES.--------------------------M = '\)
107 FORMAT (' 7.GAIN TO CONVERT VOLT TO lbf.(lbf/VOLT)-----KT = '\)
108 FORMAT (' 8.CHANNEL NUMBER.-------------------------ADCHN = '\)
109 FORMAT (' 9.CLOCK FREQUENCY.(Hz)-----------------------',/
    1          '          (1000.GE.FREQCK.LE.5000)--------------FREQCK = '\)
110 FORMAT (' ***************END OF THE DATA INPUT*****************')
C
      RETURN
      END
C
C     ***********************************************************************
C     *                          DATA DISPLAY                               *
C     ***********************************************************************
      SUBROUTINE DTDP
C
      INTEGER*2 ADCHN,ADCVAL,NP
      REAL      KT
      COMMON /PARA/D,F,DC,NT,NP,M,KT
      COMMON /ADCBL/ADCHN,ADCVAL
      COMMON /FRQ/FREQCK
      WRITE (*,201) D,F,DC,NT,NP,M,KT,ADCHN,FREQCK
C
201 FORMAT ('1*****************************************************'/
    1       ' *                    INPUT DATA                     */
    2       ' *****************************************************'/
    3       ' *                                                   */
    4       ' * 1.THE DIAMETER OF THE WORK.---',F10.4, '(in.)      */
    5       ' * 2.THE FEED.--------------------',F10.4, '(in./rev.) */
    6       ' * 3.THE DEPTH OF CUT.-----------',F10.4, '(in.)      */
    7       ' * 4.THE No. OF THE TEETH.--------',I10 , '           */
    8       ' * 5.THE No. OF THE PULSES.-------',I10 , '           */
    9       ' * 6.THE No. OF THE SAMPLES.------',I10 , '           */
    A       ' * 7.THE GAIN.-------------------',F10.4, '(lbf/VOLT) */
    B       ' * 8.THE CHANNEL NUMBER.----------',I10 , '           */
    C       ' * 9.THE CLOCK FREQUENCY.--------',F10.4, '(Hz)       */
    D       ' *                                                   */
    E       ' *****************************************************')
C
      RETURN
      END
C
```

```
C      ************************************************************
C      * MEASUREMENTS OF THE SPINDLE SPEED AND THE CUTTING FORCE  *
C      ************************************************************
       SUBROUTINE GETDAT
C
       REAL    KT,KD
       INTEGER*2 TERMCT,SOURCF,ADCHN,ADCVAL,CLKTIC,NP
       COMMON /COUNT/TERMCT,SOURCF
       COMMON /PARA/D,F,DC,NT,NP,M,KT
       COMMON /RESULT/AVRPM,AVFORC,Q,HP,HPU
       COMMON /ADCBL/ADCHN,ADCVAL
       COMMON /FRQ/FREQCK
C      ------------------- DATA INITIALIZATION ---------------------
       AVRPM=0.
       AVFORC=0.
       KD=5./127.
C      ----------------------- MEASUREMENT -------------------------
       DO 600 I=1,M
       CALL TRAIL
       CALL CLOCK
       CALL DWNCNT(NP)
       CALL CLKOUT(CLKTIC)
       DELTIM=CLKTIC/FREQCK
       RPM=(60.*NP)/(NT*DELTIM)
       AVRPM=AVRPM+RPM
       CALL ADC
       FC=KT*KD*ADCVAL
       AVFORC=AVFORC+FC
  600  CONTINUE
C
       RETURN
       END
C
C      ************************************************************
C      * CALCULATION FOR THE SPINDLE SPEED AND THE CUTTING FORCE  *
C      ************************************************************
       SUBROUTINE CALC
C
       INTEGER*2 NP
       REAL    KT
       COMMON /PARA/D,F,DC,NT,NP,M,KT
       COMMON /RESULT/AVRPM,AVFORC,Q,HP,HPU
C      ---------- Average Rotational Speed of the Workpiece --------
       AVRPM=AVRPM/FLOAT(M)
C      ------------------ Average Cutting Force --------------------
       AVFORC=AVFORC/FLOAT(M)
```

```
C        ------------------- Average Cutting Speed -------------------------------
         PAI=3.14156
         V=PAI*D*AVRPM/12.
C        -------------------- Metal Removal Rate --------------------------------
         Q=12.*V*F*DC
C        -------------------- Total Horsepower ----------------------------------
         HP=AVFORC*V/33000.
C        -------------------- Unit Horsepower -----------------------------------
         HPU=HP/Q
C
         RETURN
         END
C
C        ***********************************************************************
C        *              DISPLAY OF THE CALCULATED RESULTS                      *
C        ***********************************************************************
         SUBROUTINE RTDP
C
         COMMON /RESULT/AVRPM,AVFORC,Q,HP,HPU
         WRITE (*,700) AVRPM,AVFORC,Q,HP,HPU
C
   700   FORMAT (' *               CALCULATED RESULTS          */'
        1        ' ****************************************************'/
        2        ' *                                                 */'
        3        ' * THE SPINDLE SPEED.--------',F12.4,  '(rev./min)   */'
        4        ' * THE CUTTING FORCE.--------',F12.4,  '(lbf)        */'
        5        ' * THE MTL. RMVL. RATE.------',F12.4,  '(in3./min)   */'
        6        ' * THE HORSEPOWER.-----------',E12.4,  '(hp)         */'
        7        ' * THE UNIT HORSEPOWER.------',E12.4,  '(hp min/in.3)*/'
        8        ' *                                                 */'
        9        ' ****************************************************')
C
         RETURN
         END
C
C        ***********************************************************************
C        *           DATA ALTERATION FOR ANOTHER MEASUREMENT                   *
C        ***********************************************************************
         SUBROUTINE ANTCAL
C
         INTEGER*2 ADCHN,ADCVAL,NP
         CHARACTER AWR
         REAL      KT
         COMMON /PARA/D,F,DC,NT,NP,M,KT
         COMMON /ADCBL/ADCHN,ADCVAL
         COMMON /FRQ/FREQCK
```

252 APPENDIX A DESCRIPTION OF AN 8088-BASED SYSTEM

```
C     -------------------------------------------------------------------
C     |  selection of the parameter                                      |
C     -------------------------------------------------------------------
   90 WRITE (*,400)
      READ (*,*) N
      GO TO (1,2,3,4,5,6,7,8,9) N
C     -------------------------------------------------------------------
C     |  If the user selects the number other than 1 through 9,          |
C     |  the computer requests the user try again.                       |
C     -------------------------------------------------------------------
   13 WRITE (*,413)
      GO TO 90
    1 WRITE (*,415)
      WRITE (*,401) D
      READ (*,*) D
      GO TO 99
    2 WRITE (*,415)
      WRITE (*,402) F
      READ (*,*) F
      GO TO 99
    3 WRITE (*,415)
      WRITE (*,403) DC
      READ (*,*) DC
      GO TO 99
    4 WRITE (*,415)
      WRITE (*,404) NT
      READ (*,*) NT
      GO TO 99
    5 WRITE (*,415)
      WRITE (*,405) NP
      READ (*,*) NP
      GO TO 99
    6 WRITE (*,415)
      WRITE (*,406) M
      READ (*,*) M
      GO TO 99
    7 WRITE (*,415)
      WRITE (*,407) KT
      READ (*,*) KT
      GO TO 99
    8 WRITE (*,415)
      WRITE (*,408) ADCHN
      READ (*,*) ADCHN
      GO TO 99
```

```
    9 WRITE (*,415)
      WRITE (*,409) FREQCK
      READ  (*,*) FREQCK
   99 WRITE (*,414)
      READ (*,416) AWR
      IF (AWR .EQ. 'Y') GO TO 90
      IF (AWR .EQ. 'y') GO TO 90
      GO TO 80
C
  400 FORMAT (' INPUT THE PARAMETER NUMBER YOU WANT TO CHANGE'/
     1        '     N= '\)
  401 FORMAT ('         D = ',F10.4,'------------>      D = '\)
  402 FORMAT ('         F = ',F10.4,'------------>      F = '\)
  403 FORMAT ('        DC = ',F10.4,'------------>     DC = '\)
  404 FORMAT ('        NT = ',I10  ,'------------>     NT = '\)
  405 FORMAT ('        NP = ',I10  ,'------------>     NP = '\)
  406 FORMAT ('         M = ',I10  ,'------------>      M = '\)
  407 FORMAT ('        KT = ',F10.4,'------------>     KT = '\)
  408 FORMAT ('     ADCHN = ',I10  ,'------------>  ADCHN = '\)
  409 FORMAT ('    FREQCK = ',F10.4,'------------> FREQCK = '\)
  413 FORMAT (' YOU HAVE SELECTED A WRONG NUMBER!'/
     1        '  TRY AGAIN!')
  414 FORMAT (' DO YOU WANT TO CHANGE ANOTHER PARAMETER?(Y/N)'\)
  415 FORMAT (' PLEASE INPUT A NEW VALUE.')
  416 FORMAT (A1)
C
   80 RETURN
      END
C
C     ***********************************************************************
C     *                       CLOCK FREQUENCY INPUT                         *
C     ***********************************************************************
      SUBROUTINE CLKFRQ
C
C     THIS ROUTINE IS TO INPUT CLOCK FREQUENCY, AND THEN
C     DETERMINE TERMCT AND SOURCF FOR THE LABTENDER CLOCK.
C
      INTEGER*2 TERMCT,SOURCF
      COMMON /COUNT/TERMCT,SOURCF
      COMMON /FRQ/FREQCK
```

```
C
      SOURCF = 5
      IF(FREQCK .LE. 5000.) GOTO 9
      WRITE(*,*) 'Clock frequency must be less than 5000. Hz.'
    4 WRITE (*,100)
  100 FORMAT (' Please input the correct value.---FREQCK= '\)
      READ (*,*) FREQCK
    9 CONTINUE
C     TESTCT = REAL VALUE FOR TERMCT, NECESSARY IF TERMCT > 32767.
   10 TESTCT = AINT((1*10**SOURCF)/FREQCK)
C     CHECK TO BE SURE TESTCT AND SOURCF ARE WITHIN THE DESIRED RANGE.
      IF (TESTCT .GT. 32767.) THEN
          SOURCF = SOURCF - 1
          GOTO 10
      ELSEIF (TESTCT .LT. 2.) THEN
          GOTO 4
      ELSEIF (SOURCF .LT. 2) THEN
          GOTO 4
      ELSE
          CONTINUE
          TERMCT = INT(TESTCT)
      ENDIF
C
      RETURN
      END


TITLE    TRAIL  -  PULSE DETECTION ROUTINE

PAGE  ,132         ;Set page width to 132 characters.
;     TRAIL is a routine that returns on the trailing (falling) edge of
;     one complete pulse. Remember that our interface boxes are
;     built with reverse logic, so a high input is a 0, and a low
;     is a 1. We use pin 7 for the signal. Therefore, pin 7 low
;     is 1XXXXXXX (80H), and pin 7 high is 0XXXXXXX (0H)
;     where X= don't care.
;     FORTRAN OR ASSEMBLER ''CALL TRAIL" calls this procedure.
DATA SEGMENT PUBLIC 'DATA'
;     For local assembler program data storage
;     Not used here, but required for linking
DATA ENDS
DGROUP GROUP DATA   ;The DATA segment will be linked into the
                    ;group called DGROUP, to match Microsoft
                    ;FORTRAN convention.
```

```
CODE    SEGMENT PUBLIC 'CODE'
        ASSUME  CS:CODE,DS:DGROUP,SS:DGROUP
        PUBLIC  TRAIL       ;Make TRAIL label available to other segments.
IOPORT  EQU     828         ;Address of Digital I/O port in IBM memory
LO      EQU     80H         ;Reference pin 7 lo signal, reverse logic
HI      EQU     0H          ;Reference pin 7 hi signal, reverse logic

TRAIL   PROC FAR

        PUSH    BP          ;Save calling framepointer
        MOV     BP,SP       ;on the stack.
        MOV     AH,HI       ;Put reference HI signal in AH
        MOV     DX,IOPORT   ;Use DX for indirect addressing
                            ;required by the IN and OUT commands.

TOP:    IN      AL,DX       ;Get data from port A, input port
        AND     AL,080H     ;Mask off bits 0-6
        CMP     AL,AH       ;Bit 7 hi yet?
        JNE     TOP         ;If not, loop until it is.
        MOV     AH,LO       ;Put reference LO signal in AH.

BOTTOM:
        IN      AL,DX       ;Get data from port A
        AND     AL,080H     ;Mask off bits 0-6
        CMP     AL,AH       ;Bit 7 lo yet?
        JNE     BOTTOM      ;If not, loop until it is
        MOV     SP,BP       ;Restore framepointer
        POP     BP
        RET

TRAIL   ENDP
CODE    ENDS

        END


TITLE   DWNCNT  -   PULSE COUNTING ROUTINE

PAGE    ,132            ;Set page width to 132 characters.
;       DWNCNT is a routine that counts NP pulses.
;       FORTRAN "CALL DWNCNT(NP)" calls this procedure.
DATA SEGMENT PUBLIC 'DATA'
;       For local assembler program data storage.
;       Not used here, but required for linking.
DATA ENDS
```

```
DGROUP  GROUP  DATA    ;The DATA segment will be linked into the group
                       ;called DGROUP to match Microsoft FORTRAN convention.
CODE    SEGMENT PUBLIC 'CODE'
        ASSUME  CS:CODE,DS:DGROUP,SS:DGROUP
        PUBLIC  DWNCNT          ;Make DWNCNT label available to other segments.
        EXTRN   TRAIL:FAR       ;Tell assembler that TRAIL is an external
                                ;label (in another Code Segment).
DWNCNT PROC FAR
        PUSH    BP              ;Save calling framepointer
        MOV     BP,SP           ;on the stack.
        LES     BX,DWORD PTR [BP+6]     ;ES,BX = address of NP.
        MOV     CX,ES:[BX]      ;Move value of NP into CX from memory address in BX.
                                ;CX is now the counting register.
GO:     PUSH    CX              ;Save CX before calling TRAIL.
                                ; CX contains the present pulse count.
        CALL    TRAIL           ;Wait for a pulse.
        POP     CX              ;Restore CX
        SUB     CX,1            ;Count a pulse.
        CMP     CX,0            ;If NP pulses have not arrived
        JNE     GO              ; yet, repeat.
                                ;
                                ;The codes above can be replaced simply by
                                ;   GO:  CALL TRAIL
                                ;        LOOP GO
        MOV     SP,BP           ;Restore framepointer
        POP     BP
        RET     04H             ;Return, pop 4 bytes off stack
                                ;for the address of the 1 parameter.

DWNCNT  ENDP

CODE    ENDS

END
```

```
TITLE     CLOCK   -   ROUTINE TO INITIALIZE COUNTERS 1 AND 2

PAGE    ,132            ;Set page width to 132 characters.

;       CLOCK is a routine that initializes Counter 1 and Counter 2
;       in the Labtender. Counters can be used to count and/or
;       generate various signals. The frequency sources, signal outputs,
;       and maximum counts are all programmable. Counter 1 will be used
;       as a clock, generating pulses at a desired frequency. Counter 2
;       will count the number of pulses that Counter 1 generates.
;       FORTRAN "CALL CLOCK" calls this procedure.
;   COUNTER 1:
;       TERMCT and SOURCF control the frequency generated by Counter 1.
;       TERMCT, the terminal count for Counter 1, is the number of pulses
;       Counter 1 will count from the source frequency before generating
;       a pulse. Counter 1 counts down from TERMCT to zero repetitively.
;       SOURCF selects the source frequency for Counter 1 from the Labtender.
;       NOTE THAT SOURCF IS NOT THE ACTUAL SOURCE FREQUENCY; IT IS USED TO
;       SELECT A FREQUENCY SOURCE ON THE LABTENDER!!
;       The Labtender has a 1 MHz crystal that generates 5 possible source
;       frequencies, from 1 MHz (10 6) to 100 Hz (10 2). To choose a source
;       frequency, we define SOURCF as the power of 10 of the desired source
;       frequency from the Labtender for Counter 1. We generally need
;       frequencies below 5 kHz in Lab, and higher source frequencies have
;       occasionally caused problems, so we will limit the source frequency
;       to 100 kHz and the clock frequency to 5 kHz. SOURCF must then be
;       between 2 and 5. TERMCT must be between 2 and 32767. A program
;       called CLKFRQ.FOR is available to select TERMCT and SOURCF, and
;       can be copied into your program and modified as desired.
;       Counter 1 will generate a pulse on pin OUT1 every time it counts
;       TERMCT pulses. The output frequency of Counter 1 (the `clock'
;       frequency) is (1*10**SOURCF)/TERMCT. To generate a 500-Hz clock,
;       choose 500 = 10000/20 so SOURCF = 4 and TERMCT = 20, for example.
;       The OUT1 pin can be used to generate timer interrupts. Enabling the
;       IRQ4 interrupt will cause the timer interrupts to be sent to the
;       processor. If we do not enable IRQ4 then the interrupts will be ignored.
;   COUNTER 2:
;       Counter 2 counts the pulses generated from Counter 1. The assembler
;       routine CLKOUT(CLKTIC) can then be used to read the value in Counter 2
;       and return it to FORTRAN as CLKTIC, the number of clock counts.

DATA SEGMENT PUBLIC 'DATA'
;       For local assembler program data storage.
;       Not used, but required for linking
DATA ENDS
```

258   APPENDIX A   DESCRIPTION OF AN 8088-BASED SYSTEM

```
COUNT$A SEGMENT COMMON '$COUNT'

;       This segment is equivalent to COMMON/COUNT/TERMCT,SOURCF in FORTRAN.
;       The $ and A are required to match the segment name and class that
;       the linker uses for the FORTRAN common.  TERMCT and SOURCF must be
;       declared INTEGER*2 in the FORTRAN calling routine before the
;       common block.

TERMCT  DW      ?
SOURCF  DW      ?

COUNT$A ENDS

DGROUP GROUP DATA,COUNT$A     ;The DATA and COUNT$A segments will be
                              ;linked into the group called DGROUP,
                              ;to match Microsoft FORTRAN convention.

CODE    SEGMENT PUBLIC 'CODE'

        ASSUME CS:CODE,DS:DGROUP,SS:DGROUP
        PUBLIC CLOCK       ;Make the CLOCK label available to other segments.
;       Labtender Addresses, in IBM PC, for the programmable timer.

CLKDATA EQU     824        ;Data Port for 9513 Timer.
CLKCOM  EQU     825        ;Command Port for 9513 Timer.

CLOCK PROC FAR

        PUSH    BP         ;Save calling framepointer
        MOV     BP,SP

        MOV     DL,67      ;Print a 'C' to the screen
        MOV     AH,2       ;for debugging
        INT     21H

;       First reset all counters to zero.

        MOV     DX,CLKCOM  ;Master Reset command.
        MOV     AL,255
        OUT     DX,AL

;       Set up the Master Mode Register.

        MOV     DX,CLKCOM  ;Set Data Pointer to Master Mode Register.
        MOV     AL,23
        OUT     DX,AL
```

```
        MOV DX,CLKDATA     ;Write data to Master Mode Register, lo-byte:
        MOV AL,0           ;Use default values
        OUT DX,AL


        MOV AL,128         ;Write data to Master Mode Register, hi-byte:
        OUT DX,AL          ;Use BCD division when dividing frequencies!!
                           ;Ex: 1 MHz crystal/1000 (base 10) = 1000 Hz

;   Set up Counter 1.

        MOV DX,CLKCOM      ;Disarm Counter 1, so we can
        MOV AL,193         ;write to its control port.
        OUT DX,AL


        MOV AL,1           ;Set Data Pointer to Counter 1's Mode Register.
        OUT DX,AL


        MOV DX,CLKDATA     ;Write Data to Counter 1 Mode Register, lo-byte:
        MOV AL,33          ;Count repetitively, reload from Load Register,
        OUT DX,AL          ;Count down, Terminal Count pulse high.

;       Manipulate SOURCF to create command to select source frequency.
;       By coincidence, (17 - SOURCF) will be correct command.
        MOV CX,SOURCF          ;Put SOURCF in CX.
        MOV AX,17              ;AX = 17
        SUB AX,CX              ;AX <-- AX - CX.  Now AL has command.
        OUT DX,AL          ;Write Data to Counter 1 Mode Register, hi-byte:
                           ;Count on rising edge, source frequency = 10**SOURCF.

;       Put TERMCT in Counter 1's Load Register.
;       TERMCT is the value from which Counter 1 will start counting down.

        MOV DX,CLKCOM      ;Set Data Pointer to Counter 1's Load Register.
        MOV AL,9
        OUT DX,AL


        MOV DX,CLKDATA     ;Write data to Counter 1's Load register.
        MOV AX,TERMCT      ;Put TERMCT in AX.
        OUT DX,AL          ;Send lo-byte to Load Register.
        MOV AL,AH          ;Put hi-byte of TERMCT into AL.
        OUT DX,AL          ;Send hi-byte to Load Register.
```

260 APPENDIX A DESCRIPTION OF AN 8088-BASED SYSTEM

```
;       Set up Counter 2.

        MOV DX,CLKCOM       ;Disarm Counter 2, so we can
        MOV AL,194          ;write to its control port.
        OUT DX,AL

        MOV AL,2            ;Set Data Pointer to Counter 2's Mode Register.
        OUT DX,AL

        MOV DX,CLKDATA      ;Write Data to Counter 2 Mode Register, lo-byte:
        MOV AL,42           ;Count repetitively, reload from Load Register,
        OUT DX,AL           ;Count up, square-wave output.

        MOV AL,0            ;Write Data to Counter 2 Mode Register, hi-byte:
        OUT DX,AL           ;Count on rising edge, source frequency = Counter 1.
                            ;Each time Counter 1 = 0 we get a pulse.

;       Put zero in Counter 2's Load Register.

        MOV DX,CLKCOM       ;Set Data Pointer to Counter 1's Load Register.
        MOV AL,10
        OUT DX,AL

        MOV AX,0            ;Put zero in AX.
        OUT DX,AL           ;Send lo-byte to Load Register.
        MOV AL,AH           ;Put hi-byte of AX into AL.
        OUT DX,AL           ;Send hi-byte to Load Register.

;       Turn on Counters 1 & 2.

        MOV AL,99           ;Command to Load and Arm Counters 1 & 2.
        OUT DX,AL

        MOV     SP,BP       ;Restore framepointer
        POP     BP
        RET                 ;Return

CLOCK ENDP

CODE ENDS

END
```

A.3 Program Listings

```
TITLE    CLKOUT   -   ROUTINE TO READ COUNTER 2

PAGE   ,132            ;Set page width to 132 characters.

;        CLKOUT is a procedure to read Counter 2 and return the value to
;        FORTRAN as CLKTIC, the number of clock counts.  To do this we will
;        save the present value of Counter 2 in the Counter 2 Hold Register,
;        and then put the value of the Hold Register in CLKTIC. We need to
;        use the Hold Register because we cannot directly read a Counter.
;        Note that calling CLKOUT does not stop Counter 2, so we could
;        read CLKTIC repeatedly, as Counter 2 counts pulses, if desired.
;        The FORTRAN program must be designed to make sure that CLKTIC
;        never overflows its 16-bit size (CLKTIC < 32767).  Otherwise,
;        the correct count would be lost.
;        FORTRAN ``CALL CLKOUT(CLKTIC)" calls this procedure.
;        Counter 2 counts the pulses generated from Counter 1. CLKTIC can
;        be used in FORTRAN to calculate the elapsed time, DELTIM, since
;        calling CLOCK to start Counters 1 and 2.
;        DELTIM = CLKTIC/FREQCK = pulses/(pulses/sec) = seconds, where
;        FREQCK = (1*10**SOURCF)/TERMCT.

DATA SEGMENT PUBLIC 'DATA'
;        For local assembler program data storage.
;        Not used, but required for linking
DATA ENDS

DGROUP GROUP DATA       ;The DATA segment will be linked into the group
                        ;called DGROUP to match Microsoft FORTRAN convention.

CODE    SEGMENT PUBLIC 'CODE'

        ASSUME CS:CODE,DS:DGROUP,SS:DGROUP
        PUBLIC CLKOUT     ;Make the CLKOUT label available to other segments.

;       Labtender Addresses in IBM PC for the programmable timer.
CLKDATA EQU     824      ;Data Port for 9513 Timer.
CLKCOM  EQU     825      ;Command Port for 9513 Timer.

CLKOUT  PROC FAR

        PUSH    BP       ;Save calling framepointer
        MOV     BP,SP
```

262 APPENDIX A DESCRIPTION OF AN 8088-BASED SYSTEM

```
        MOV     DX,CLKCOM  ;Save the value of Counter 2 in the
        MOV     AL,162     ;Counter 2 Hold Register.
        OUT     DX,AL

        MOV     AL,18      ;Set the Data Pointer register at
        OUT     DX,AL      ;the Counter 2 Hold Register.

        MOV     DX,CLKDATA ;Set DX at the clock data port.
        IN      AL,DX      ;Read the lo-byte of Counter 2.
        MOV     CL,AL      ;Save in CL.
        IN      AL,DX      ;Read the hi-byte of Counter 2.
        MOV     CH,AL      ;Save in CH.

        LES     BX,DWORD PTR [BP+6] ;
        MOV     ES:[BX],CX          ;Put value of Counter 2 in CLKTIC

        MOV     SP,BP      ;Restore framepointer
        POP     BP
        RET     04H        ;Return, pop 4 bytes (address of 1 parameter).

CLKOUT  ENDP

CODE    ENDS

END


TITLE  ADC - ANALOG-TO-DIGITAL CONVERSION ROUTINE

;       ADC is a routine that converts the analog input signal to
;       digital integer number. This routine is written for the
;       LABTENDER board that supports 8 bits analog to digital
;       conversion. The input voltage range is from +5 V to -5 V.

DATA    SEGMENT PUBLIC 'DATA'
;       For local assembler program data storage
;       Not used here, but required for linking
DATA    ENDS

ADCBL$A SEGMENT  COMMON  '$ADCBL'

;       This segment is equivalent to COMMON/ADCBL/ADCHN,ADCVAL
;       in FORTRAN. The $ and A are required to match the segment
;       name and class that the linker uses for the FORTRAN common.
;       ADCHN and ADCVAL must be declared INTEGER*2 in the FORTRAN
;       calling routine before the common block.
```

```
ADCHN       DW      ?
ADCVAL      DW      ?

ADCBL$A ENDS

DGROUP GROUP DATA,ADCBL$A  ;The DATA and ADCBL$A segments will be
                           ;linked into the group called DGROUP,
                           ;to match Microsoft FORTRAN convention.

CODE    SEGMENT PUBLIC 'CODE'

        ASSUME  CS:CODE,DS:DGROUP,SS:DGROUP
        PUBLIC  ADC         ;Make ADC label available to other segments.

;-----------------------------------------------------------------------
                        SUBTTL  A/D CONVERSION
;-----------------------------------------------------------------------
;   ADCHN is passed to ADC from the calling program, and ADCVAL
;   is returned through the COMMON/ADCBL/ADCHN,ADCVAL block.
;   FORTRAN "CALL ADC" calls this A/D procedure

LABTDR  EQU     816         ;Address of Labtender in IBM-PC memory
ADDATA  EQU     817         ;A/D Converter data register

ADC PROC FAR

        PUSH    BP          ;Save calling framepointer
        MOV     BP,SP       ;on the stack.

        MOV     DX,LABTDR   ;Indirect addressing using DX will have to
                            ;be used when using IN and OUT commands.
        MOV     AX,ADCHN    ;Get channel number
        MOV     AH,0        ;Mask off high byte, we only use AL

;       Now create the proper command byte for the Labtender's A/D routine,
;       based on the channel number.  Use the AL register.
        SHL     AL,1        ;Shift ADCHN left 1 bit (from bits 0-2 to 1-3)
        AND     AL,3FH      ;Mask off bits 6 and 7
        OR      AL,1        ;Put a 1 in bit 0 to start conversion
        OUT     DX,AL       ;Send command to ADC command port
```

```
;       Now create a byte that will match the value in the Labtender's
;       status port when the A/D conversion is complete.
        AND     AL,0FEH     ;Mask off error bit (bit 0), leave bits 1-6
        OR      AL,80H      ;the same so we are still looking at the same
                            ;channel, and put a 1 in bit 7 (status bit -
                            ;a 1 indicates that conversion is complete).
        MOV     AH,AL;Save the reference for `conversion complete' in AH.

CYCLE:
        IN      AL,DX       ;Read status port.
        CMP     AL,AH       ;Conversion complete?
        JNE     CYCLE       ;Wait until it is.

        MOV     DX,ADDATA   ;Point to data port.
        IN      AL,DX       ;Read data.
        MOV     AH,0        ;Clear hi-byte of AX
        SUB     AX,128      ;CONVERT FROM UNSIGNED TO SIGNED
                            ;0 to 255 becomes -128 to 127
        MOV     ADCVAL,AX   ;Return A/D value to calling program.

        MOV     SP,BP       ;Restore framepointer
        POP     BP
        RET                 ;Return

ADC ENDP

CODE ENDS

END
```


```
C       LAB3
C       ###########################################################
C       #                 STEPPING MOTOR CONTROL                  #
C       ###########################################################
C       ###########################################################
C       #                                                         #
C       #                      MAIN PROGRAM                       #
C       #                                                         #
C       ###########################################################
C
```

```
C       ------------------------- NOMENCLATURE ---------------------------------
C
C       AMAX  : PERMISSIBLE MAX. ACCELERATION.(IN./SEC.2)
C       AMIN  : PERMISSIBLE MIN. ACCELERATION.(IN./SEC.2)
C       ARCON : CONSTANT ACCELERATION.(BINARY WORD)
C       ARX   : ACCELERATION FOR X-AXIS.(BINARY WORD)
C       ARY   : ACCELERATION FOR Y-AXIS.(BINARY WORD)
C       AX    : ACCELERATION FOR X-AXIS. (in./sec.2)
C       DIRX  : DIRECTION OF MOTION ALONG X-AXIS.
C       DIRY  : DIRECTION OF MOTION ALONG Y-AXIS.
C       DRX   : DISPLACEMENT ALONG X-AXIS.(BINARY WORD)
C       DRY   : DISPLACEMENT ALONG Y-AXIS.(BINARY WORD)
C       DX    : DISPLACEMENT ALONG X-AXIS.(IN.) (SINGLE AXIS MOTION)
C       DXX   : DISPLACEMENT ALONG X-AXIS.(IN.) (TWO AXIS MOTION)
C       DYY   : DISPLACEMENT ALONG Y-AXIS.(IN.) (TWO AXIS MOTION)
C       FREQ  : CLOCK FREQUENCY.(Hz)
C       NLEFT : No. OF STEPS REQUIRED FOR ACCEL. AND DECEL.(BINARY WORD)
C       OUTWRD: CONTENT OF THE SIGNAL TO THE MOTOR.(BINARY WORD)
C       P1    : START POINT OF THE CIRCULAR MOTION.
C       P2    : STOP POINT OF THE CIRCULAR MOTION.
C       PITCH : PITCH OF THE LEAD SCREW.(REV./IN.)
C       R     : RADIUS.(IN.)
C       SLWRT : SLEW RATE OF THE MOTOR.(Hz)
C       STEP  : RESOLUTION OF THE MOTOR.(STEPS/REV.)
C       VC    : TANGENTIAL VELOCITY.      (in./sec)(CIRCULAR MOTION)
C       VCMAX : MAX. OF VC.(IN./SEC)
C       VCMIN : MIN. OF VC.(IN./SEC)
C       VRCON : CONSTANT VELOCITY.(BINARY WORD)
C       VX    : VELOCITY FOR X-AXIS.     (in./sec)
C       VY    : VELOCITY FOR Y-AXIS.     (in./sec)
C       XCOUNT: DISPLACEMENT ALONG X-AXIS.(BINARY WORD)
C       YCOUNT: DISPLACEMENT ALONG Y-AXIS.(BINARY WORD)
C       -------------------------------------------------------------------------
        CHARACTER       ANS
        INTEGER         CHOICE,P1,P2
        INTEGER*2       TERMCT,SOURCF
        INTEGER*2       ARX,VRX,DRX,XCOUNT,ARCON,VRCON,NLEFT,DIRX,OUTWRD
        INTEGER*2       ARY,VRY,DRY,YCOUNT,DIRY
        COMMON /COUNT / TERMCT,SOURCF
        COMMON /XBLK/ ARX,VRX,DRX,XCOUNT,ARCON,VRCON,NLEFT,
       1 DIRX,OUTWRD
```

```
      COMMON /YBLK  / ARY,VRY,DRY,YCOUNT,DIRY
      COMMON /SLCT  / CHOICE
      COMMON /XDATA / AX,VX,DX,VMIN,VMAX,AMIN,AMAX
      COMMON /XYDATA/ VT,DXX,DYY,VMINY,VMAXY
      COMMON /CDATA / VC,R,P1,P2,VCMIN,VCMAX
      COMMON /PARA  / PITCH,STEP,SLWRT,FREQ,MAXNUM
C     ----------------------------------------------------------------
      CALL DTIN
  100 CALL DTVD
      CALL CLKFRQ
      CALL CLOCK
      CALL SETDDA
  200 CALL SELECT
      GO TO (1,2,3,4) CHOICE
C     -------------------SINGLE-AXIS MOTION---------------------------
    1 CALL DTINX
      CALL DTVDX
      CALL CALCX
      WRITE (*,600)
      READ  (*,610) ANS
      CALL ENABLE
      CALL CHECKX
      CALL DISABL
      GO TO 900
C     -------------------TWO-AXIS MOTION------------------------------
    2 CALL DTINXY
      CALL DTVDXY
      CALL CALCXY
      WRITE (*,600)
      READ  (*,610) ANS
      CALL ENABLE
      CALL CHEKXY
      CALL DISABL
      CALL YCLEAR
      GO TO 900
C     -------------------CIRCULAR MOTION------------------------------
    3 CALL DTINCR
      CALL DTVDCR
      CALL CALCCR
      WRITE (*,600)
      READ  (*,610) ANS
      CALL ENABLE
      CALL CR
      CALL DISABL
      CALL YCLEAR
```

```
C       -------------------CONTINUE---------------------------------------
  900 WRITE (*,630)
      READ  (*,610) ANS
      IF (ANS.EQ.'1') GO TO 300
      IF (ANS.EQ.'2') GO TO 200
      GO TO 4
  300 CALL DTALT
      GO TO 100
C       -------------------QUIT-------------------------------------------
    4 WRITE (*,620)
      STOP
C
  600 FORMAT (
     1'        ATTENTION!!'/
     2'        Have you set up the table?'/
     3'        If you have finished setting up the table, '/
     4'        then hit the return key to start the motion.')
  610 FORMAT (A1)
  620 FORMAT ('0 This is the end of the program. Good by!')
  630 FORMAT ('0Which one do you wish to do?'//
     1'0      1. Alter the parameter.'/
     2'       2. Another motion.'/
     3'       3. Quit the program.'//
     4'0 Please enter the number. No. = '\)
C
      END
C
C     ###############################################################
C     #                      DATA INPUT                             #
C     ###############################################################
      SUBROUTINE DTIN
C
      COMMON /PARA/PITCH,STEP,SLWRT,FREQ,MAXNUM
C
      BIT=16
      MAXNUM=2**(BIT-1)
      WRITE (*,600)
      WRITE (*,610)
      READ (*,*) PITCH
      WRITE (*,620)
      READ (*,*) STEP
```

268 APPENDIX A DESCRIPTION OF AN 8088-BASED SYSTEM

```
        WRITE (*,630)
        READ (*,*) SLWRT
        WRITE (*,640)
        READ (*,*) FREQ
        WRITE (*,650)
        RETURN
C
 600 FORMAT (
       #'1****************************************************'/
       #' *                                                   *'/
       #' *              STEPPING MOTOR CONTROL               *'/
       #' *                                                   *'/
       #' ****************************************************'/
       #'                                                      '/
       #'           The following data are required.           '/
       #'                                                      '/
       #'              1. Pitch of the lead screw.             '/
       #'              2. Resolution of the motor.             '/
       #'              3. Slew rate of the motor.              '/
       #'              4. Clock frequency.                     '/
       #'                                                      '/
       #'           Please enter the above values              '/
       #'           according to the instruction.              ')
 610 FORMAT (
       #'0 1.Pitch of the lead screw.(rev./in.)               '/
       #'   (Typical value, 40 (rev./in.)------------PITCH = '       )
 620 FORMAT (
       #'0 2.Resolution of the motor.(steps/rev.)             '/
       #'   (Typical value, 200 (steps/rev.)----------STEP = '       )
 630 FORMAT (
       #'0 3.Slew rate of the motor. (Hz)                     '/
       #'   (Typical value, 1000 (Hz)----------------SLWRT = '       )
 640 FORMAT (
       #'0 4.Clock frequency.(Hz)                             '/
       #'   (Frequency must be less or equal to the slew rate !!)   '/
       #'                                    ---------FREQ = '\      )
 650 FORMAT (' ****************END OF THE DATA INPUT*****************')
C
        END
C
```

```fortran
C     ******************************************************************
C     *                        DATA DISPLAY                            *
C     ******************************************************************
      SUBROUTINE DTDP
C
      COMMON /PARA/PITCH,STEP,SLWRT,FREQ,MAXNUM
C
      WRITE (*,600) PITCH,STEP,SLWRT,FREQ
      RETURN
C
  600 FORMAT (
     #'1****************************************************************'/
     #' *                                                              */
     #' *                         INPUT DATA                           */
     #' ****************************************************************'/
     #' *                                                              */
     #' * 1.The pitch of the lead screw.--------',F8.0,' (rev./in.)    */
     #' * 2.The resolution of the motor.--------',F8.0,' (steps/rev.)  */
     #' * 3.The slew rate of the motor.---------',F8.0,' (Hz)          */
     #' * 4.The clock frequency.----------------',F8.0,' (Hz)          */
     #' *                                                              */
     #' ****************************************************************')
C
      END
C
C     ******************************************************************
C     *                       DATA VALIDATION                          *
C     ******************************************************************
      SUBROUTINE DTVD
C
      CHARACTER ANS
      COMMON /PARA/PITCH,STEP,SLWRT,FREQ,MAXNUM
C
  810 CALL DTDP
      WRITE (*,600)
      READ  (*,610) ANS
      IF (ANS.EQ.'Y') GO TO 900
      IF (ANS.EQ.'y') GO TO 900
  820 WRITE (*,620)
  800 READ  (*, *) N
      GO TO (1,2,3,4) N
      WRITE (*,630)
      GO TO 800
```

```
      1 WRITE (*,640)
        WRITE (*,650) PITCH
        READ  (*,  *) PITCH
        GO TO 810
      2 WRITE (*,640)
        WRITE (*,660) STEP
        READ  (*,  *) STEP
        GO TO 810
      3 WRITE (*,640)
        WRITE (*,670) SLWRT
        READ  (*,  *) SLWRT
        GO TO 810
      4 WRITE (*,640)
        WRITE (*,680) FREQ
        READ  (*,  *) FREQ
        GO TO 810
    900 IF (FREQ.LE.SLWRT) GO TO 999
        WRITE (*,690)
        GO TO 820
    999 RETURN
C
    600 FORMAT ('0Are all data correct? (Y/N)= '\)
    610 FORMAT (A1)
    620 FORMAT ('0Which parameter is wrong?'/
       #         ' Enter the parameter number. No. = '\)
    630 FORMAT ('0You have entered a wrong number. Try again!! No. = '\)
    640 FORMAT ('0Please enter the correct value.')
    650 FORMAT ('0PITCH = ',F8.0,'(rev./in.)------->PITCH = '\)
    660 FORMAT ('0STEP  = ',F8.0,'(steps/rev.)----->STEP  = '\)
    670 FORMAT ('0SLWRT = ',F8.0,'(Hz)------------>SLWRT = '\)
    680 FORMAT ('0FREQ  = ',F8.0,'(Hz)------------>FREQ  = '\)
    690 FORMAT ('0ATTENTION!'/
       #'        The clock frequency must be less or equal to the'/
       #'        slew rate!'/
       #'        Please correct the frequency and/or the slew rate. ')
C
        END
C
C    ####################################################################
C    #                    MOTION TYPE SELECTION                        #
C    ####################################################################
        SUBROUTINE SELECT
C
        INTEGER CHOICE
        COMMON /SLCT/ CHOICE
```

```
C
         WRITE (*,600)
      2  READ  (*, *) CHOICE
         GO TO (1,1,1,1) CHOICE
         WRITE (*,610)
         GO TO 2
      1  RETURN
C
     600 FORMAT (
                 #'0         What type of motion do you desire?    '/
                 #'                                                '/
                 #'          Please enter the number of the motion. '/
                 #'                                                '/
                 #'          1. Single axis motion with acceleration.  '/
                 #'          2. Two axis motion with constant velocity. '/
                 #'          3. Circular motion with constant velocity. '/
                 #'          4. Quit the program.                  '/
                 #'                                                '/
                 #'          No. = '\)
     610 FORMAT ('0 You have entered a wrong number. Try again!! No. = '\)
C
         END
C
C
C    ****************************************************************
C    *            DATA INPUT FOR SINGLE-AXIS MOTION                 *
C    ****************************************************************
         SUBROUTINE DTINX
C
         COMMON /PARA / PITCH,STEP,SLWRT,FREQ,MAXNUM
         COMMON /XDATA/ AX,VX,DX,VMIN,VMAX,AMIN,AMAX
C        ---------------CALCULATION FOR VMAX AND VMIN----------------
         VMAX=FREQ/(PITCH*STEP)
         VMIN=VMAX/MAXNUM
         WRITE (*,600)
         WRITE (*,610)
         READ  (*, *) DX
         WRITE (*,620) VMIN,VMAX
         READ  (*, *) VX
         AMIN1=VX*VX*FREQ/(2.*ABS(DX)*FREQ-VX)
         AMIN2=FREQ**2/(STEP*PITCH*MAXNUM)
         IF (AMIN1.GT.AMIN2) GO TO 10
         AMIN=AMIN2
         GO TO 20
      10 AMIN=AMIN1
```

```
      20 AMAX=VX
         WRITE (*,630) AMIN, AMAX
         READ  (*,  *) AX
         RETURN
C
     600 FORMAT (
        #'0*****************************************************************'/
        #' *                                                               * '/
        #' *           SINGLE-AXIS MOTION WITH ACCELERATION                * '/
        #' *                                                               * '/
        #' *****************************************************************'/
        #'                                                                   '/
        #'              Please enter the following values.                   '/
        #'                                                                   ')
     620 FORMAT (
        #'0 2. Constant velocity. (in./sec.)'/
        #'    (The range is,',F10.7,' < VX < ',F10.7,' (in./sec))'/
        #'    ------- VX = '\)
     610 FORMAT (
        #'0 1. Displacement. (in.)'/
        #'    (Displacement must be within -2.0=< DX =< 2.0.)'/
        #'    (Show minus sign, if negative direction.)'/
        #'    ------- DX = '\)
     630 FORMAT (
        #'0 3. Acceleration. (in./sec.2)'/
        #'    (The range is ,',F10.7,' < AX < ',F8.4,' (in./sec.2)'/
        #'    Otherwise, velocity will not reach the desired constant'/
        #'    velocity.)'/
        #'    ------- AX = '\)
C
         END
C
C      *************************************************************************
C      *            DATA DISPLAY FOR SINGLE-AXIS MOTION                        *
C      *************************************************************************
         SUBROUTINE DTDPX
C
         COMMON /XDATA/ AX,VX,DX,VMIN,VMAX,AMIN,AMAX
C
         WRITE (*,600) AX,VX,DX
         RETURN
```

```fortran
C
  600 FORMAT (
     #'0***************************************************************'/
     #' *                                                             * '/
     #' *                      INPUT DATA                             * '/
     #' *                                                             * '/
     #' ***************************************************************'/
     #' *                                                             */
     #' *         1.Acceleration      = ',F10.7,' (in./sec.2)         */
     #' *         2.Constant velocity= ',F10.7,' (in./sec)            */
     #' *         3.Displacement      = ',F8.5,' (in.)                */
     #' *                                                             */
     #' ***************************************************************')
C
      END
C
C     ############################################################################
C     #          DATA VALIDATION FOR SINGLE-AXIS MOTION                         #
C     ############################################################################
      SUBROUTINE DTVDX
C
      CHARACTER       ANS
      COMMON /PARA / PITCH,STEP,SLWRT,FREQ,MAXNUM
      COMMON /XDATA/ AX,VX,DX,VMIN,VMAX,AMIN,AMAX
C
   10 CALL DTDPX
      WRITE (*,500)
      READ  (*,510) ANS
      IF (ANS.EQ.'Y') GO TO 100
      IF (ANS.EQ.'y') GO TO 100
      WRITE (*,520)
   20 READ  (*,  *) N
      GO TO (1,2,3) N
      WRITE (*,530)
      GO TO 20
    1 WRITE (*,540)
      WRITE (*,550) DX
      READ  (*,  *) DX
      GO TO 10
    2 WRITE (*,540)
      WRITE (*,560) VX
      READ  (*,  *) VX
      GO TO 10
```

```
      3 WRITE (*,540)
        WRITE (*,570) AX
        READ  (*,  *) AX
        GO TO 10
C
    100 IF (DX.LE.2.0.AND.DX.GE.-2.0) GO TO 110
        WRITE (*,600) DX
        READ  (*,  *) DX
        GO TO 100
    110 IF (VX.GT.VMIN.AND.VX.LT.VMAX) GO TO 120
        WRITE (*,610) VX,VMIN,VMAX
        READ  (*,  *) VX
        GO TO 110
C
    120 AMIN1=VX*VX*FREQ/(2.*ABS(DX)*FREQ-VX)
        AMIN2=FREQ**2/(STEP*PITCH*MAXNUM)
        IF (AMIN1.GT.AMIN2) GO TO 30
        AMIN=AMIN2
        GO TO 40
     30 AMIN=AMIN1
     40 AMAX=VX
C
        IF (AX.GT.AMIN.AND.AX.LT.AMAX) GO TO 999
        WRITE (*,620) AX,AMIN,AMAX
        READ  (*,  *) AX
        GO TO 120
    999 RETURN
C
    500 FORMAT ('0 Are all data correct? (Y/N)= '\)
    510 FORMAT (A1)
    520 FORMAT ('0 Which data is wrong?   Enter the data number. No. = '\)
    530 FORMAT ('0 You have entered a wrong number. Try again! No. = '\)
    540 FORMAT ('  Please enter the correct value.')
    550 FORMAT (' DX =',F8.5,' (in.)----------------> DX = '\)
    560 FORMAT (' VX =',F8.5,' (in./sec.)-----------> VX = '\)
    570 FORMAT (' AX =',F8.5,' (in./sec.2)----------> AX = '\)
    600 FORMAT (
       #'0Displacement = ',F10.7,' is out of the range.'/
       #' The range is +-2.0,        Please correct the data.'/
       #' Displacement -------- DX  = '\)
    610 FORMAT (
       #'0Constant velocity = ',F8.5,' is out of range.'/
       #' The range is ',F10.7,' < VX < ',F10.7/
       #' Please correct the data.'/
       #' Constant velocity --- VX  = '\)
```

```
      620 FORMAT (
             #'0Acceleration = ',F10.7,' is out of range.'/
             #' The range is ',F10.7,' < AX   <',F10.7/
             #' Please correct the data.'/
             #' Acceleration -------- AX   = '\)
C
          END
C
C     ##########################################################################
C     #           CALCULATION FOR SINGLE-AXIS MOTION                           #
C     ##########################################################################
          SUBROUTINE CALCX
C
          CHARACTER ANS
          INTEGER*2      ARX,VRX,DRX,XCOUNT,ARCON,VRCON,NLEFT,DIRX,OUTWRD
          COMMON /XBLK / ARX,VRX,DRX,XCOUNT,ARCON,VRCON,NLEFT,
         1 DIRX,OUTWRD
          COMMON /XDATA/ AX,VX,DX,VMIN,VMAX,AMIN,AMAX
          COMMON /PARA / PITCH,STEP,SLWRT,FREQ,MAXNUM
C     ------------------Acceleration conversion--------------------------------
C     10.5 is added, so that the ARX will be the nearest round No.1
C     -------------------------------------------------------------------------
          ARX=(MAXNUM*AX*PITCH*STEP/FREQ/FREQ)+0.5
C     ----------Initial value for velocity register----------------------------
          VRX=0
C     ----------Initial value for displacement register------------------------
          DRX=MAXNUM-1
C     ----------Total steps for displacement-----------------------------------
          XCOUNT=ABS(DX)*PITCH*STEP
C     ----------Constant acceleration------------------------------------------
          ARCON=ARX
C     ----------Value for constant velocity------------------------------------
          VRCON=(MAXNUM*VX*PITCH*STEP/FREQ)+0.5
C     ----------Value to calculate deceleration point--------------------------
          NLEFT=XCOUNT
C     ----------Setting up the direction register------------------------------
          IF (DX.LT.0.) GO TO 100
          DIRX=1
          GO TO 200
      100 DIRX=0
      200 CONTINUE
C
          RETURN
          END
```

276 APPENDIX A DESCRIPTION OF AN 8088-BASED SYSTEM

```fortran
C
C     ##########################################################################
C     #                   DATA INPUT FOR TWO-AXIS MOTION                        #
C     ##########################################################################
      SUBROUTINE DTINXY
C
      COMMON /PARA   / PITCH,STEP,SLWRT,FREQ,MAXNUM
      COMMON /XYDATA/ VT,DXX,DYY,VMINY,VMAXY
C     --------------- data input for displacements ----------------------------
      WRITE (*,600)
      READ  (*, *) DXX
      WRITE (*,610)
      READ  (*, *) DYY
C     -------- calculation for max. and min. tangential velocity --------------
      VMAX=FREQ/(PITCH*STEP)
      IF(ABS(DXX).LT.ABS(DYY)) GO TO 100
      VMAXY=VMAX*SQRT(1+(DYY/DXX)**2)
      VMINY=VMAXY/MAXNUM
      GO TO 200
  100 VMAXY=VMAX*SQRT(1+(DXX/DYY)**2)
      VMINY=VMAXY/MAXNUM
C     ------------- data input for tangential velocity ------------------------
  200 WRITE (*,620) VMINY,VMAXY
      READ  (*, *) VT
      RETURN
C
  600 FORMAT (
     #'0**********************************************************'/
     #' *                                                        *'/
     #' *         TWO-AXIS MOTION WITH CONSTANT VELOCITY          *'/
     #' *                                                        *'/
     #' **********************************************************'/
     #'                                                           '/
     #'            Please enter the following values.             '/
     #'                                                           '/
     #' 1. Displacement along X-axis. (in.)                       '/
     #' (Show minus sign if negative direction.)--------DXX = '   )
  610 FORMAT (
     #'02. Displacement along Y-axis. (in.)              '/
     #' (Show minus sign if negative direction.)--------DYY = '   )
  620 FORMAT (
     #'03. Constant tangential velocity. (in./sec)   '/
     #' (The range is ,',F8.5,' < VT < ',F8.5,' (in./sec) '/
     #' ---------------------------------------------- VT = '     )
C
      END
```

```
C
C      ****************************************************************
C      *              DATA VALIDATION FOR TWO-AXIS MOTION              *
C      ****************************************************************
       SUBROUTINE DTVDXY
C
       CHARACTER        ANS
       COMMON /XYDATA/ VT,DXX,DYY,VMINY,VMAXY
       COMMON /PARA  / PITCH,STEP,SLWRT,FREQ,MAXNUM
C
       I=0
    20 CALL DTDPXY
       WRITE (*,500)
       READ  (*,510) ANS
       IF (ANS.EQ.'Y') GO TO 70
       IF (ANS.EQ.'y') GO TO 70
       WRITE (*,520)
    10 READ  (*, *) N
       GO TO (1,2,3) N
       WRITE (*,530)
       GO TO 10
     1 WRITE (*,540)
       WRITE (*,550) DXX
       READ  (*, *) DXX
       GO TO 20
     2 WRITE (*,540)
       WRITE (*,560) DYY
       READ  (*, *) DYY
       GO TO 20
     3 WRITE (*,540)
       WRITE (*,570) VT
       READ  (*, *) VT
       GO TO 20
C
    70 IF (DXX.LE.2.0.AND.DXX.GE.-2.0) GO TO 80
       WRITE (*,580) DXX
       READ  (*, *) DXX
       I=I+1
       GO TO 70
    80 IF (DYY.LE.2.0.AND.DYY.GE.-2.0) GO TO 90
       WRITE (*,590) DYY
       READ  (*, *) DYY
       I=I+1
       GO TO 80
```

```
C     ------------------------------------------------------------------
C        If the user altered the displacement, then check the max.       |
C        and min. velocity again                                         |
C     ------------------------------------------------------------------
   90 IF (I.EQ.0) GO TO 310
C
      VMAX=FREQ/(PITCH*STEP)
      IF (ABS(DXX).GT.ABS(DYY)) GO TO 300
      VMAXY=VMAX*SQRT(1.+(DYY/DXX)**2)
      VMINY=VMAXY/MAXNUM
      GO TO 310
  300 VMAXY=VMAX*SQRT(1.+(DXX/DYY)**2)
      VMINY=VMAXY/MAXNUM
C
  310 IF(VT.GT.VMINY.AND.VT.LT.VMAXY) GO TO 100
      WRITE (*,600) VT,VMINY,VMAXY
      READ  (*, *) VT
      GO TO 90
  100 RETURN
C
  500 FORMAT ('0Are all data correct? (Y/N)= '\)
  510 FORMAT (A1)
  520 FORMAT ('0Which data is wrong? Enter the data number. No.  = '\)
  530 FORMAT ('0You have entered a wrong number. Try again. No.  = '\)
  540 FORMAT ('0Please enter the correct value.')
  550 FORMAT ('  DXX =',F8.5,' (in.)--------------> DXX = '\)
  560 FORMAT ('  DYY =',F8.5,' (in.)--------------> DYY = '\)
  570 FORMAT ('   VT =',F8.5,' (in./sec)---------> VT  = '\)
  580 FORMAT (
     #'0DXX= ',F8.5,' is out of the range. The range of DXX = +-2.0'/
     #' Please correct the input data.  DXX = '\) 590 FORMAT (
     #'0DYY= ',F8.5,' is out of the range. The range of DYY = +-2.0'/
     #' Please correct the input data.  DYY = '\)
  600 FORMAT (
     #'0VT = ',F10.7, ' is out of the range. '/
     #' The range is ,'F10.7, ' < VT < ',F10.7/
     #' Please correct the input data.   VT = '\)
C
      END
C
```

```
C     ##############################################################
C     #            DATA DISPLAY FOR TWO-AXIS MOTION                 #
C     ##############################################################
      SUBROUTINE DTDPXY
C
      COMMON /XYDATA/ VT,DXX,DYY,VMINY,VMAXY
C
      WRITE (*,600) DXX,DYY,VT
      RETURN
C
  600 FORMAT (
     #'0**************************************************************'/
     #' *                                                            *'/
     #' *                       INPUT DATA                           *'/
     #' *                                                            *'/
     #' **************************************************************'/
     #' *                                                            *'/
     #' *       1. Displacement along X-axis =',F8.5,' (in.)         *'/
     #' *       2. Displacement along Y-axis =',F8.5,' (in.)         *'/
     #' *       3. Tangential velocity       =',F10.7,'(in./sec)     *'/
     #' *                                                            *'/
     #' **************************************************************')
C
      END
C
C     ##############################################################
C     #            CALCULATION FOR TWO-AXIS MOTION                  #
C     ##############################################################
      SUBROUTINE CALCXY
C
      CHARACTER     ANS
      INTEGER*2     ARX,VRX,DRX,XCOUNT,ARCON,VRCON,NLEFT,DIRX,OUTWRD
      INTEGER*2     ARY,VRY,DRY,YCOUNT,DIRY
      COMMON /XBLK / ARX,VRX,DRX,XCOUNT,ARCON,VRCON,NLEFT,
     1 DIRX,OUTWRD
      COMMON /YBLK / ARY,VRY,DRY,YCOUNT,DIRY
      COMMON /XYDATA/ VT,DXX,DYY,VMINY,VMAXY
      COMMON /PARA / PITCH,STEP,SLWRT,FREQ,MAXNUM
C     ----------------------------------------------------------------
C     to avoid computational error, DXX has been checked.
C     ----------------------------------------------------------------
      IF (DXX.NE.0.) GO TO 10
      VX=0.
      VY=VT
      GO TO 20
```

```
      10 THETA=ATAN(ABS(DYY/DXX))
         VX=VT*COS(THETA)
         VY=VT*SIN(THETA)
      20 CONTINUE
C        ----------Setting for X-data------------------------------------
C        ----------Acceleration------------------------------------------
         ARX=0
C        ----------Velocity----------------------------------------------
         VRX=(MAXNUM*VX*PITCH*STEP/FREQ)+0.5
C        ----------Displacement------------------------------------------
         DRX=MAXNUM-1
C        ----------Counter-----------------------------------------------
         XCOUNT=ABS(DXX)*PITCH*STEP
C        ----------Direction---------------------------------------------
         IF(DXX.LT.0.) GO TO 100
         DIRX=1
         GO TO 200
     100 DIRX=0
C        ----------Setting for Y-data------------------------------------
C        ----------Acceleration------------------------------------------
     200 ARY=0
C        ----------Velocity----------------------------------------------
         VRY=(MAXNUM*VY*PITCH*STEP/FREQ)+0.5
C        ----------Displacement------------------------------------------
         DRY=MAXNUM-1
C        ----------Counter-----------------------------------------------
         YCOUNT=ABS(DYY)*PITCH*STEP
C        ----------Direction---------------------------------------------
         IF(DYY.LT.0.) GO TO 300
         DIRY=1
         GO TO 400
     300 DIRY=0
     400 CONTINUE
C
         RETURN
         END
C
C        ****************************************************************
C        *              DATA INPUT FOR CIRCULAR MOTION                  *
C        ****************************************************************
         SUBROUTINE DTINCR
C
         INTEGER      P1,P2
         COMMON /PARA / PITCH,STEP,SLWRT,FREQ,MAXNUM
         COMMON /CDATA/ VC,R,P1,P2,VCMIN,VCMAX
```

```
C
C       ---------------Calculation for VCMAX and VCMIN---------------------------
        VCMAX=FREQ/(PITCH*STEP)
        VCMIN=SQRT(2.)*VCMAX/MAXNUM
C
        WRITE (*,600)
        WRITE (*,610) VCMIN,VCMAX
        READ  (*,  *) VC
        WRITE (*,620)
        READ  (*,  *) R
        WRITE (*,630)
        WRITE (*,640)
        READ  (*,  *) P1
        WRITE (*,650)
        READ  (*,  *) P2
        RETURN
C
  600 FORMAT (
       #'0****************************************************************'/
       #' *                                                              *'/
       #' *                      CIRCULAR MOTION                         *'/
       #' *                                                              *'/
       #' ****************************************************************'/
       #'                                                                 '/
       #'                 Please enter the following values.              '/
       #'                                                                 ')
  610 FORMAT (
       #'0 1. Tangential velocity. (in./sec)'/
       #'    (The range is,',F10.7, ' < VC < ',F10.7, ' (in./sec))'/
       #'    -------- VC = '\)
  620 FORMAT (
       #'02. Radius. (IN.)'/
       #'    (Be sure that the max. is 2.0 (in.))'/
       #'    --------  R = '\)
```

```
      630 FORMAT (
             #'0    Select the start point and the stop point.'/
             #'                      POS.Y                      '/
             #'                        2                        '/
             #'                      **O**                      '/
             #'   PLAN VIEW      *    |    * R                  '/
             #'                *      |   / *                   '/
             #'              *        |  /    *                 '/
             #'            *          | /       *               '/
             #'          *            |/          *             '/
             #'    NEG.X 3 o--------------------------o 1 POS.X ')
      640 FORMAT (
             #'          *            |           *             '/
             #'            *          |         *               '/
             #'              *        |       *                 '/
             #'                *      |     *                   '/
             #'                  *    |   *                     '/
             #'                      **O**                      '/
             #'                        4                        '/
             #'                      NEG.Y                      '/
             #'    Please enter the point number.               '/
             #'3.Start point P1 = '\)
      650 FORMAT (
             #'4.Stop point P2 = '\)
C
          END
C
C      ****************************************************************
C      *           DATA VALIDATION FOR CIRCULAR MOTION                *
C      ****************************************************************
          SUBROUTINE DTVDCR
C
          CHARACTER     ANS
          INTEGER       P1,P2
          COMMON /PARA /PITCH,STEP,SLWRT,FREQ,MAXNUM
          COMMON /CDATA/VC,R,P1,P2,VCMIN,VCMAX
       20 CALL DTDPCR
          WRITE (*,500)
          READ  (*,510) ANS
          IF (ANS.EQ.'Y') GO TO 100
          IF (ANS.EQ.'y') GO TO 100
          WRITE (*,520)
       10 READ  (*, *) N
          GO TO (1,2,3,4) N
          WRITE (*,530)
          GO TO 10
```

```
    1 WRITE (*,540)
      WRITE (*,550) VC
      READ  (*,  *) VC
      GO TO 20
    2 WRITE (*,540)
      WRITE (*,560) R
      READ  (*,  *) R
      GO TO 20
    3 WRITE (*,540)
      WRITE (*,570) P1
      READ  (*,  *) P1
      GO TO 20
    4 WRITE (*,540)
      WRITE (*,580) P2
      READ  (*,  *) P2
      GO TO 20
  100 IF (VC.GT.VCMIN.AND.VC.LT.VCMAX) GO TO 200
      WRITE (*,600) VC,VCMIN,VCMAX
      READ  (*,  *) VC
      GO TO 100
  200 IF (R.LE.2.0) GO TO 300
      WRITE (*,610) R
      READ  (*,  *) R
      GO TO 200
  300 P=ABS((P1-P2)/2.-(P1-P2)/2)
      IF (P.NE.0.) GO TO 400
      WRITE (*,620)
      READ  (*,  *) P1
      WRITE (*,630)
      READ  (*,  *) P2
      GO TO 300
  400 RETURN
C
  500 FORMAT ('0Are all data correct? (Y/N)= '\)
  510 FORMAT (A1)
  520 FORMAT ('0Which data is wrong?'/
     #'      Enter the number of the data No. = '\)
  530 FORMAT ('0You have entered a wrong number. Try again! No.  = '\)
  540 FORMAT ('0Please enter the correct data.')
  550 FORMAT (' VC =',F10.7,' (in./sec)----------> VC = '\)
  560 FORMAT ('  R =',F8.5,' (in.)--------------->  R = '\)
  570 FORMAT (' P1 =',I8  ,' -------------------> P1 = '\)
  580 FORMAT (' P2 =',I8  ,' -------------------> P2 = '\)
```

```
      600 FORMAT (
          #'0Tangential velocity = ,',F10.7, ' is out of the range.'/
          #' The range is ,',F10.7, ' < VC < ',F10.7, ' (in./sec)'/
          #' Please correct the input data. ------------- VC = '\)
      610 FORMAT (
          #'0Radius =',F8.5,' is too large. Max. radius = 2.0 (in.)'/
          #' Please correct the input data. ------------- R = '\)
      620 FORMAT (
          #'0Your selected points are not correct.'/
          #' Please correct your data.   '/
          #' Start point, P1 = '\)
      630 FORMAT (' Stop point, P2 = '\)
C
          END
C
C     *****************************************************************
C     #           DATA DISPLAY FOR CIRCULAR MOTION                    #
C     *****************************************************************
          SUBROUTINE DTDPCR
C
          INTEGER    P1,P2
          COMMON /CDATA/ VC,R,P1,P2,VCMIN,VCMAX
          WRITE (*,600) VC,R,P1,P2
C
      600 FORMAT (
          #'0****************************************************'/
          #' *                                                  */
          #' *                   INPUT DATA                     */
          #' *                                                  */
          #' ****************************************************/
          #'                                                    */
          #'     1. Tangential velocity = ',F10.7, ' (in./sec)  */
          #'     2. Radius               = ',F8.5,'   (in.)     */
          #'     3. Start                = ',I8  ,'             */
          #'     4. Stop                 = ',I8  ,'             */
          #'                                                    */
          #' ****************************************************)
C
          RETURN
          END
C
```

```
C       ################################################################
C       #              CALCULATION FOR CIRCULAR MOTION                 #
C       ################################################################
        SUBROUTINE CALCCR
C
        INTEGER         P1,P2
        INTEGER*2       ARX,VRX,DRX,XCOUNT,ARCON,VRCON,NLEFT,DIRX,OUTWRD
        INTEGER*2       ARY,VRY,DRY,YCOUNT,DIRY
        COMMON /XBLK / ARX,VRX,DRX,XCOUNT,ARCON,VRCON,NLEFT,
       1 DIRX,OUTWRD
        COMMON /YBLK / ARY,VRY,DRY,YCOUNT,DIRY
        COMMON /CDATA/ VC,R,P1,P2,VCMIN,VCMAX
        COMMON /PARA / PITCH,STEP,SLWRT,FREQ,MAXNUM
C       ----------Setting for X-data------------------------------------
C       ----------Acceleration------------------------------------------
        ARX=0
C       ----------Velocity----------------------------------------------
        VRX=0
C       ----------Displacement------------------------------------------
        DRX=MAXNUM-1
C       ----------Counter-----------------------------------------------
        XCOUNT=R*PITCH*STEP
C       ----------Constant velocity-------------------------------------
        VRCON=(VC*MAXNUM*PITCH*STEP/FREQ)+0.5
C       ----------NLEFT-------------------------------------------------
        NLEFT=XCOUNT
C       ----------Setting for Y-data------------------------------------
C       ----------Acceleration------------------------------------------
        ARY=0
C       ----------Velocity----------------------------------------------
        VRY=0
C       ----------Displacement------------------------------------------
        DRY=DRX
C       ----------Counter-----------------------------------------------
        YCOUNT=XCOUNT
C       ----------Start and Stop----------------------------------------
C       ----------Counterclockwise--------------------------------------
        IF (P1 .EQ. 1 .AND. P2 .EQ. 2) GO TO 100
        IF (P1 .EQ. 2 .AND. P2 .EQ. 3) GO TO 110
        IF (P1 .EQ. 3 .AND. P2 .EQ. 4) GO TO 120
        IF (P1 .EQ. 4 .AND. P2 .EQ. 1) GO TO 130
C       ----------Clockwise---------------------------------------------
        IF (P1 .EQ. 1 .AND. P2 .EQ. 4) GO TO 140
        IF (P1 .EQ. 4 .AND. P2 .EQ. 3) GO TO 150
        IF (P1 .EQ. 3 .AND. P2 .EQ. 2) GO TO 160
        IF (P1 .EQ. 2 .AND. P2 .EQ. 1) GO TO 170
```

```
C
  100 DIRX=0
      DIRY=1
      ARCON=1
      GO TO 200
  110 DIRX=0
      DIRY=0
      ARCON=0
      GO TO 200
  120 DIRX=1
      DIRY=0
      ARCON=1
      GO TO 200
  130 DIRX=1
      DIRY=1
      ARCON=0
      GO TO 200
C
  140 DIRX=0
      DIRY=0
      ARCON=1
      GO TO 200
  150 DIRX=0
      DIRY=1
      ARCON=0
      GO TO 200
  160 DIRX=1
      DIRY=1
      ARCON=1
      GO TO 200
  170 DIRX=1
      DIRY=0
      ARCON=0
C
  200 CONTINUE
      RETURN
      END
```

```
C
C      ###########################################################################
C      #                    CLEARING Y-AXIS PARAMETERS                           #
C      ###########################################################################
       SUBROUTINE YCLEAR
C
       INTEGER*2    ARY,VRY,DRY,YCOUNT,DIRY
       COMMON /YBLK/ ARY,VRY,DRY,YCOUNT,DIRY
       ARY=0
       VRY=0
       DRY=0
       YCOUNT=0
       RETURN
       END
C
C      ###########################################################################
C      #                    ALTERATION FOR THE PARAMETERS                        #
C      ###########################################################################
       SUBROUTINE DTALT
C
       CHARACTER ANS
       COMMON /PARA/ PITCH,STEP,SLWRT,FREQ,MAXNUM
C
    30 WRITE (*,600) PITCH,STEP,SLWRT,FREQ
    10 READ  (*,  *) N
       GO TO (1,2,3,4) N
       WRITE (*,610)
       GO TO 10
     1 WRITE (*,620)
       WRITE (*,630) PITCH
       READ  (*,  *) PITCH
       GO TO 20
     2 WRITE (*,620)
       WRITE (*,640) STEP
       READ  (*,  *) STEP
       GO TO 20
     3 WRITE (*,620)
       WRITE (*,650) SLWRT
       READ  (*,  *) SLWRT
       GO TO 20
     4 WRITE (*,620)
       WRITE (*,660) FREQ
       READ  (*,  *) FREQ
```

288   APPENDIX A   DESCRIPTION OF AN 8088-BASED SYSTEM

```fortran
   20 WRITE (*,670)
      READ  (*, *) ANS
      IF (ANS.EQ.'Y') GO TO 30
      IF (ANS.EQ.'y') GO TO 30
      RETURN
C
  600 FORMAT ('0Which parameter do you want to change?       '/
     #'                                                      '/
     #'                 1. PITCH = ',F8.2,' (rev./in.)       '/
     #'                 2. STEP  = ',F8.2,' (steps/rev.)     '/
     #'                 3. SLWRT = ',F8.2,' (Hz)             '/
     #'                 4. FREQ  = ',F8.2,' (Hz)             '/
     #'                                                      '/
     #' Enter the parameter number. No. = '\)
  610 FORMAT ('0You have entered a wrong number. Try again! No. = '\)
  620 FORMAT ('0Please enter a new value.')
  630 FORMAT (' PITCH =',F8.2,' (rev./in.)----------> PITCH = '\)
  640 FORMAT (' STEP  =',F8.2,' (steps/rev.)--------> STEP  = '\)
  650 FORMAT (' SLERT =',F8.2,' (Hz)----------------> SLWRT = '\)
  660 FORMAT (' FREQ  =',F8.2,' (Hz)----------------> FREQ  = '\)
  670 FORMAT ('0Do you want to change another parameter? (Y/N)= '\)
C
      END
C
C ****************************************************************
C *                  CLOCK FREQUENCY INPUT                       *
C ****************************************************************
      SUBROUTINE CLKFRQ
C
C     THIS ROUTINE IS TO INPUT CLOCK FREQUENCY, AND THEN
C     DETERMINE TERMCT AND SOURCF FOR THE LABTENDER CLOCK.
C
      INTEGER*2 TERMCT,SOURCF
      COMMON /COUNT/TERMCT,SOURCF
      COMMON /PARA/PITCH,STEP,SLWRT,FREQ,MAXNUM
C
      SOURCF = 5
      IF (FREQ.LE.5000.) GO TO 9
    4 WRITE (*,100)
  100 FORMAT (' Please input the correct value.---FREQ= '\)
      READ (*, *) FREQ
    9 CONTINUE
C     TESTCT = REAL VALUE FOR TERMCT, NECESSARY IF TERMCT > 32767.
   10 TESTCT = AINT((1*10**SOURCF)/FREQ)
```

```
C       CHECK TO BE SURE TESTCT AND SOURCF ARE WITHIN THE DESIRED RANGE.
        IF (TESTCT .GT. 32767.) THEN
            SOURCF = SOURCF - 1
            GOTO 10
          ELSEIF (TESTCT .LT. 2.) THEN
            GOTO 4
          ELSEIF (SOURCF .LT. 2) THEN
            GOTO 4
          ELSE
            CONTINUE
            TERMCT = INT(TESTCT)
        ENDIF
C
        RETURN
        END


TITLE    ENABLE  -  ROUTINE TO ENABLE INTERRUPTS

PAGE   ,132           ;Set page width to 132 characters.
;      FORTRAN ``CALL ENABLE" calls this procedure.
DATA SEGMENT PUBLIC 'DATA'
;      For local assembler program data storage.
;      Not used, but required for linking
DATA ENDS
DGROUP GROUP DATA         ;The DATA segment will be linked into the group
                          ;called DGROUP to match Microsoft FORTRAN convention.

CODE   SEGMENT PUBLIC 'CODE'
       ASSUME CS:CODE,DS:DGROUP,SS:DGROUP
       PUBLIC ENABLE      ;Make the ENABLE label available to other segments.

;      Labtender Address, in IBM PC.
INTCLR EQU     818        ;Timer Interrupt Clear address.

;      8259 Programmable Interrupt Controller Addresses, in IBM PC.
IMR    EQU     021H       ;Interrupt Mask Register.

ENABLE PROC FAR

       PUSH    BP         ;Save calling framepointer
       MOV     BP,SP

       MOV     DX,INTCLR  ;Clear timer interrupt,
       MOV     AL,0       ;just in case it is not already.
       OUT     DX,AL      ;Write to clear the interrupt
```

290 APPENDIX A   DESCRIPTION OF AN 8088-BASED SYSTEM

```
        MOV  DL,69           ;Print an `E' to the screen
        MOV  AH,2            ;for debugging
        INT  21H


        MOV  DX,IMR
        IN   AL,DX           ;Get contents of IMR.
        AND  AL,11101111B    ;Mask to include bit 4. 0 = enable.
        CLI                  ;Clear interrupt flag
        OUT  DX,AL           ;Reset the IMR.
        STI                  ;Reset the interrupt flag

        MOV      SP,BP       ;Restore framepointer
        POP      BP
        RET                  ;Return

ENABLE  ENDP

CODE    ENDS

END



TITLE   DISABL  -  ROUTINE TO DISABLE INTERRUPTS

PAGE   ,132            ;Set page width to 132 characters.
;       FORTRAN ``CALL DISABL" calls this procedure.
DATA SEGMENT PUBLIC 'DATA'
;       For local assembler program data storage.
;       Not used, but required for linking
DATA ENDS
DGROUP GROUP DATA      ;The DATA segment will be linked into the group
                       ;called DGROUP to match Microsoft FORTRAN convention.
CODE    SEGMENT PUBLIC 'CODE'
        ASSUME CS:CODE,DS:DGROUP,SS:DGROUP
        PUBLIC DISABL  ;Make the DISABL label available to other segments.
;       Labtender Addresses, in IBM PC, for initialization routine.
INTCLR  EQU      818   ;Timer Interrupt Clear Address.
IOPORTB EQU      829   ;Parallel Port B Data.

;       8259 Programmable Interrupt Controller Addresses, in IBM PC.
IMR     EQU  021H      ;Interrupt Mask Register.

;       Constants:
TRAIL   EQU      0H
```

```
DISABL  PROC FAR

        PUSH    BP              ;Save calling framepointer
        MOV     BP,SP

        MOV DX,IMR
        IN  AL,DX               ;Get contents of IMR.
        OR  AL,00010000B        ;Mask to block bit 4. 1 = disable.
        OUT DX,AL               ;Reset the IMR.

        MOV DX,INTCLR           ;Clear timer interrupt,
        MOV AL,0                ;just in case it is not already.
        CLI                     ;Clear interrupt flag
        OUT DX,AL               ;Write to clear the interrupt
        STI                     ;Reset the interrupt flag

        MOV     DX,IOPORTB      ;To allow the stepper motor to be

                                ;manually reset, we must leave the
        MOV     AL,TRAIL        ;output to it high, so we do not
        OUT     DX,AL           ;complement (NOT) the lo trail, here.

        MOV DL,68               ;Print a 'D' to the screen
        MOV AH,2                ;for debugging
        INT 21H

        MOV     SP,BP           ;Restore framepointer
        POP     BP
        RET                     ;Return

DISABL  ENDP

CODE ENDS

END


TITLE   SETDDA - LAB THREE ASSEMBLY CODE

PAGE    ,132
;
;                       USER NOTES
;
;       To link the assembly language modules to a FORTRAN driver the
;       following common blocks must be included in the calling program's
;       main and subroutines (only needed where they are used):
```

292 APPENDIX A DESCRIPTION OF AN 8088-BASED SYSTEM

```
;
;       COMMON/COUNT/TERMCT,SOURCF
;       COMMON/XBLK/ARX,VRX,DRX,XCOUNT,ARCON,VRCON,NLEFT,DIRX,OUTWRD
;       COMMON/YBLK/ARY,VRY,DRY,YCOUNT,DIRY
;
;       All variables in common blocks must be declared INTEGER*2.
;
;       Routines are called as follows (not necessarily in this order):
;
;       CALL CLOCK, CALL SETDDA, CALL ENABLE, CALL CHECKX, CALL DISABL
;       CALL CHEKXY
;
;       CLOCK and SETDDA must be called before attempting to run the rest
;       of the routines.  CLOCK sets up the timer.  SETDDA sets up the
;       interrupt vector and the I/O port.
;
;       DIRX and DIRY are set in FORTRAN to indicate the direction of travel.
;       DIRX or DIRY = 1 = POSIT in assy --> POSITIVE MOVE
;                    = 0 = NEGAT in assy --> NEGATIVE MOVE
;       POSIT and/or NEGAT can be used as constants to compare to DIR,
;       for branching control in the DDA macro.
;
;-----------------------------------------------------------------------------
        SUBTTL  LAB3 ROUTINES - DDA16 MACRO, SETDDA, AND DDASER.
;-----------------------------------------------------------------------------

DATA SEGMENT PUBLIC 'DATA'
;       For local assembler program data storage.
;       Not used, but required for linking DATA ENDS

XBLK$A SEGMENT COMMON '$XBLK'

;       This segment is equivalent to FORTRAN's
;       COMMON/XBLK/ARX,VRX,XRX,XCOUNT,ARCON,VRCON,NLEFT,DIRX,OUTWRD.
;       The $ and A are required to match the segment name and class that
;       the linker uses for the FORTRAN common.  All variables must be
;       declared INTEGER*2 in the FORTRAN calling routine before the
;       common block.

ARX     DW      ?
VRX     DW      ?
DRX     DW      ?
XCOUNT  DW      ?
ARCON   DW      ?
```

```
VRCON     DW      ?
NLEFT     DW      ?
DIRX      DW      ?
OUTWRD    DW      ?

XBLK$A    ENDS


YBLK$A    SEGMENT COMMON '$YBLK'

;        This segment is equivalent to COMMON/YBLK/ARY,VRY,DRY,YCOUNT,DIRY.
;        The $ and A are required to match the segment name and class that
;        the linker uses for the FORTRAN common.  All variables must be
;        declared INTEGER*2 in the FORTRAN calling routine before the
;        common block.

ARY       DW      ?
VRY       DW      ?
DRY       DW      ?
YCOUNT    DW      ?
DIRY      DW      ?

YBLK$A    ENDS

DGROUP    GROUP   DATA,XBLK$A,YBLK$A
          ;The DATA,XBLK$A, and YBLK$A segments will be linked into the group
          ;called DGROUP to match Microsoft FORTRAN convention.

CODE      SEGMENT PUBLIC 'CODE'

          ASSUME  CS:CODE,DS:DGROUP,SS:DGROUP
          PUBLIC  SETDDA      ;Make the SETDDA label available to other segments.

;------------------------------------------------------------------------------
;         Macro for LAB #3:  DDA16
;------------------------------------------------------------------------------

;         Constants for DDA macro:
PLUSX     EQU     01H
MINUSX    EQU     02H
PLUSY     EQU     04H
MINUSY    EQU     08H
POSIT     EQU     1H
NEGAT     EQU     0H
```

294   APPENDIX A   DESCRIPTION OF AN 8088-BASED SYSTEM

```
DDA16 MACRO    AR,VR,DR,COUNT,PLUS,MINUS,DIR
      LOCAL    SNDPLS,NEGA,CT,FINISH
; notice the LOCAL statement comes right after the MACRO statement

        MOV AX,AR       ;GET AR.
        ADD VR,AX       ;VR<--VR+AR.
        MOV AX,VR       ;GET VR.
        ADD DR,AX       ;DR<--DR+VR.
        JO  SNDPLS      ;JUMP TO SNDPLS IF OVERFLOW.
        JMP FINISH      ;JUMP TO FINISH IF NOT.

SNDPLS:
        MOV AX,POSIT    ;GET POSIT.
        CMP DIR,AX      ;POSITIVE MOVE?
        JNE NEGA        ;IF NOT, JUMP TO NEGA.
        MOV AX,PLUS     ;IF IT IS,
        OR  OUTWRD,AX   ;OUTWRD<--PLUS.
        JMP CT          ;JUMP TO CT.

NEGA:
        MOV AX,MINUS    ;GET MINUS.
        OR  OUTWRD,AX   ;OUTWRD<--MINUS

CT:
        DEC COUNT       ;COUNT<--COUNT-1
        MOV AX,7FFFH    ;CLEAR SIGN BIT OF DR
        AND DR,AX       ;
FINISH:
        NOP             ;THIS IS LIKE continue STATEMENT IN FORTRAN

        ENDM

;--------------------------------------------------------------------------------
        SUBTTL  SETDDA ROUTINE
;--------------------------------------------------------------------------------
;       This routine initializes the digital I/O port and
;       sets up the IRQ4 interrupt vector.
;       FORTRAN ``CALL SETDDA" calls this procedure.

;       Labtender Address, in IBM PC, for SETDDA routine.
IOCOM   EQU     831     ;Parallel Port Control Register.

SETDDA  PROC FAR

        PUSH    BP              ;Save calling framepointer
        MOV     BP,SP
```

```
;       Set up the Digital I/O Port

        MOV     DX,IOCOM        ;Indirect addressing, address
                                ;of I/O command port to DX
        MOV     AL,90H          ;A input, B output, mode 0.
        OUT     DX,AL           ;Set up the I/O chip.

;       Load the IRQ4 interrupt vector (address to go to on interrupt 4).

        CLI                     ;Disable interrupts while we are modifying them
                                ;to prevent any castastrophes.
        PUSH    ES              ;Save Extra Segment
        MOV     AX,0000H
        MOV     ES,AX           ;Point Extra Segment at the ISR address table.
        MOV     DI,30H          ;Offset of entry for IRQ4 in the address table.
        MOV     AX,OFFSET DDASER ;Get offset of DDASER, our interrupt
                                ;service routine.
        STOSW                   ;Load offset in AX into address in DI,

                                ;and add 2 to DI to prepare to store CS.
        MOV     AX,CS           ;Get Code Segment for assembly code.
        STOSW                   ;Load code segment in AX into address in DI.
                                ;Now the 4-byte address of our DDASER begins at memory
                                ;location 30H. An IRQ4 interrupt (if enabled) will
                                ;cause the the microprocessor to immediately go to
                                ;that address, and start performing our DDASER.
        POP     ES              ;Restore Extra Segment
        STI                     ;Enable interrupts

        MOV     SP,BP           ;Restore framepointer
        POP     BP
        RET                     ;Return

SETDDA ENDP

;-------------------------------------------------------------------------
        SUBTTL  INTERRUPT SERVICE ROUTINE
;-------------------------------------------------------------------------

;       Labtender Addresses, in IBM PC, for DDASER routine.
INTCLR  EQU     818     ;Timer Interrupt Clear Address.
IOPORTB EQU     829     ;Parallel Port B Data.
```

```
;       8259 Programmable Interrupt Controller Addresses, in IBM
PC.
ICR     EQU     20H     ;Interrupt Command Register.
IMR     EQU     21H     ;Interrupt Mask Register.

;       Constants for DDASER routine.
TRAIL   EQU     0H      ;Value for a low signal to OUTWRD.
EOI     EQU     20H     ;End of Interrupt signal for 8259 Interrupt
Controller.

DDASER  PROC    FAR
        PUSH    BP              ;Save calling framepointer
        MOV     BP,SP

        PUSH    AX              ;Save all the registers we may be
        PUSH    BX              ;modifying. This is done automatically
        PUSH    CX              ;only during FORTRAN CALLs, not interrupts.
        PUSH    DX

        PUSH    DS              ;Save the data segment register
        MOV     AX,DGROUP       ;Store the segment DGROUP into the data segment
        MOV     DS,AX           ;so that we assure the addressing is correct

        MOV     AX,TRAIL        ;Initialize OUTWRD to zero.
        MOV     OUTWRD,AX


;       Check XCOUNT. If zero, bypass macro call for x DDA

DDAX:   CMP     XCOUNT,0
        JE      DDAY

;       Otherwise call DDA macro for x.

        DDA16   ARX,VRX,DRX,XCOUNT,PLUSX,MINUSX,DIRX

;       Check YCOUNT. If zero, bypass macro call for y DDA.

DDAY:   CMP     YCOUNT,0
        JE      CONTIN

;       Otherwise, call DDA macro for y.

        DDA16   ARY,VRY,DRY,YCOUNT,PLUSY,MINUSY,DIRY
```

```
CONTIN:  MOV      AX,OUTWRD
         CMP      AL,TRAIL
         JE       CLKEOI      ;Do not bother to send it if it is zero.

;        Otherwise send pulse.
;             translator logic requires that we complement:

             NOT      AL

         MOV      DX,IOPORTB  ;Must use indirect addressing with
                              ;address in DX.
         OUT      DX,AL       ;Send OUTWRD
         MOV      AL,TRAIL    ;Load TRAIL into AL.
         NOT      AL          ;Complement TRAIL for translator.

         MOV      CX,25
PLSE:    NOP                  ;Hold pulse for 25 counts, so we get a
         LOOP     PLSE        ;nice pulse, not just an impulse.
         OUT      DX,AL       ;Now send the TRAIL (low) signal in AL
                              ;to end the OUTWRD pulse.
CLKEOI:
;        First indicate to the timer that the interrupt it sent
;        has been serviced. This will allow it to send another one.
         MOV DX,INTCLR         ;Timer Interrupt Clear addr.
         MOV AL,0              ;Data could be anything
         OUT DX,AL             ;Write to clear the interrupt

;        Now tell the 8259 that the interrupt has been serviced
         MOV      AL,EOI       ;Get end-of interrupt signal
         MOV      DX,ICR       ;Send it to the 8259's
         OUT      DX,AL        ;Interrupt Command Register

         POP      DS

         POP      DX           ;Restore registers.
         POP      CX
         POP      BX
         POP      AX

         MOV      SP,BP        ;Restore framepointer
         POP      BP
```

298 APPENDIX A DESCRIPTION OF AN 8088-BASED SYSTEM

```
        IRET                    ;Interrupt return.
DDASER  ENDP


CODE ENDS

END



TITLE   CHECKX  -  SINGLE AXIS MOTION

PAGE    ,132            ;Set page width to 132 characters.
;       FORTRAN "CALL CHECKX" calls this procedure.
DATA SEGMENT PUBLIC 'DATA'
;       For local assembler program data storage.
;       Not used, but required for linking
DATA ENDS
XBLK$A SEGMENT COMMON '$XBLK'
;       This segment is equivalent to FORTRAN's
;       COMMON/XBLK/ARX,VRX,XRX,XCOUNT,ARCON,VRCON,NLEFT,DIRX,OUTWRD.
;       The $ and A are required to match the segment name and class that
;       the linker uses for the FORTRAN common.  All variables must be
;       declared INTEGER*2 in the FORTRAN calling routine before the
;       common block.
ARX     DW      ?
VRX     DW      ?
DRX     DW      ?
XCOUNT  DW      ?
ARCON   DW      ?
VRCON   DW      ?
NLEFT   DW      ?
DIRX    DW      ?
OUTWRD  DW      ?
XBLK$A ENDS
DGROUP GROUP DATA,XBLK$A ;The DATA and XBLK$A segments will be linked into the
group
        ;called DGROUP to match Microsoft FORTRAN convention.

CODE    SEGMENT PUBLIC 'CODE'
        ASSUME  CS:CODE,DS:DGROUP,SS:DGROUP
        PUBLIC  CHECKX          ;Make the CHECKX label available to other segments.

CHECKX PROC FAR

        PUSH    BP              ;Save calling framepointer
        MOV     BP,SP
```

```
            MOV     DL,88           ;Print an X to the screen
            MOV     AH,2
            INT     21H

PART1:      MOV     AX,VRX
            CMP     VRCON,AX        ;Wait till VRX = VRCON
            JG      PART1
            MOV     ARX,0           ;Turn off acceleration
            MOV     CX,XCOUNT       ;Determine number of pulses left to
                                    ;finish move.
            SUB     NLEFT,CX        ;NLEFT <---- NLEFT - XCOUNT

PART2:      MOV     AX,NLEFT
            CMP     XCOUNT,AX       ;Loop till ready to
            JG      PART2           ;start deceleration
            NEG     ARCON           ;Negate ARCON
            MOV     CX,ARCON
            MOV     ARX,CX          ;and load it to ARX

PART3:      CMP     XCOUNT,0        ;Loop till move
            JG      PART3           ;is complete.

BOTTOM:
            MOV     DL,70           ;Print an F to the screen for ``finished"
            MOV     AH,2
            INT     21H

            MOV     SP,BP           ;Restore framepointer
            POP     BP
            RET                     ;Return

CHECKX ENDP

CODE ENDS

END



TITLE   CHEKXY  -  TWO AXIS MOTION

PAGE    ,132            ;Set page width to 132 characters.
;       FORTRAN ``CALL CHEKXY" calls this procedure.  DATA SEGMENT PUBLIC 'DATA'
;       For local assembler program data storage.
;       Not used, but required for linking
DATA ENDS
```

300 APPENDIX A DESCRIPTION OF AN 8088-BASED SYSTEM

```
XBLK$A SEGMENT COMMON '$XBLK'
;       This segment is equivalent to FORTRAN's
;       COMMON/XBLK/ARX,VRX,XRX,XCOUNT,ARCON,VRCON,NLEFT,DIRX,OUTWRD.
;       The $ and A are required to match the segment name and class that
;       the linker uses for the FORTRAN common. All variables must be
;       declared INTEGER*2 in the FORTRAN calling routine before the
;       common block.
ARX        DW      ?
VRX        DW      ?
DRX        DW      ?
XCOUNT     DW      ?
ARCON      DW      ?
VRCON      DW      ?
NLEFT      DW      ?
DIRX       DW      ?
OUTWRD     DW      ?
XBLK$A ENDS
YBLK$A SEGMENT COMMON '$YBLK'
;       This segment is equivalent to COMMON/YBLK/ARY,VRY,DRY,YCOUNT,DIRY.
;       The $ and A are required to match the segment name and class that
;       the linker uses for the FORTRAN common. All variables must be
;       declared INTEGER*2 in the FORTRAN calling routine before the
;       common block.
ARY        DW      ?
VRY        DW      ?
DRY        DW      ?
YCOUNT     DW      ?
DIRY       DW      ?
YBLK$A ENDS
DGROUP GROUP DATA,XBLK$A,YBLK$A
           ;The DATA,XBLK$A, and YBLK$A segments will be linked into the group
           ;called DGROUP to match Microsoft FORTRAN convention.
CODE   SEGMENT PUBLIC 'CODE'
           ASSUME CS:CODE,DS:DGROUP,SS:DGROUP

           PUBLIC  CHEKXY     ;Make the CHEKXY label available to other segments.
           EXTRN   ENABLE:FAR,DISABL:FAR
                        ;Tell assembler that ENABLE and DISABL are labels in
                        ;another code segment, so we can call them if we want.
                        ;Normally we call them from the FORTRAN program.
CHEKXY PROC   FAR
       PUSH   BP          ;Save calling framepointer
       MOV    BP,SP
```

```
GOX:    MOV     AX,XCOUNT   ;Get XCOUNT
        CMP     AX,0        ;XCOUNT=0?
        JG      GOX         ;If not 0, jump to GOX

GOY:    MOV     AX,YCOUNT   ;Get YCOUNT
        CMP     AX,0        ;YCOUNT=0?
        JG      GOY         ;If not 0, jump to GOY

        MOV     SP,BP       ;Restore framepointer
        POP     BP
        RET                 ;Return

CHEKXY  ENDP

CODE ENDS

END


TITLE   CR  -  CIRCULAR MOTION

PAGE    ,132            ;Set page width to 132 characters.
;       FORTRAN "CALL CR" calls this procedure.
DATA SEGMENT PUBLIC 'DATA'
;       For local assembler program data storage.
;       Not used, but required for linking
DATA ENDS
XBLK$A SEGMENT COMMON '$XBLK'
;       This segment is equivalent to FORTRAN's
;       COMMON/XBLK/ARX,VRX,DRX,XCOUNT,ARCON,VRCON,NLEFT,DIRX,OUTWRD.
;       The $ and A are required to match the segment name and class that
;       the linker uses for the FORTRAN common.  All variables must be
;       declared INTEGER*2 in the FORTRAN calling routine before the
;       common block.
ARX     DW      ?
VRX     DW      ?
DRX     DW      ?
XCOUNT  DW      ?
ARCON   DW      ?
VRCON   DW      ?
NLEFT   DW      ?
DIRX    DW      ?
OUTWRD  DW      ?
XBLK$A ENDS
YBLK$A SEGMENT COMMON '$YBLK'
```

```
;       This segment is equivalent to COMMON/YBLK/ARY,VRY,DRY,YCOUNT,DIRY.
;       The $ and A are required to match the segment name and class that
;       the linker uses for the FORTRAN common.  All variables must be
;       declared INTEGER*2 in the FORTRAN calling routine before the
;       common block.
ARY     DW      ?
VRY     DW      ?
DRY     DW      ?
YCOUNT  DW      ?
DIRY    DW      ?
YBLK$A  ENDS
DGROUP  GROUP   DATA,XBLK$A,YBLK$A
;The DATA,XBLK$A, and YBLK$A segments will be linked into the
;group called DGROUP to match Microsoft FORTRAN convention.

CODE    SEGMENT PUBLIC 'CODE'

        ASSUME  CS:CODE,DS:DGROUP,SS:DGROUP
        PUBLIC  CR      ;Make the CR label available to other segments.
        EXTRN   ENABLE:FAR,DISABL:FAR
                        ;Tell assembler that ENABLE and DISABL are labels in
                        ;another code segment, so we can call them if we want.
                        ;Normally we call them from the FORTRAN program.

CR      PROC    FAR

        PUSH    BP              ;Save calling framepointer
        MOV     BP,SP

        MOV     DL,67           ;Print a C to the screen
        MOV     AH,2
        INT     21H

CMOVE:  MOV     CX,XCOUNT ;
        CMP     CX,0      ;
        JG      START           ;If XCOUNT>0, then calculate velocity.

GO:     MOV     CX,YCOUNT ;
        CMP     CX,0      ;
        JG      START           ;If YCOUNT>0, then calculate velocity.
        JMP     FIN             ;If it is not, jump to FIN.

START:  MOV     AX,ARCON        ;distinguish the starting point.
        CMP     AX,0            ;if ARCON=1, then VRX inc. and VRY dec.
        JE      YSTART          ;if ARCON=0, then VRX dec. and VRY inc.
```

```
XSTART:
        MOV     BX,NLEFT    ;Get NLEFT.
        SUB     BX,YCOUNT   ;BX <-- NLEFT-YCOUNT, BX is Y position.
        MOV     AX,VRCON    ;Get VRCON=VC.
        MUL     BX          ;DX+AX <-- VC*(NLEFT-YCOUNT).
        MOV     BX,NLEFT    ;BX <-- NLEFT.
        DIV     BX          ;AX <-- DX+AX/BX=VC*(NLEFT-YCOUNT)/NLEFT.
        MOV     VRX,AX      ;VRX <-- AX.

        MOV     BX,XCOUNT   ;Get XCOUNT.
        MOV     AX,VRCON    ;Get VRCON=VC.
        MUL     BX          ;DX+AX <-- VC*XCOUNT.
        MOV     BX,NLEFT    ;BX <-- NLEFT.
        DIV     BX          ;AX <-- DX+AX/BX=VC*XCOUNT/NLEFT.
        MOV     VRY,AX      ;VRY <--AX.
        JMP     CMOVE

YSTART:
        MOV     BX,YCOUNT   ;Get YCOUNT
        MOV     AX,VRCON    ;Get VRCON=VC.
        MUL     BX          ;DX+AX <-- VC*YCOUNT.
        MOV     BX,NLEFT    ;BX <-- NLEFT.
        DIV     BX          ;AX <-- DX+AX/BX=VC*YCOUNT/NLEFT.
        MOV     VRX,AX      ;VRX <-- AX.

        MOV     BX,NLEFT    ;Get NLEFT.
        SUB     BX,XCOUNT   ;BX <-- NLEFT-XCOUNT, BX IS X POSITION.
        MOV     AX,VRCON    ;Get VRCON=VC.
        MUL     BX          ;DX+AX <-- VC*(NLEFT-XCOUNT).
        MOV     BX,NLEFT    ;BX <-- NLEFT.
        DIV     BX          ;AX <-- DX+AX/BX=VC*(NLEFT-XCOUNT)/NLEFT
        MOV     VRY,AX      ;VRY <-- AX.
        JMP     CMOVE       ;

FIN:    NOP

        MOV     SP,BP       ;Restore framepointer
        POP     BP
        RET                 ;Return

CHEKCR  ENDP

CODE ENDS

END
```

```
C     LAB4
C     ************************************************************************
C     *                            DC MOTOR CONTROL                          *
C     ************************************************************************
C     ************************************************************************
C     *                                                                      *
C     *                            MAIN PROGRAM                              *
C     *                                                                      *
C     ************************************************************************
C     ----------------------- NOMENCLATURE ---------------------------------
C     GAIN   : OPEN LOOP GAIN.
C     TAU    : TIME CONSTANT.
C     KP     : PROPORTIONAL CONTROL GAIN.
C     KI     : INTEGRAL CONTROL GAIN.
C     TSAMPL : SAMPLING TIME.(sec)
C     NSAMPL : No. OF SAMPLES.
C     OLREF  : STEP INPUT VALUE.(OPEN LOOP)
C     CLREF  : DESIRED VALUE.(CLOSED LOOP)
C     ADCVAL : VALUE FROM A/D CONVERTER.
C     DACVAL : VALUE SEND TO D/A CONVERTER.
C     ADCHN  : ADC CHANNEL NO.
C     DACHN  : DAC CHANNEL NO.
C     FREQCK : CLOCK FREQUENCY.(Hz)
C     Y(I)   : ADC SAMPLING VALUE.
C     E(I)   : ERROR BETWEEN CLREF AND Y(I).
C     M(I)   : CONTROLLER VALUE.
C     ----------------------------------------------------------------------
      CHARACTER          ANS
      INTEGER*2          TERMCT,SOURCF,K
      INTEGER*2          ADCHN,ADCVAL,DACHN,DACVAL
      INTEGER            YSS,OLREF,CLREF,M(1000),Y(1000),E(1000)
      REAL               KP,KI
      COMMON /COUNT /    TERMCT,SOURCF
      COMMON /ADCBL /    ADCHN,ADCVAL
      COMMON /DACBL /    DACHN,DACVAL
      COMMON /CBLK  /    K
      COMMON /PARA  /    RUNTIM,FREQCK,OLREF,CLREF,TSAMPL,NSAMPL
      COMMON /RESULT/    YSS,GAIN,TAU,OLRISE,CLRISE,OVERSH
      COMMON /CONST /    C0,C1,KP,KI
      COMMON /CONST1/    A,B,P,Q
      COMMON /ARRAY /    M,Y,E
```

```
C     ------------------------------------------------------------
      CALL DTIN
100   CALL DTVD
      CALL CLKFRQ
      CALL CLOCK
      CALL SETISR
      CALL MTONOL
      CALL CALCOL
      CALL GRAFOL
      CALL DTINCL
210   CALL DTVDCL
      CALL CLKFRQ
      CALL CLOCK
      CALL CALCCL
      CALL MTONCL
      CALL GRAFCL
      WRITE (*,600)
      READ (*,620) ANS
      IF (ANS.EQ.'Y') GO TO 200
      IF (ANS.EQ.'y') GO TO 200
      WRITE (*,610)
      READ (*,620) ANS
      IF (ANS.EQ.'Y') GO TO 300
      IF (ANS.EQ.'y') GO TO 300
      GO TO 999
200   CALL ALTCL
      CALL CLKFRQ
      CALL CLOCK
      GO TO 210
300   CALL ALTOL
      GO TO 100
C
600   FORMAT ('0Do you desire another C/L run? (Y/N) = '\)
610   FORMAT ('0Do you desire another O/L run? (Y/N) = '\)
620   FORMAT (A1)
C
999   STOP
      END
```

```
C
C       ************************************************************
C       *                       DATA INPUT                         *
C       ************************************************************
        SUBROUTINE DTIN
C
        INTEGER*2    ADCHN,ADCVAL,DACHN,DACVAL
        INTEGER      OLREF,CLREF
        COMMON /ADCBL / ADCHN,ADCVAL
        COMMON /DACBL / DACHN,DACVAL
        COMMON /PARA  / RUNTIM,FREQCK,OLREF,CLREF,TSAMPL,NSAMPL
C
        WRITE (*,600)
   10   WRITE (*,610)
        READ  (*, *) ADCHN
        IF (ADCHN.GE.0.AND.ADCHN.LE.5) GO TO 20
        WRITE (*,510)
        GO TO 10
   20   WRITE (*,620)
        READ  (*, *) DACHN
        IF (DACHN.GE.0.AND.DACHN.LE.5) GO TO 30
        WRITE (*,510)
        GO TO 20
   30   WRITE (*,630)
        READ  (*, *) RUNTIM
        IF (RUNTIM.GE.4..AND.RUNTIM.LE.10.) GO TO 40
        WRITE (*,510)
        GO TO 30
   40   WRITE (*,640)
        READ  (*, *) FREQCK
        IF (FREQCK.GE.1..AND.FREQCK.LE.100.) GO TO 50
        WRITE (*,510)
        GO TO 40
   50   WRITE (*,650)
        READ  (*, *) OLREF
        IF (OLREF.GE.70.AND.OLREF.LE.127) GO TO 999
        WRITE (*,510)
        GO TO 50
  999   WRITE (*,670)
        RETURN
C
  510   FORMAT ('0 Your input data is not correct. Try again.')
```

```
  600 FORMAT (
       #'1****************************************************'/
       #' *                                                   *'/
       #' *                 DC MOTOR CONTROL                  *'/
       #' *                                                   *'/
       #' ****************************************************'/
       #'                                                      '/
       #'      The following data are required to open loop control.   '/
       #'                                                      '/
       #'             1. ADC channel No.                       '/
       #'             2. DAC channel No.                       '/
       #'             3. Running time.(sec)                    '/
       #'             4. Clock frequency.(Hz)                  '/
       #'             5. Open loop reference value.            '/
       #'                                                      '/
       #'             Please enter the above values            '/
       #'             according to the instruction.            ')
  610 FORMAT (
       #'0 1.ADC channel No.(0-5)--------------------ADCHN = '       )
  620 FORMAT (
       #'0 2.DAC channel No.(0-5)--------------------DACHN = '       )
  630 FORMAT (
       #'0 3.Running time.(sec.) (4.-10.)------------RUNTIM = '      )
  640 FORMAT (
       #'0 4.Clock frequency.(Hz) (1.-100.)----------FREQCK = '      )
  650 FORMAT (
       #'0 5.Open loop reference value.(70-127)-------OLREF = '      )
  670 FORMAT (
       #'0********************END OF THE DATA INPUT********************')
C
      END
C
C     ############################################################################
C     #                           DATA DISPLAY                                   #
C     ############################################################################
      SUBROUTINE DTDP
C
      INTEGER*2      ADCHN,ADCVAL,DACHN,DACVAL
      INTEGER        OLREF,CLREF
      COMMON /ADCBL / ADCHN,ADCVAL
      COMMON /DACBL / DACHN,DACVAL
      COMMON /PARA  / RUNTIM,FREQCK,OLREF,CLREF,TSAMPL,NSAMPL
C
      WRITE (*,600) ADCHN,DACHN,RUNTIM,FREQCK,OLREF
      RETURN
```

```fortran
C
  600 FORMAT (
     #'1**************************************************************'/
     #' *                       INPUT DATA                           *'/
     #' **************************************************************'/
     #' *                                                            *'/
     #' * 1.ADC channel------------------------',I8  ,' *'/
     #' * 2.DAC channel------------------------',I8  ,' *'/
     #' * 3.Running time-----------------------',F8.0,' (sec)        *'/
     #' * 4.Clock frequency.-------------------',F8.0,' (Hz)         *'/
     #' * 5.Open-loop reference value.---------',I8  ,' *'/
     #' *                                                            *'/
     #' **************************************************************')
C
      END
C
C     ############################################################################
C     #                          DATA VALIDATION                                 #
C     ############################################################################
      SUBROUTINE DTVD
C
      CHARACTER        ANS
      INTEGER*2        ADCHN,ADCVAL,DACHN,DACVAL
      INTEGER          OLREF,CLREF
      COMMON /ADCBL /  ADCHN,ADCVAL
      COMMON /DACBL /  DACHN,DACVAL
      COMMON /PARA  /  RUNTIM,FREQCK,OLREF,CLREF,TSAMPL,NSAMPL
C
  810 CALL DTDP
      WRITE (*,600)
      READ  (*,610) ANS
      IF (ANS.EQ.'Y') GO TO 900
      IF (ANS.EQ.'y') GO TO 900
  820 WRITE (*,620)
  800 READ  (*, *) N
      GO TO (1,2,3,4,5) N
      WRITE (*,630)
      GO TO 800
    1 WRITE (*,640)
      WRITE (*,650) ADCHN
      READ  (*, *) ADCHN
      IF (ADCHN.GE.0.AND.ADCHN.LE.5) GO TO 810
      WRITE (*,500)
      GO TO 1
```

```
      2 WRITE (*,640)
        WRITE (*,660) DACHN
        READ  (*, *) DACHN
        IF (DACHN.GE.0.AND.DACHN.LE.5) GO TO 810
        WRITE (*,500)
        GO TO 2
      3 WRITE (*,640)
        WRITE (*,670) RUNTIM
        READ  (*, *) RUNTIM
        IF (RUNTIM.GE.4..AND.RUNTIM.LE.10.) GO TO 810
        WRITE (*,500)
        GO TO 3
      4 WRITE (*,640)
        WRITE (*,680) FREQCK
        READ  (*, *) FREQCK
        IF (FREQCK.GE.1..AND.FREQCK.LE.100.) GO TO 810
        WRITE (*,500)
        GO TO 4
      5 WRITE (*,640)
        WRITE (*,690) OLREF
        READ  (*, *) OLREF
        IF (OLREF.GE.70.AND.OLREF.LE.127) GO TO 810
        WRITE (*,500)
        GO TO 5
    900 RETURN
C
    500 FORMAT ('0Your input data is not correct. Try again!')
    600 FORMAT ('0Are all data correct? (Y/N) = '\)
    610 FORMAT (A1)
    620 FORMAT ('0Which data is wrong?'/
       #             ' Enter the data number. No. = '\)
    630 FORMAT ('0You have entered a wrong number. Try again!! No. = '\)
    640 FORMAT ('0Please enter the correct value.')
    650 FORMAT ('0ADCHN = ',I8 ,'----------------->ADCHN = '\)
    660 FORMAT ('0DACHN = ',I8 ,'----------------->DACHN = '\)
    670 FORMAT ('0RUNTIM= ',F8.0,'(sec)----------->RUNTIM= '\)
    680 FORMAT ('0FREQCK= ',F8.0,'(Hz)------------>FREQCK= '\)
    690 FORMAT ('0OLREF = ',I8 ,'----------------->OLREF = '\)
C
        END
C
C
C   **************************************************************************
C   #                           MOTOR-OFF ROUTINE                            #
C   **************************************************************************
        SUBROUTINE OFF
```

```
C
      INTEGER*2      ADCHN,ADCVAL,DACHN,DACVAL
      COMMON /ADCBL / ADCHN,ADCVAL
      COMMON /DACBL / DACHN,DACVAL
  100 DACVAL=0
      CALL DAC
      CALL ADC
      ABADC=ABS(ADCVAL)
      IF(ABADC.GT.2) GO TO 100
      RETURN
      END
C
C     ########################################################################
C     #                    RUN THE MOTOR (OPEN LOOP)                         #
C     ########################################################################
      SUBROUTINE MTONOL
C
      INTEGER*2      ADCHN,ADCVAL,DACHN,DACVAL,K
      INTEGER        YSS,OLREF,CLREF,M(1000),Y(1000),E(1000)
      COMMON /ADCBL / ADCHN,ADCVAL
      COMMON /DACBL / DACHN,DACVAL
      COMMON /PARA  / RUNTIM,FREQCK,OLREF,CLREF,TSAMPL,NSAMPL
      COMMON /RESULT/ YSS,GAIN,TAU,OLRISE,CLRISE,OVERSH
      COMMON /ARRAY / M,Y,E
      COMMON /CBLK  / K
C
      CALL OFF
      NSAMPL=FREQCK*RUNTIM
      DACVAL=OLREF
      TSAMPL=1./FREQCK
      CALL DAC
      CALL ENABLE
      K=0
  120 KSAVE=K
  130 IF(KSAVE.EQ.K) GO TO 130
      CALL ADC
      Y(K)=ADCVAL
      IF(K.LT.NSAMPL) GO TO 120
      CALL DISABL
      CALL OFF
C
      RETURN
      END
C
```

```
C      ##########################################################################
C      #                    CALCULATION FOR OPEN-LOOP CONTROL                   #
C      ##########################################################################
       SUBROUTINE CALCOL
C
       INTEGER        YSS,OLREF,CLREF,M(1000),Y(1000),E(1000)
       COMMON /RESULT/ YSS,GAIN,TAU,OLRISE,CLRISE,OVERSH
       COMMON /PARA  / RUNTIM,FREQCK,OLREF,CLREF,TSAMPL,NSAMPL
       COMMON /ARRAY / M,Y,E
C      -----------------calculation for the gain--------------------------------
       SUMY=0.
       DO 100 I=NSAMPL-9,NSAMPL
       SUMY=SUMY+Y(I)
  100  CONTINUE
       YSS=SUMY/10.+0.5
       GAIN=FLOAT(YSS)/OLREF
C      -----------------calculation for the time constant----------------------
       YTAU=0.632*YSS
       DO 200 I=1,NSAMPL
       IF (Y(I).GE.YTAU) GO TO 300
  200  CONTINUE
  300  TAU=(((YTAU-Y(I-1))/(Y(I)-Y(I-1)))+(I-2))*TSAMPL
C      -----------------calculation for OLRISE----------------------------------
       DO 700 I=1,NSAMPL
       IF(Y(I).GE.YSS) GO TO 800
  700  CONTINUE
  800  OLRISE=(((YSS-Y(I-1))/(Y(I)-Y(I-1)))+(I-2))*TSAMPL
       RETURN
       END
C
C      ##########################################################################
C      #                    DATA INPUT FOR CLOSED LOOP                          #
C      ##########################################################################
       SUBROUTINE DTINCL
C
       INTEGER        CLREF,OLREF
       REAL           KP,KI
       COMMON /PARA  / RUNTIM,FREQCK,OLREF,CLREF,TSAMPL,NSAMPL
       COMMON /CONST / C0,C1,KP,KI
C
   10  WRITE (*,600)
       READ  (*, *) CLREF
       IF (CLREF.GE.0.AND.CLREF.LE.127) GO TO 20
       WRITE (*,500)
       GO TO 10
```

```
   20 WRITE (*,610)
      READ  (*, *) KP
      WRITE (*,620)
      READ  (*, *) KI
C
  500 FORMAT ('0 Your input data is not correct. Try again!')
  600 FORMAT (
     #'0****************************************************'/
     #'0Please enter the following values.'/
     #'                                    '/
     #'0    1.Closed-loop reference value.(0-127)-----CLREF= '\)
  610 FORMAT (
     #'0    2.Gain of the P control.--------------KP= '\)
  620 FORMAT (
     #'0    3.Gain of the I control.--------------KI= '\)
C
      RETURN
      END
C
C     ************************************************************
C     *            DATA VALIDATION FOR CLOSED LOOP               *
C     ************************************************************
      SUBROUTINE DTVDCL
C
      CHARACTER ANS
      INTEGER CLREF,OLREF
      REAL    KP,KI
      COMMON /PARA/ RUNTIM,FREQCK,OLREF,CLREF,TSAMPL,NSAMPL
      COMMON /CONST/ C0,C1,KP,KI
C
  100 CALL DTDPCL
      WRITE (*,600)
      READ  (*,610) ANS
      IF (ANS.EQ.'Y') GO TO 900
      IF (ANS.EQ.'y') GO TO 900
      WRITE (*,620)
   10 READ  (*, *) N
      GO TO (1,2,3,4,5) N
      WRITE (*,630)
      GO TO 10
    1 WRITE (*,640)
      WRITE (*,650) CLREF
      READ  (*, *) CLREF
      IF (CLREF.GE.0.AND.CLREF.LE.127) GO TO 100
      WRITE (*,500)
      GO TO 1
```

```
    2 WRITE (*,640)
      WRITE (*,660) KP
      READ  (*,  *) KP
      GO TO 100
    3 WRITE (*,640)
      WRITE (*,670) KI
      READ  (*,  *) KI
      GO TO 100
    4 WRITE (*,640)
      WRITE (*,700) TSAMPL
      READ  (*,  *) TSAMPL
      IF (TSAMPL.GE.0.01.AND.TSAMPL.LE.1.) GO TO 100
      WRITE (*,500)
      GO TO 4
    5 WRITE (*,640)
      WRITE (*,710) RUNTIM
      READ  (*,  *) RUNTIM
      IF (RUNTIM.GE.4..AND.RUNTIM.LE.10.) GO TO 100
      WRITE (*,500)
      GO TO 5
  900 FREQCK=1./TSAMPL
      RETURN
C
  500 FORMAT (' Your input data is not correct. Try again!')
  600 FORMAT ('0Are all data correct? (Y/N) = '\)
  610 FORMAT (A1)
  620 FORMAT ('0Which data is wrong? Enter the data number. No.  = '\)
  630 FORMAT ('0You have entered a wrong number. Try again! No.  = '\)
  640 FORMAT ('0Please enter the new value.')
  650 FORMAT ('0CLREF= ',I8,'------>CLREF = '\)
  660 FORMAT ('0KP    = ',F10.4,'----> KP = '\)
  670 FORMAT ('0KI    = ',F10.4,'----> KI = '\)
  700 FORMAT ('0TSAMPL= ',F10.4,'------>TSAMPL = '\)
  710 FORMAT ('0RUNTIM= ',F10.4,'------>RUNTIM = '\)
C
      END
C
C     *************************************************************************
C     *                 DATA DISPLAY FOR CLOSED LOOP                          *
C     *************************************************************************
      SUBROUTINE DTDPCL
C
      INTEGER CLREF,OLREF
      REAL KP,KI
      COMMON /PARA  / RUNTIM,FREQCK,OLREF,CLREF,TSAMPL,NSAMPL
```

```
      COMMON /CONST/  C0,C1,KP,KI
      WRITE (*,600) CLREF,KP,KI,TSAMPL,RUNTIM
 600  FORMAT (
     #'0****************************************************'/
     #'0                        1. CLREF = ',I8  ,'          '/
     #'                         2. KP    = ',F10.4,'         '/
     #'                         3. KI    = ',F10.4,'         '/
     #'                         4. TSAMPL= ',F10.4,'         '/
     #'                         5. RUNTIM= ',F10.4,'         '/
     #'0****************************************************')
      RETURN
      END
C
C     ###################################################################
C     #         CALCULATION FOR CLOSED-LOOP CONTROL                    #
C     ###################################################################
      SUBROUTINE CALCCL
C
      CHARACTER      ANS
      INTEGER        YSS,OLREF,CLREF,M(1000),Y(1000),E(1000)
      REAL           KP,KI
      COMMON /PARA  / RUNTIM,FREQCK,OLREF,CLREF,TSAMPL,NSAMPL
      COMMON /RESULT/ YSS,GAIN,TAU,OLRISE,CLRISE,OVERSH
      COMMON /CONST / C0,C1,KP,KI
      COMMON /CONST1/ A,B,P,Q
      COMMON /ARRAY / M,Y,E
C     -----------------calculation for constants----------------------
      A=-1./TAU
      B=-1.*A*GAIN
      P=EXP(A*TSAMPL)
      Q=(P-1.)*B/A
      WRITE (*,610) A,B,P,Q,KP,KI
 610  FORMAT (
     #'0        A= ',E10.4, ',     B= ',E10.4/
     #'         P= ',E10.4, ',     Q= ',E10.4/
     #'         KP= ',E10.4, ',    KI= ',E10.4)
C     ----calculation for the roots of the characteristic equation------------
 3000 CA=1.
      CB=Q*(KP+KI*TSAMPL)-1.-P
      CC=P-Q*KP
      D=CB**2-4.*CA*CC
      IF(D.LT.0.) GO TO 100
C     ----------REAL POLE--------------------------------------------
      Z1=(-1.*CB+SQRT(D))/(2.*CA)
      Z2=(-1.*CB-SQRT(D))/(2.*CA)
```

```
      WRITE (*,650) Z1,Z2
      IF (ABS(Z1).GT.ABS(Z2)) GO TO 900
      ABZ=ABS(Z2)
      GO TO 1000
  900 ABZ=ABS(Z1)
      GO TO 1000
C     ----------COMPLEX POLE------------------------------------------
  100 ZR=(-1.*CB)/(2.*CA)
      ZI=SQRT(-D)/(2.*CA)
      WRITE (*,600) ZR,ZI
      ABZ=SQRT(ZR**2+ZI**2)
 1000 IF(ABZ.GT.1.) GO TO 710
      IF(ABZ.EQ.1.) GO TO 720
      WRITE (*,500)
      GO TO 800
  710 WRITE (*,510)
      GO TO 800
  720 WRITE (*,520)
C
  500 FORMAT ('0THE SYSTEM IS STABLE.')
  510 FORMAT ('0THE SYSTEM IS UNSTABLE.')
  520 FORMAT ('0THE SYSTEM IS LIMITEDLY STABLE.')
  600 FORMAT (
     #'0The poles are located as follows.'/
     #'    REAL Z= ',E10.4,'   IMAGI. Z= ',E10.4)
  650 FORMAT (
     #'0The poles are located as follows.'/
     #'        Z1= ',E10.4,'        Z2= ',E10.4)
C     ------------------calculation for C0 and C1-------------------
  800 C0=KP+KI*TSAMPL
      C1=-1.*KP
C
      RETURN
      END
C
C     ################################################################
C     #            RUN THE MOTOR (CLOSED LOOP)                       #
C     ################################################################
      SUBROUTINE MTONCL
C
      INTEGER*2    ADCHN,ADCVAL,DACHN,DACVAL,K
      INTEGER      YSS,OLREF,CLREF,M(1000),Y(1000),E(1000)
      REAL         KP,KI
      COMMON /ADCBL / ADCHN,ADCVAL
      COMMON /DACBL / DACHN,DACVAL
```

```
      COMMON /PARA   / RUNTIM,FREQCK,OLREF,CLREF,TSAMPL,NSAMPL
      COMMON /RESULT/ YSS,GAIN,TAU,OLRISE,CLRISE,OVERSH
      COMMON /CONST / C0,C1,KP,KI
      COMMON /ARRAY / M,Y,E
      COMMON /CBLK  / K
C
      CALL OFF
      NSAMPL=RUNTIM/TSAMPL
      M(1)=0
      Y(1)=0
      E(1)=CLREF
      CALL ENABLE
      K=1
  100 KSAVE=K
  200 IF(KSAVE.EQ.K) GO TO 200
      CALL ADC
      Y(K)=ADCVAL
      E(K)=CLREF-Y(K)
      M(K)=C0*E(K)+C1*E(K-1)+M(K-1)
      IF (M(K).GT.127) M(K)=127
      IF (M(K).LT.-128) M(K)=-128
      DACVAL=M(K)
      CALL DAC
      IF (K.LT.NSAMPL) GO TO 100
      CALL DISABL
      CALL OFF
C     ----------CALCULATION FOR OVERSHOOT-------------------------------------
      YMAX=0.
      DO 300 I=1,NSAMPL
      IF (Y(I).GT.YMAX) YMAX=Y(I)
  300 CONTINUE
      IF(YMAX.GT.CLREF) GO TO 400
      OVERSH=0.
      GO TO 900
  400 OVERSH=(YMAX-CLREF)*100./CLREF
C     ----------CALCULATION FOR CLOSED-LOOP RISE TIME------------------------
      DO 777 I=1, NSAMPL
      IF(Y(I).GE.CLREF) GO TO 888
  777 CONTINUE
      CLRISE=9999.99
      GO TO 900
  888 CLRISE=(((CLREF-Y(I-1))/(Y(I)-Y(I-1)))+(I-2))*TSAMPL
  900 RETURN
      END
C
```

```
C     ##############################################################
C     #              DATA ALTERATION FOR CLOSED LOOP               #
C     ##############################################################
      SUBROUTINE ALTCL
C
      CHARACTER         ANS
      INTEGER           OLREF,CLREF
      REAL              KP,KI
      COMMON /CONST /   C0,C1,KP,KI
      COMMON /PARA  /   RUNTIM,FREQCK,OLREF,CLREF,TSAMPL,NSAMPL
C
   30 WRITE (*,600) KI,KP,CLREF,TSAMPL,RUNTIM
   20 READ  (*,  *) N
      GO TO (1,2,3,4,5) N
      WRITE (*,610)
      GO TO 20
    1 WRITE (*,620)
      WRITE (*,630) KI
      READ  (*,  *) KI
      GO TO 10
    2 WRITE (*,620)
      WRITE (*,640) KP
      READ  (*,  *) KP
      GO TO 10
    3 WRITE (*,620)
      WRITE (*,650) CLREF
      READ  (*,  *) CLREF
      IF (CLREF.GE.0.AND.CLREF.LE.127) GO TO 10
      WRITE (*,500)
      GO TO 3
    4 WRITE (*,620)
      WRITE (*,680) TSAMPL
      READ  (*,  *) TSAMPL
      IF (TSAMPL.GE.0.01.AND.TSAMPL.LE.1.) GO TO 700
      WRITE (*,500)
      GO TO 4
  700 FREQCK=1./TSAMPL
      GO TO 10
    5 WRITE (*,620)
      WRITE (*,690) RUNTIM
      READ  (*,  *) RUNTIM
      IF (RUNTIM.GE.4..AND.RUNTIM.LE.10.) GO TO 10
      WRITE (*,500)
      GO TO 5
```

```
      10 WRITE (*,660)
         READ  (*,670) ANS
         IF (ANS.EQ.'Y') GO TO 30
         IF (ANS.EQ.'y') GO TO 30
         RETURN
C
     500 FORMAT ('0 Your input data is not correct. Try again!')
     600 FORMAT ('0Which parameter do you want to change?
        '/
        #' If you want to change the poles, you must select 1., 2.or 4.'/
        #'                                                              '/
        #'                   1.  KI    = ',F10.4,'              '/
        #'                   2.  KP    = ',F10.4,'              '/
        #'                   3.  CLREF = ',I8   ,' (0-127) '/
        #'                   4.  TSAMPL= ',F10.4,'(0.01-1.) (sec)         '/
        #'                   5.  RUNTIM= ',F10.4,'(4-10) (sec)            '/
        #'                                                              '/
        #' Enter the parameter number. No. = '\)
     610 FORMAT ('0You have entered a wrong number. Try again! No. = '\)
     620 FORMAT ('0Please enter a new value.')
     630 FORMAT ('0KI    =',F10.4,' --------------------> KI    = '\)
     640 FORMAT ('0KP    =',F10.4,' --------------------> KP    = '\)
     650 FORMAT ('0CLREF =',I8   ,' --------------------> CLREF = '\)
     660 FORMAT ('0Do you want to change another parameter? (Y/N)= '\)
     670 FORMAT (A1)
     680 FORMAT ('0TSAMPL=',F10.4,' -------------------->TSAMPL = '\)
     690 FORMAT ('0RUNTIM=',F10.4,' -------------------->RUNTIM = '\)
C
         END
C
C
C    ###########################################################################
C    #               DATA ALTERATION FOR OPEN LOOP                             #
C    ###########################################################################
         SUBROUTINE ALTOL
C
         INTEGER        OLREF,CLREF
         COMMON /PARA / RUNTIM,FREQCK,OLREF,CLREF,TSAMPL,NSAMPL
C
      30 WRITE (*,600) RUNTIM,FREQCK,OLREF
      20 READ  (*, *) N
         GO TO (1,2,3) N
         WRITE (*,610)
         GO TO 20
```

```
    1 WRITE (*,620)
      WRITE (*,630) RUNTIM
      READ  (*,  *) RUNTIM
      IF (RUNTIM.GE.4..AND.RUNTIM.LE.20.) GO TO 10
      WRITE (*,500)
      GO TO 1
    2 WRITE (*,620)
      WRITE (*,640) FREQCK
      READ  (*,  *) FREQCK
      IF(FREQCK.GE.1..AND.FREQCK.LE.100.) GO TO 10
      WRITE (*,500)
      GO TO 2
    3 WRITE (*,620)
      WRITE (*,650) OLREF
      READ  (*,  *) OLREF
      IF(OLREF.GE.70.AND.OLREF.LE.127) GO TO 10
      WRITE (*,500)
      GO TO 3
   10 WRITE (*,660)
      READ  (*,670) ANS
      IF (ANS.EQ.'Y') GO TO 30
      IF (ANS.EQ.'y') GO TO 30
      RETURN
C
  600 FORMAT ('0Which parameter do you want to change?        '/
     #'                                                       '/
     #'                 1.  RUNTIM = ',F10.4, '(4-10) (sec)   '/
     #'                 2.  FREQCK = ',F10.4, '(1.-100.) (Hz) '/
     #'                 3.  OLREF  = ',I8   ,' (70-127) '/
     #'                                                       '/
     #' Enter the parameter number. No. = '\)
  500 FORMAT ('0Your input data is not correct. Try again!')
  610 FORMAT ('0You have entered a wrong number. Try again! No. = '\)
  620 FORMAT ('0Please enter a new value. ')
  630 FORMAT ('0 RUNTIM = ',F8.4,'------------------> RUNTIM = '\)
  640 FORMAT ('0 FREQCK = ',F8.4,'------------------> FREQCK = '\)
  650 FORMAT ('0 OLREF  = ',I8  ,'------------------> OLREF  = '\)
  660 FORMAT ('0Do you want to change another parameter? (Y/N)= '\)
  670 FORMAT (A1)
C
      END
C
```

320 APPENDIX A   DESCRIPTION OF AN 8088-BASED SYSTEM

```
C      ****************************************************************
C      *                    CLOCK FREQUENCY INPUT                     *
C      ****************************************************************
       SUBROUTINE CLKFRQ
C
C      THIS ROUTINE IS TO INPUT CLOCK FREQUENCY, AND THEN
C      DETERMINE TERMCT AND SOURCF FOR THE LABTENDER CLOCK.
C
       INTEGER*2        TERMCT,SOURCF
       COMMON /COUNT /  TERMCT,SOURCF
       COMMON /PARA  /  RUNTIM,FREQCK,OLREF,CLREF,TSAMPL,NSAMPL
C
       SOURCF = 5
       IF (FREQCK.LE.5000.) GO TO 9
     4 WRITE (*,100)
   100 FORMAT (' Please input the correct value.---FREQCK= '\)
       READ (*, *) FREQCK
     9 CONTINUE
C      TESTCT = REAL VALUE FOR TERMCT, NECESSARY IF TERMCT > 32767.
    10 TESTCT = AINT((1*10**SOURCF)/FREQCK)
C      CHECK TO BE SURE TESTCT AND SOURCF ARE WITHIN THE DESIRED RANGE.
       IF (TESTCT .GT. 32767.) THEN
          SOURCF = SOURCF - 1
          GOTO 10
        ELSEIF (TESTCT .LT. 2.) THEN
          GOTO 4
        ELSEIF (SOURCF .LT. 2) THEN
          GOTO 4
        ELSE
          CONTINUE
          TERMCT = INT(TESTCT)
       ENDIF
C
       RETURN
       END
C
C      ****************************************************************
C      *                    GRAFCL SUBROUTINE                         *
C      ****************************************************************
       SUBROUTINE GRAFCL
C
       INTEGER          CLREF,OLREF,YSS
       INTEGER          M(1000),Y(1000),E(1000)
       REAL             KP,KI,KIDT
       CHARACTER        ANSWER,BEGFMT*15,ENDFMT*7,FRMAT*40
```

```
      COMMON /PARA   / RUNTIM,FREQCK,OLREF,CLREF,TSAMPL,NSAMPL
      COMMON /RESULT/ YSS,GAIN,TAU,OLRISE,CLRISE,OVERSH
      COMMON /CONST / C0,C1,KP,KI
      COMMON /CONST1/ A,B,P,Q
      COMMON /ARRAY / M,Y,E
C
      DATA BEGFMT / '(1X,F7.3,2X,' /
      DATA ENDFMT / 'X, ''*'')' /
C
      KIDT=KI*TSAMPL
      WRITE(*,'(A\)') '0Do you want to print C/L run data (Y/N) ? '
      READ (*,'(A1)') ANSWER
      IF (ANSWER .EQ. 'Y') GO TO 900
      IF (ANSWER .EQ. 'y') GO TO 900
      GO TO 100
C     PRINT HEADER FOR RUN DATA
  900 WRITE(*,130) CLREF,KP,KIDT,TSAMPL,CLRISE,OVERSH,A,B,P,Q
      WRITE (*,50)
      DO 10 I = 1,NSAMPL
      TIME = (I-1)/FREQCK
      WRITE(*,70) TIME,M(I),Y(I),E(I),CLREF
   10 CONTINUE
C
   50 FORMAT ('0TIME',3X,'OUTPUT, M',3X,'RESPONSE, Y',3X,'ERROR, E',
     *3X,'C/L REFEFENCE')
   70 FORMAT (F7.3,5X,I4,8X,I4,7X,I4,9X,I4)
C
  100 WRITE(*,'(A\)') '0Do you want to plot response (Y/N) ? '
      READ (*,'(A1)') ANSWER
      IF (ANSWER .EQ. 'Y') GO TO 910
      IF (ANSWER .EQ. 'y') GO TO 910
      GO TO 200
C
  910 WRITE(*,130) CLREF,KP,KIDT,TSAMPL,CLRISE,OVERSH,A,B,P,Q
  130 FORMAT('1'/
     *'         CLOSED LOOP RESPONSE OF DC SERVO MOTOR, Y(T)'/
     *'              CLOSED LOOP REFERENCE = ',I4,/
     *'    KP = ',F7.2,',   KIDT = ',F7.2,',    SAMPLING TIME = ',F7.3/
     *'    RISE TIME = ',F7.3,' (SEC),',8X,'% OVERSHOOT = ',F7.2/
     *'      A = ',E12.6,',   B = ',E12.6/
     *'      P = ',E12.6,',   Q = ',E12.6)
```

```
C       CHECK IF THERE ARE TOO MANY POINTS FOR 60 LINE PLOT.
C       SET PRINTER TO 8 LINES/INCH (INSTEAD OF 6) TO SHORTEN PLOTS.
C       DOS COMMAND IS MODE LPT1:80,8
C       KSTEP IS THE STEP SIZE FOR THE OUTPUT K COUNT FOR PLOTTING.
C       KSTEP = 1 UNLESS THERE ARE TOO MANY POINTS, THEN KSTEP > 1.
        KSTEP = 1
        NLINE = 60
        IF (NLINE .GE. NSAMPL) THEN
          NLINE = NSAMPL
          GOTO 150
        ELSE
          KSTEP = NINT(REAL(NSAMPL)/(REAL(NLINE)))
          WRITE(*,140) KSTEP
          NLINE = NSAMPL/KSTEP
        ENDIF
C
  140 FORMAT(' Output plot shows one  *  for every ',I3,' data points'/)
  150 CONTINUE
C       PRINT RESPONSE AXIS FOR Y(K)
        WRITE(*,135)
  135 FORMAT(10X,7('+',9('-')),'+'/
       *10X,'0',9X,'1',9X,'2',9X,'3',9X,'4',9X,'5',9X,'6',9X,'7'/
       *10X,7('0',9X),'0'/)
        DO 170 I = 1,NLINE
        K = I * KSTEP
        TIME = (K-KSTEP)/FREQCK
        NBLANK = Y(K)
C       NBLANK IS THE # OF BLANKS BEFORE THE ASTERISK (PLOT TRACE).
C
C       OUR FORMAT REQUIRES 0 <= NBLANK <= 70, FOR 80 CHARACTERS/LINE.
C       ANY NEGATIVE Y(K) WILL BE CLIPPED AT ZERO FOR PLOT.
        IF(NBLANK .LE. 0) NBLANK = 0
        IF(NBLANK .GT. 70) NBLANK = 70
C       FIRST DEFINE THE VARIABLE FORMAT (``FRMAT") TO LOCATE THE *.
        WRITE(FRMAT,180) BEGFMT,NBLANK,ENDFMT
C       NOW USE THIS FORMAT TO PRINT THE NEXT DATA LINE.
  170 WRITE(*,FRMAT) TIME
C       DEFINE THE FORMAT FOR OUR VARIABLE FORMAT, ``FRMAT"
  180 FORMAT(A15,I3,A7)
C       FORM FEED
        WRITE(*,'(''1'')')
  200 CONTINUE
        RETURN
        END
C
```

```
C     ************************************************************************
C     *                         GRAFOL SUBROUTINE                            *
C     ************************************************************************
      SUBROUTINE GRAFOL
C
      INTEGER       OLREF,CLREF,YSS
      INTEGER       M(1000),Y(1000),E(1000)
      REAL          KI,KP
      CHARACTER     ANSWER,BEGFMT*15,ENDFMT*7,FRMAT*40
      COMMON /PARA  / RUNTIM,FREQCK,OLREF,CLREF,TSAMPL,NSAMPL
      COMMON /RESULT/ YSS,GAIN,TAU,OLRISE,CLRISE,OVERSH
      COMMON /ARRAY / M,Y,E
C
      DATA BEGFMT / '(1X,F7.3,2X,' /
      DATA ENDFMT / 'X, ''*'')' /
C
      WRITE(*,'(A\)') '0Do you want to print O/L run data (Y/N)
     ? '
      READ (*,'(A1)') ANSWER
      IF (ANSWER .EQ. 'Y') GO TO 900
      IF (ANSWER .EQ. 'y') GO TO 900
      GO TO 100
C     PRINT HEADER FOR RUN DATA
  900 WRITE(*,130) OLREF,GAIN,TAU,YSS,TSAMPL
      WRITE (*,50)
      DO 10 I = 1,NSAMPL
      TIME = (I-1)/FREQCK
      WRITE(*,70) TIME,Y(I)
   10 CONTINUE
   50 FORMAT ('0 TIME',3X,'RESPONSE, Y')
   70 FORMAT (F7.3,5X,I4)
  100 WRITE(*,'(A\)') '0Do you want to plot response (Y/N) ? '
      READ (*,'(A1)') ANSWER
      IF (ANSWER .EQ. 'Y') GO TO 910
      IF (ANSWER .EQ. 'y') GO TO 910
      GO TO 200
C
  910 WRITE(*,130) OLREF,GAIN,TAU,YSS,TSAMPL
  130 FORMAT('1'//
     *       '         OPEN LOOP RESPONSE OF D.C. SERVO MOTOR, Y(T)'//
     *       '               OPEN LOOP REFERENCE = ',I4,/
     *       '     GAIN = ',F7.2,',',15X,'TIME CONSTANT = ',F7.2,/
     *       '     STEADY STATE RESPONSE = ',I4,',    SAMPLING TIME = ',
     *       F7.3/) .
```

```
C         CHECK IF THERE ARE TOO MANY POINTS FOR 60 LINE PLOT.
C         SET PRINTER TO 8 LINES/INCH (INSTEAD OF 6) TO SHORTEN PLOTS.
C         DOS COMMAND IS MODE LPT1:80,8
C         KSTEP IS THE STEP SIZE FOR THE OUTPUT K COUNT FOR PLOTTING.
C         KSTEP = 1 UNLESS THERE ARE TOO MANY POINTS, THEN KSTEP > 1.
          KSTEP = 1
          NLINE = 60
          IF (NLINE .GE. NSAMPL) THEN
              NLINE = NSAMPL
              GOTO 150
          ELSE
              KSTEP = NINT(REAL(NSAMPL)/(REAL(NLINE)))
              WRITE(*,140) KSTEP
              NLINE = NSAMPL/KSTEP
          ENDIF
C
  140 FORMAT(' Output plot shows one  *  for every ',I3,' data points'/)
  150 CONTINUE
C
C         PRINT RESPONSE AXIS FOR Y(K)
          WRITE(*,135)
  135 FORMAT(10X,7('+',9('-')),'+'/
     *10X,'0',9X,'1',9X,'2',9X,'3',9X,'4',9X,'5',9X,'6',9X,'7'/
     *10X,7('0',9X),'0'/)
C
          DO 170 I = 1,NLINE
          K = I * KSTEP
          TIME = (K-KSTEP)/FREQCK
          NBLANK = Y(K)
C         NBLANK IS THE # OF BLANKS BEFORE THE ASTERISK (PLOT TRACE).
C
C         OUR FORMAT REQUIRES 0 <= NBLANK <= 70, FOR 80 CHARACTERS/LINE.
C         ANY NEGATIVE Y(K) WILL BE CLIPPED AT ZERO FOR PLOT.
          IF(NBLANK .LE. 0) NBLANK = 0
          IF(NBLANK .GT. 70) NBLANK = 70
C         FIRST DEFINE THE VARIABLE FORMAT (``FRMAT") TO LOCATE THE
*.
          WRITE(FRMAT,180) BEGFMT,NBLANK,ENDFMT
C         NOW USE THIS FORMAT TO PRINT THE NEXT DATA LINE.
  170 WRITE(*,FRMAT) TIME
C         DEFINE THE FORMAT FOR OUR VARIABLE FORMAT, ``FRMAT"
  180 FORMAT(A15,I3,A7)
```

```
C       FORM FEED
        WRITE(*,'(''1'')')
  200 CONTINUE
        RETURN
        END


        TITLESETISR - LAB FOUR ASSEMBLY CODE

        PAGE    ,132            ;Set page width to 132 characters.
        ;       This routine initializes the digital I/O port and
        ;       sets up the IRQ4 interrupt vector.
        ;       FORTRAN "CALL SETISR" calls this procedure.
        DATA SEGMENT PUBLIC 'DATA'
        ;       For local assembler program data storage.
        ;       Not used, but required for linking
        DATA ENDS
        CBLK$A SEGMENT COMMON '$CBLK'
        ;       This segment is equivalent to FORTRAN's COMMON/CBLK/K
        ;       The $ and A are required to match the segment name and class that
        ;       the linker uses for the FORTRAN common.  All variables must be
        ;       declared INTEGER*2 in the FORTRAN calling routine before the
        ;       common block.
        K       DW      ?
        CBLK$A ENDS
        DGROUP GROUP DATA,CBLK$A ;The DATA,CBLK$A segment will be linked into the group
                ;called DGROUP to match Microsoft FORTRAN convention.

        CODE    SEGMENT PUBLIC 'CODE'

                ASSUME CS:CODE,DS:DGROUP,SS:DGROUP
                PUBLIC  SETISR  ;Make the INIT label available to other segments.

        ;       Labtender Address, in IBM PC, for SETISR routine.
        IOCOM   EQU     831     ;Parallel Port Control Register.

        SETISR  PROC FAR

                PUSH    BP              ;Save calling framepointer
                MOV     BP,SP

                MOV DL,73               ;Print an 'I' to the screen
                MOV AH,2                ;for debugging
                INT 21H
```

Load the IRQ4 interrupt vector (address to go to upon interrupt 4).

```
        CLI                 ;Disable interrupts while we're modifying them
                            ;to prevent any castastrophes.
        PUSH ES             ;Save Extra Segment
        MOV AX,0000H
        MOV ES,AX           ;Point Extra Segment at the ISR address table.
        MOV DI,30H          ;Offset of entry for IRQ4 in the address table.
        MOV AX,OFFSET ISR   ;Get offset of DDASER, our interrupt
                            ;service routine.
        STOSW               ;Load offset in AX into address in DI,
                            ;and add 2 to DI to prepare to store cs.
        MOV AX,CS           ;Get Code Segment for assembly code.
        STOSW               ;Load code segment in AX into address in DI.
                            ;Now the 4-byte address of our DDASER begins at memory
                            ;location 30H. An IRQ4 interrupt (if enabled) will
                            ;cause the the microprocessor to immediately go to
                            ;that address, and start performing our
DDASER.
        POP ES              ;Restore Extra Segment
        STI                 ;Enable interrupts

        MOV     SP,BP       ;Restore framepointer
        POP     BP
        RET                 ;Return

SETISR ENDP

;--------------------------------------------------------------------------------
        SUBTTL  INTERRUPT SERVICE ROUTINE
;--------------------------------------------------------------------------------
;       Labtender Addresses, in IBM PC, for DDASER routine.
INTCLR  EQU     818     ;Timer Interrupt Clear Address.

;       8259 Programmable Interrupt Controller Addresses, in IBM PC.
ICR     EQU     20H     ;Interrupt Command Register.
IMR     EQU     21H     ;Interrupt Mask Register.

;       Constants for DDASER routine.
EOI     EQU     20H     ;End of Interrupt signal for 8259 Interrupt Controller.
```

```
ISR    PROC FAR
       PUSH    BP              ;Save calling framepointer
       MOV     BP,SP

       PUSH AX                 ;Save all the registers we'll be
       PUSH BX
       PUSH CX                 ;modifying.  This is done automatically
       PUSH DX                 ;only during FORTRAN CALLs, not interrupts.

       PUSH DS                 ;Save the data segment register
       MOV   AX,DGROUP         ;Store the segment DGROUP into the data segment
       MOV   DS,AX             ;so that we assure the addressing is correct

       INC     K               ;K = K + 1, count another time step.

CLKEOI:
       MOV DX,INTCLR           ;Timer Interrupt Clear addr.
       MOV AL,0                ;Data could be anything
       OUT DX,AL               ;Write to clear the interrupt

       MOV     AL,EOI          ;Get end-of interrupt signal
       MOV     DX,ICR          ;Send it to the 8259's
       OUT     DX,AL           ;Interrupt Command Register

       POP     DS

       POP     DX              ;Restore registers.
       POP     CX
       POP     BX
       POP     AX

       MOV     SP,BP           ;Restore framepointer
       POP     BP

       IRET                    ;Interrupt return.
ISR    ENDP

CODE ENDS

END
```

```
TITLE   DAC - DIGITAL-TO-ANALOG CONVERSION ROUTINE

PAGE    ,132    ;Set page width to 132 characters.
;       This routine sends the digital-to-analog conversion value
(DACVAL)
;       for a specified channel (DACHN).
;       FORTRAN "CALL DAC" calls this D/A procedure.
DATA SEGMENT PUBLIC 'DATA'
;       For local assembler program data storage
;       Not used here, but required for linking
DATA ENDS
DACBL$A SEGMENT COMMON '$DACBL'
;       This segment is equivalent to COMMON/DACBL/DACHN,DACVAL
;       in FORTRAN. The $ and A are required to match the segment
;       name and class that the linker uses for the FORTRAN common.
;       DACHN and DACVAL must be declared INTEGER*2 in the FORTRAN
;       calling routine before the common block.
DACHN   DW      ?
DACVAL  DW      ?
DACBL$A ENDS

DGROUP GROUP DATA,DACBL$A ;The DATA and DACBL$A segments will be
                          ;linked into the group called DGROUP,
                          ;to match Microsoft FORTRAN convention.

CODE    SEGMENT PUBLIC 'CODE'

        ASSUME  CS:CODE,DS:DGROUP,SS:DGROUP
        PUBLIC  DAC             ;Make DAC label available to other segments.

DACNTL  EQU     820     ;D/A control port location in memory
DADATA  EQU     821     ;D/A converter data register

DAC PROC FAR

        PUSH    BP              ;Save calling framepointer
        MOV     BP,SP           ;on the stack.

        MOV     DX,DADATA       ;Indirect addressing using DX will have to
                                ;be used when using IN and OUT commands.

        MOV     BX,DACVAL       ;Get D/A value
        MOV     BH,0            ;Mask off hi byte
        ADD     BX,128          ;CONVERT FROM SIGNED TO UNSIGNED
                                ;-128 to 127 becomes 0 to 255.
```

```
        MOV     AX,DACHN    ;Get channel number
        MOV     AH,0        ;Mask off high byte, we only use AL

;       Now create the proper command byte for the Labtender's D/A routine,
;       based on the channel number.  Use the AL register.
        AND     AL,0FH      ;Mask of bits 4-7
        OR      AL,8        ;Set bit 3 hi
        MOV     AH,AL       ;Save D/A command byte in AH

        MOV     AL,BL       ;Get DACVAL (only lower byte)
        OUT     DX,AL       ;Load value into D/A data register
        MOV     DX,DACNTL   ;Point to D/A control word
        MOV     AL,AH       ;Get D/A command byte
        OUT     DX,AL       ;Start conversion

        MOV     SP,BP       ;Restore framepointer
        POP     BP
        RET                 ;Return

DAC ENDP

CODE ENDS

END
```

APPENDIX B
DESCRIPTION OF A Z80-BASED SYSTEM

This appendix provides an overview of a laboratory system based on Xycom 3800B Development Systems and Xycom 3935 Target Systems. We will primarily be describing the development systems, which consist of

1. A Z80-based microcomputer
2. A CRT terminal
3. Dual floppy disk drives
4. ADC, DAC, digital I/O port, and clock

Some of the development systems also have a printer. A complete set of Xycom manuals should be available in the laboratory for your reference. A book such as [Zaks80], describing the Z80 hardware and assembly language programming, should also be available as a reference. This appendix just provides a brief overview.

Section B.1 provides an overview of the development system software and describes methods for passing parameters between FORTRAN and assembly language routines. Section B.2 provides necessary information for programming the ADCs and DACs. Section B.3 describes how a program developed and tested on a development system can be down-loaded to the target system. Section B.4 contains program listings for the case studies described in Chapters 9 through 12.

B.1 DEVELOPMENT SOFTWARE FOR THE XYCOM SYSTEM

In order to develop and run programs on the Xycom processor, several pieces of software common to most small processors must be used: the operating system (SPDOS and FOS), editor, FORTRAN compiler, assembler, and linker. The purpose of this section is to (1) briefly describe each of these various types of development software in terms of its purpose and relation to other software; (2) describe the type of files that serve as input and output; and (3) briefly describe several of the more useful commands. Further information concerning the development software can be found in the Xycom SPDS, FORTRAN, and assembler manuals.

SPDS Operating System (SPDOS)

The SPDOS contains the I/O handling software (i.e., assigning a logical to a physical device) as well as providing features that allow a user to debug his or her assembly language programs. The debugger allows for the possibility of suspending a program; resuming execution; and examining, displaying, and modifying the contents of registers or memory. The SPDOS is the program loaded by the bootstrap (a "read only memory" hardware program that enables the processor to load disk files into memory) so that when the processor is turned on, the SPDOS is the first development program encountered. For example, when the processor is first turned on, the query

```
Load Y/N?
```

appears. If the answer is yes, the processor prompts with an exclamation mark to indicate to the user that he or she is in the operating system.

SPDOS Commands

A   Assign Logical Device to Physical Device (see Table B.1). The format of the A command is:

```
!A logdevice,phydevice,file name:disk drive no.<CR>
```

For example, to assign the printer as the list device:

```
!A L,P<CR>
```

For example, to assign a file on a disk drive 1 as the list device:

Table B.1 Logical I/O Devices

Letter	Number	Logic Device	Description
C	2	Console Input	A character-oriented device used for input operations
E	3	Console ("Echo") Output	A character-oriented device, output only.
L	1	List Output	A character-oriented, output-only device that accepts a character from the program and records it on some external medium in human readable form.
P	5	Punch Output	A character-oriented, output-only device that accepts a character from the program and records it on some external medium.
R	4	Reader Input	A character-oriented, input-only device that transfers data on command and signals the program when there is no more data (an end-of-file condition).
S	0	Console Status	A status of the console input device to determine if a character is ready.

Note: These are the logical unit numbers used in FORTRAN READ or WRITE statements, for example,
```
READ(2,10) X    read from terminal
WRITE(3,10) X   write to terminal
```

　　　　　　　　!A L,D,FILE:1<CR>

Q　　　　　Run FOS (see next section). The format of the Q command is: !Q This causes loading and transfer to the FOS disk operating system, if present, on a System Diskette located in drive 0.

C　　　　　Copy. *Warning*: This is a command that copies disks not files! For example typing !C will produce a copy of the system disk in drive 0 onto your disk in drive 1, wiping out any programs that you had. (For debugging commands, see Chapter 1 of the SPDS manual.)

File Operating System (FOS)

The FOS is a software development tool for the control and data management of the dual floppy disk drives. Once in FOS, the user is able to call other development software, such as the editor, FORTRAN compiler, assembler, and linker, necessary to create and run programs.

FOS Commands

COPY The format of the COPY command is:

?COPY<CR>

The purpose of this command is to copy the entire contents of the diskette in drive unit 0 to the diskette in drive unit 1.

CREAT The format of this command is

?CREAT,new filename, new filesize<CR>

The purpose of this command is to create a user-designated file-directory entry with the specified file name and file size in sectors.

DELET The format of the DELET command is:

?DELET:unit number, filename 1, filename 2,
..., filename n

The purpose of this command is to delete the designated, nonpermanent files from the diskette, in specified drive unit, and then to replace the contents of that diskette's user file area and file directory area, thus making the disk space available for additional files.

EXIT The format of the EXIT command is

?EXIT,<CR>

The purpose of this command is to return control to SPDOS. SPDOS will print out the start address of FOS and then return an exclamation mark (!).

INIT Initialize a Diskette. The format of the INIT command is

?INIT,unit number <CR>

The purpose of this command is to initialize the file directory area on the specified drive unit. Warning: This command erases the entire diskette.

LIST The format of the LIST command is

?LIST,unit number <CR>

The purpose of this command is to list the contents of the file directory of the diskette in the specified drive unit. This command will list the file names, attributes, file's starting track and sectors, and the file's size in sectors on the designated list device.

MERGE The format for the MERGE command is

```
?MERGE,new filename, filename 1, filename 2,
..., filename n<CR>
```

The purpose of this command is to create a new file whose contents is the concatenation of the contents of the specified files, in the order in which they appear in the command.

PRINT The format of the PRINT command is

```
?PRINT,filename, first line #, last line #<CR>
```

The purpose of this command is to print the contents of the specified file to the designated list device. This command may be aborted by typing any key. With the additional two variables, the user may print as much or as little of a file as needed. First line # and last line # are inserted as decimal values. If no numbers appear in the line number area, the entire file will be printed.

RENAM The format of the RENAM command is

```
?RENAM,old filename, new filename<CR>
```

The purpose of this command is to modify the specified file's file-directory entry by replacing its existing file name with a new file name.

RUN The format of the RUN command is

```
?RUN,object filename<CR>
```

The purpose of this command is to load the contents of the object file into RAM memory for execution. The data is loaded into memory at the location specified in the object file.

VIEW The format of the VIEW command is

```
?VIEW,filename, lines per frame, first line<CR>
```

The purpose on this command is to display, on the console device, the contents of the specified file one frame at a time. The number of lines per displayed frame, if not specified, is 14 by default. The first line displayed is line 0, if not specified otherwise. Lines per frame and/or first line may be omitted and, if so, are assumed to be 14 and 0, respectively. All numbers are in decimal. When in the VIEW command, the following five keys may be used:

N	Causes the next frame to be displayed
P	Causes the previous frame to be displayed
F	Causes the first frame to be displayed (i.e., that frame whose first line is "first line")
B	Causes the beginning frame to be displayed (i.e., that frame whose first line is 1)
CR	(Carriage return) returns to FOS executive.

Editor

The editor is the program designed to facilitate the entry and the modification of FORTRAN or assembly language files. It allows the user to enter characters, append them, insert them, add lines, remove lines, search for characters or strings. Refer to the manual for the editor for a list of editor commands.

FORTRAN Compiler

The FORTRAN Compiler is the development program that translates a FORTRAN code file into machine language (i.e., instructions in their binary form) and places the binary instructions in a new file (object file). The compiler is used after the editor and prior to the linker when writing software. The FORTRAN Compiler is called, while in FOS, by the command:

```
?FTN,fortfil:1,objfil:1,option number
```

Valid option numbers can be in the range 1 through 4. They select the listing and object file options as follows:

1. No listing or object file is produced. This option would be selected when one wishes only to see if the compiler detects any syntax errors in the source file. The object file need not be specified in this case.
2. Only a listing is produced. The listing is directed to the SPDOS list device.
3. Only an object file is produced.
4. Both a listing and an object file are produced.

For example:

```
?FTN,FILE1:1,FILE2:1,3
```

This command will take the FORTRAN code in FILE1 on disk 1, compile it, and put the object code generated on FILE2 on disk 1.

Universal Assembler

The universal assembler is the program that translates the mnemonic representation of instructions into their binary equivalent. It normally translates one symbolic instruction into one binary instruction (which may occupy 1, 2, or 3 bytes). The resulting binary code is called object code and is directly executable by the microcomputer. As a side effect, the assembler will also produce a complete symbolic listing of the program, as well as the equivalence tables to be used by the programmer and the symbol occurrence list in the program.

In addition, the assembler will list syntax errors such as instructions misspelled or illegal, branching errors, duplicate labels, or missing labels. It will not detect logical errors (this is your problem). The assembler is called, while in FOS, by the command:

?UASMB,asblfil:1,objfil:1,option number.

Valid option numbers can be in the range 1 through 4, interpreted as follows:

1. No listing or object file is produced. This option would be selected when one wishes only to see if the assembler detects any syntax errors in the source file. The object file need not be specified in this case.
2. Only a listing is produced. The listing is directed to the SPDOS list device. The object file need not be specified in this case.
3. Only an object file is produced.
4. Both a listing and an object file are produced.

As an example, the command

?UASMB,FILE1,FILE2,3

would cause the assembler UASMB to assemble the assembly language statements of file FILE1 and write the corresponding relocatable object code into file FILE2. No listing would be produced, but any errors would be noted on the console.

Linker

The linker resolves internal and external references between the object files, performs library searches for system subroutines, and generates a load map of memory showing the locations of the main program, subroutines, and common areas. The linker is used after assembling or compiling a file. The linker is called, while in FOS, by the command:

?RLINK,,OUTFILE:1

Linker Commands

E(XIT) Returns to SPDOS
H(ELP) Prints a list of available commands
L(INK) Links all programs in the object files listed
M(AP) Produces a listing of all declared symbols found in the linked and searched files and physical addresses associated with each symbol
W(RITE) Writes all the object files previously linked and searched to a disk file
Q(UIT) Returns to FOS
S(EARCH) This command is essentially the same as the LINK command, the difference being not all the modules found in the files are used. If a module contains an entry point that is required by a previous module, that module will be read and used. This is used for library files, where many utility programs are stored. Only the routines that are needed at the time the file is read will be used
U(REF) Lists all externally referenced symbols that are undefined at the time the command is issued.

Example B.1 Creating and Running a Program

Turn on the computer, and wait for the message:

LOAD Y/N?

Respond by placing the system disk in drive 0 and your disk in drive 1, then typing **Y**. The computer will display the SPDOS prompt character "!". Respond by typing **Q** to use FOS. In response to the FOS prompt character "?", type

?INIT,1

to initialize your disk (*Warning:* Do this the first time only!). Now use the editor to create a file by typing

?EDIT,,FCHP9:1

This will create the file named FCHP9 on your disk. You will now see the editor prompt character ">", and can use editor commands to type a FORTRAN program. Typing **STOP** while in the editor returns you to FOS. You can now compile your FORTRAN program by typing,

?FTN,FCHP9:1,OCHP9:1,3

Now invoke the editor again, as needed, to correct your FORTRAN program

```
?EDIT,FCHP9:1,FCH9:1
```

After your editing session, FCH9 will contain the corrected FORTRAN program. You can recompile it as follows:

```
?FTN,FCH9:1,OCH9:1,3
```

You can now invoke the linker to link your program with any required system routines,

```
?RLINK,,@ CH9:1
>LINK OCH9:1
>SEARCH F4LIB
>WRITE
```

The file @CH9 is now ready to be executed:

```
? @ CH9:1
```

Sharing Data between FORTRAN and Assembly Routines

We will frequently link together routines written in FORTRAN and routines written in assembly language. In many cases data must be shared by both types of routines. There are two basic approaches that can be used to accomplish this:

1. Use of common blocks
2. Passing parameters in subroutine calls

Passing Parameters via Common Blocks

The use of common blocks for passing parameters is very efficient. It is very important to consider the type of the data being shared. This method is illustrated by the program segments that follow, and by the program listings in Section B.4. In the FORTRAN program:

```
INTEGER*2 ARM,LEG
C    These are 16-bit words
LOGICAL*1 FOOT,HAND
C    These are 8-bit words
COMMON/BLK1/ARM,LEG,FOOT,HAND
ARM  =  32000
```

```
LEG  =  0
FOOT =  0
HAND =  .TRUE.
C True is -1, False is 0
```

In the Assembly Program:

```
COMMON /BLK 1/
ARM:DEFS 2    ;Reserve 2 bytes
LEG:DEFS 2    ;Reserve 2 bytes
FOOT:DEFS 1   ;Reserve 1 byte
HAND:DEFS 1   ;Reserve 1 byte
CSEG
```

Passing Parameters via Subroutine Calls

A subprogram reference with no parameters generates a simple "CALL" instruction. The corresponding subprogram should return via a simple "RET". CALL and RET are Z80 opcodes. A subprogram reference with parameters results in a somewhat more complex calling sequence. Parameters are always passed by reference (i. e., the thing passed is actually the address of the low byte of the actual argument). Therefore, parameters always occupy 2 bytes each, regardless of type. The method of passing the parameters depends on the number of parameters to pass:

1. If the number of parameters is less than or equal to 3, they are passed in the registers. Parameter 1 will be in HL, 2 in DE (if present), and 3 in BC (if present).
2. If the number of parameters is greater than 3, they are passed as follows:
   a. Parameter 1 in HL
   b. Parameter 2 in DE
   c. Parameters 3 through *n* in a contiguous data block. BC will point to the low byte of this data block (i.e., to the low byte of parameter 3).

Note that, with this scheme, the subprogram must know how many parameters to expect in order to find them. Conversely, the calling program is responsible for passing the correct number of parameters. Neither the compiler nor the run-time system checks for the correct number of parameters. When accessing parameters in a subprogram, do not forget that they are pointers to the actual arguments passed. It is entirely up to the programmer to see to it that the arguments in the calling program match in number, type, and length with parameters expected by the subprogram. This applies to FORTRAN subprograms, as well as those written in assembly language. An example is given next for one parameter:

In the FORTRAN program:

```
LOGICAL*1 COUNT
COUNT = 100
CALL COUNTR(COUNT)
```

In the Assembly Program:

```
COUNTR:   PUSH BC    ; Save BC registers
                     ; load B from location in HL
          LD B, (HL)

                     ; decrement B until zero

LOOP;          NOP   ; do nothing

          DJNZ LOOP
                     ; restore BC registers, and
                       return

          POP BC
          RET
```

The value passed to the assembly language program (COUNT) is decremented in the B register until zero. The B register is loaded with the value whose address is in the HL register pair.

B.2 PROGRAMMING THE DACs, ADCs, AND DIGITAL I/O PORTS

Both analog-to-digital converters (ADCs) and digital-to-analog converters (DACs) are 12-bit converters that are jumpered to convert voltage signals in the range of $\pm$ 10 V. The data registers used for both are set up for two's complement integer values. This means that,

Integer values		Analog signal
2047_{10}	=	10 V
0_{10}	=	0 V
-2048_{10}	=	-10 V

The ADC is set up to convert a maximum of 16 channels of analog data, with each of these channels switched to the converter by multiplexing. There are only 2 DACs that can send analog signals. For testing the programs listed in Section B.4, a buffer box was used to ensure that the computer interface was not damaged by excessively high levels of the input signals. This buffer box contained connections for two ADC and two DAC channels as well as digital I/O (8-bits), as illustrated in Figure B.1.

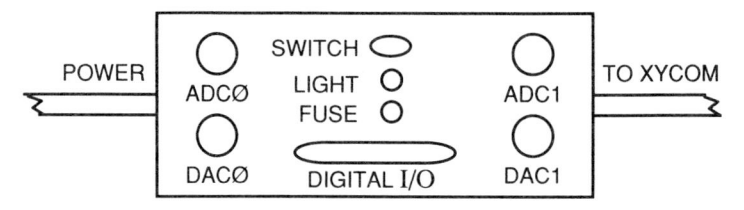

342   APPENDIX B   DESCRIPTION OF A Z80-BASED SYSTEM

Figure B.1 Buffer box layout

Programming the ADCs

The functions to be performed are

1. Select which channel to convert (multiplexing) and start the conversion.
2. Wait for the conversion to be completed.
3. Properly store the converted value.

On the Xycom machines, the functions just listed are performed by loading and storing in specific locations in memory that can be thought of as registers:

1. When the MULIPLEX register at address FF71 is loaded with a channel number, the conversion starts on that channel. The lowest 4 bits (0 to 3) are used to specify channels 0 to 15.
2. Bit 7 of the STATUS register at address FF70 is 1 when a conversion is complete. It is automatically reset to 0 when the least significant byte of the data register is stored.
3. The data is stored in DATAREG at locations FF72 and FF73 and must be read starting with the MS byte and then the LS byte. Otherwise, a conversion could start before all the data is stored. Another detail that must be taken care of is the order in which the converted data is stored. Integer values are stored in memory with the LS byte first and the MS byte next; however, this converter returns values in the DATAREG in the opposite order.

Programming the DACs

The functions to be performed are

1. Select which DAC channel to use.
2. Load the integer value to be converted.

On the Xycom, these two functions are done using the same memory locations as were used for storing data from the ADC; the DATAREG at addresses FF72 and FF73:

1. Selecting the DAC is done by putting a 0 or 1, corresponding to channel 0 or channel 1, in Bit 7, along with the MS Byte of the data to be converted.
2. This information is converted as soon as the LS byte is loaded.

B.3 DOWN-LOADING FROM THE XYCOM 3800B TO THE XYCOM 3935

In some applications the final implementation of your program will be done on a target system. This will involve down-loading the program you developed on the development system to a target-system processor. Here we describe the down-loading to the Xycom 3935 of programs developed and tested on the Xycom 3800B. To do this you will have to know about the hardware differences between these systems, linking procedures, preparation for down-loading, and, finally, how to execute your program on the 3935.

Since the 3935 target system is designed to be a dedicated processor that does only one thing, it does not support all the hardware that the 3800B system does. As a result, you should check for hardware differences between the two systems, taking particular note of differences in device addresses, for example, the parallel I/O port address. Table B.2 gives a summary of these differences.

Table B.2 Comparison of Xycom 3800B and 3935 Systems

Hardware	3800B	3935
Parallel I/O port	Address 0FF22H	Address 0FF32H
ADC and DAC	Address 0FF70H	Address 0FF70H
Real-time clock	Same	Same
Keyboard	Hazeltine 1420	Small alphanumeric keypad with default lowercase characters
Hardcopy	Printer	None
Disk drives	2–8 in. floppies	None

These differences in hardware may necessitate reassembling the Z80 code that refers to these specific addresses. The other steps involved will be linking, actually down-loading, and finally executing the program on the 3935 target.

Linking Procedure

Because of the differences in hardware between the 3800B and 3935 system, the routines in the system library are slightly different. This means that you will have to relink your object files with this new library as a first step in down-loading your program. In the example that follows, it is assumed that two relocatable object files have been created, namely, .OBJ1:1 and . OBJ2:1, and they will be linked with routines in MELIB. The linked object file will be stored in @CH9:1. The difference in linking will only be in the name of the library of service routines that are searched, so that instead of F4LIB the library will be SCLIB for those routines that are to be run on the 3935 target system. In the example given here, all user entries are in bold characters:

```
?RLINK,,@ CH9:1
>L .OBJ1:1, .OBJ2:1
>S MELIB
>S SCLIB
>W
?
```

The result is a linked binary object file.

Binary to Hexadecimal Conversion

The program is transmitted to the target system using a 20-milliamp current loop, and the communications protocol assumes that, rather than binary data, the transmitted data is in the form of hexadecimal characters. This means that the linked file must be transformed from binary to hexadecimal; to do this the program BNHX must be run as follows:

```
?BNHX,@ LAB2:1,XLAB2:1
?
```

where the hexidecimal file is stored in XLAB2:1, and can be transmitted at any convenient time.

Down-loading Procedure

The down-loading procedure can be expedited if two people are present: one to handle the 3800B and the other the 3935 target system. The steps are as follows:

1. Turn on the power for both systems and make sure they are reset. Note that on the 3935 a key is used for this purpose; for down-loading it should be in the "D" position.
2. At the 3800B, establish the link by forcing the system into the bootstrap monitor where the console prompt is "!", and enter

    ```
    !L C
    !
    ```

    to link the CPUs. Note that if a "*" appears, something is wrong, so reset both systems and try again.
3. Assign the hexidecimal file to be down-loaded to the reader device and initialize it using the commands

    ```
    !A R,D,XLAB2:1
    ! U R,0
    !
    ```

    Note that the spaces after A and U will be generated by the computer, and if the name of your file is 5 characters long, the computer will automatically insert the ":". Also, if at this point you get a "*", it means you made an error, but there is no need to start over, only repeat the command.
4. Transmit the file to the target using the command

    ```
    !T R
    ```

    After the R is entered, the disk drive will operate from time to time until the entire file has been transmitted, and then it will respond with a ".", indicating you have command of the 3935 at the 3800B console.
5. Disconnect the 3800B from the 3935 target with the command.

    ```
    .<CTRL-D>
    !
    ```

    and the two systems are operating independently.

So to review, the following keyboard sequence is used at the 3800B, assuming you started in the File Operating System (FOS)

```
?EXIT
!L C
!A R,D,XLAB2:1
!U R,0
!T R
.<CRTL-D>
!
```

and the action moves to the target system.

Execution on the 3935

Once the program is down-loaded, the first thing to do is to press the reset button; your program remains intact in memory. The resulting screen prompt is a ". ". Turn the key to position "R" for Run, then set the small character size by entering

```
.C2
```

Small characters should then appear.

Finally, to execute your program, you must indicate to the processor where it is to go to start executing instructions. This is set at the linking operation, and the default starting address is always 04800H. The procedure to start is the entry

```
.G4800
```

Then your program should start executing in the same way it did when you were writing and debugging it on the 3800B system.

B.4 PROGRAM LISTINGS

In this section are complete program listing for each of the case studies described in Chapter 9 through 12 as implemented on the Xycom-based systems described in Sections B.1 through B.3. The programs are written in FORTRAN and Z80 Assembly language. They have all been successfully tested on actual laboratory systems. Real-time microcomputer systems, as we have seen in this textbook, are complex, so it is unlikely that these programs would run without modification on a similar system. These listings are intended to provide detailed examples

B.4 Program Listings 347

that can help you to develop your own programs on similar systems or in similar applications. We encourage you to gain experience by actually applying the material covered in the book, and close with some words of advice:

- Back up all your programs on a diskette; it is only a matter of time before you will need these backups.
- Develop methodical procedures for troubleshooting your system to find malfunctioning (hardware or software) components.
- Just as in software development, develop and test all modules (hardware and software) of your system independently (e.g., use a function generator to test parallel I/O and ADC systems).

```
0001    C....  PROOGRAM ROLMIL ....
0002    C
0003    C....  ME488 LAB ASSIGNMENT NO.1: COMPUTATION OF THE
0004    C....  STAND CLEARENCE FOR A ROLLING MILL
0005    C....          THIS IS AN INTERACTVE PROGRAM WHICH CALCULATES
0006    C....  THE FREE LENGTH (LO) TO BE SET ON THE LEAD SCREWS OF A
0007    C....  ROLLING MILL WITHOUT A LOAD BY ACCOUNTING FOR THE
0008    C....  ELASTIC DEFORMATION OF THE LEAD SCREWS DUE TO THE ROLL
0009    C....  SEPARATING FORCE. REFERENCE: ME488 HANDOUT FOR
0010    C....  LAB NO. 1      -     A.G. ULSOY        12/81
0011    C
0012           INTEGER FLAG,FLAG1,FLAG2,FLAG3,FLAG4
0013           REAL L,LO,L1,L2,K,N,MILMOD
0014           COMMON/INOUT/NI,NO,IOSW
0015           DATA FLAG1,FLAG2,FLAG3,FLAG4/4*'Y'/,PI/3.141593/
0016           NI = 2
0017           NO = 3
0018    C....  USE IOSW=0 OR 1 FOR DEBUGGING AND 2 FOR FINAL RUNS ....
0019           IOSW = 0
0020           NRUN = 1
0021           WRITE(NO,200)
0022    C....  BEGIN MAIN PROGRAM LOOP ....
0023         1 IF(FLAG1.NE.'Y')GO TO 2
0024           WRITE(NO,210)
0025           READ(NI,110) L1,L2,R,D,E
0026           IF(IOSW.LT.1)WRITE(NO,340)L1,L2,R,D,E
0027         2 IF(FLAG2.NE.'Y')GO TO 3
0028           WRITE(NO,220)
0029           READ(NI,110)HO,W
0030           IF(IOSW.LT.1)WRITE(NO,340)HO,W
0031         3 IF(FLAG3.NE.'Y')GO TO 4
0032           WRITE(NO,230)
0033           READ(NI,110)K,N,CW
0034           IF(IOSW.LT.1)WRITE(NO,340)K,N,CW
0035         4 IF(FLAG4.NE.'Y')GO TO 5
0036           WRITE(NO,240)
0037           READ(NI,110)HF,U
0038           IF(IOSW.LT.1)WRITE(NO,340)HF,U
0039         5 CONTINUE
0040           DH = (HO - HF)
0041           H = (HO + HF)/2.0
0042           CL = SQRT((R**2)-((R-(DH/2.0))**2))
0043           L = L1-HF-2.0*R
0044    C....  NO=1 AT THIS POINT WILL ROUTE RESULTS TO PRINTER ....
0045           IF(IOSW.LT.1)WRITE(NO,250)CL,L
0046    C....  CALCULATE FLOW STRAINS/STRESSES ....
0047           CALL EPSSIG(K,N,HO,HF,CW,E1,E2,S0,S1,S2)
0048           IF(IOSW.LT.1)WRITE(NO,260)E1,S1,E2,S2,S0
```

348 APPENDIX B DESCRIPTION OF A Z80-BASED SYSTEM

```
0049    C.... CALCULATE ROLL SEPARATING FORCE ....
0050          F = FSW(H,S0,U,CL)
0051          IF(IOSW.LT.2)WRITE(NO,270)F
0052          MILMOD = (2.0*E*PI*(D**2))/(4.0*(L-L2))
0053          IF(IOSW.LT.2)WRITE(NO,280)MILMOD
0054          DL = F*W/MILMOD
0055          LO = L + DL
0056          WRITE(NO,290)NRUN,DL,LO
0057    C.... IF NO=1 WAS SET ABOVE THEN SET NO=3 HERE ....
0058          WRITE(NO,295)
0059          READ(NI,120)FLAG
0060          IF(FLAG.EQ.'Y')STOP
0061          WRITE(NO,300)
0062          READ(NI,120)FLAG1
0063          WRITE(NO,310)
0064          READ(NI,120)FLAG2
0065          WRITE(NO,320)
0066          READ(NI,120)FLAG3
0067          WRITE(NO,330)
0068          READ(NI,120)FLAG4
0069          NRUN = NRUN + 1
0070          GO TO 1
0071    C
0072    C.... FORMAT STATEMENTS ....
0073      110 FORMAT(8F10.0)
0074      120 FORMAT(A1)
0075      200 FORMAT('1THIS PROGRAM CALCULATES THE REQUIRED FREE LENGTH OF
0076         $A ROLLING'/1X,'MILL BY ACCOUNTING FOR THE ELASTIC DEFORMATION
0077         $OF THE LEAD SCREWS'/1X,'DUE TO THE ROLL SEPARATING FORCE.'/
0078         $1X,'(ANSWER QUESTIONS WITH Y OR N FOLLOWED BY RETURN)')
0079      210 FORMAT('0INPUT   MILL PARAMETERS: L1,L2,R,D,E'/)
0080      220 FORMAT('0INPUT   WORKPIECE GEOMETRIC PARAMETERS: H0,W'/)
0081      230 FORMAT('0INPUT WORKPIECE MATERIAL PROPERTIES: K,N,CW'/)
0082      240 FORMAT('0INPUT PROCESS PARAMETERS:HF,U'/)
0083      250 FORMAT('0CONTACT LENGTH =            ',G12.4/
0084         $         ' LOADED LENGTH =            ',G12.4)
0085      260 FORMAT('0FLOW STRAIN AT ENTRY =      ',G12.4/
0086         $         ' FLOW STRESS AT ENTRY =     ',G12.4/
0087         $         ' FLOW STRAIN AT EXIT =      ',G12.4/
0088         $         ' FLOW STRESS AT EXIT =      ',G12.4/
0089         $         ' AVERAGE FLOW STRESS =      ',G12.4)
0090      270 FORMAT('0ROLL SEPARATING FORCE/UNIT LENGTH = ',G12.4)
0091      280 FORMAT('0MILL MODULUS =            ',G12.4)
0092      290 FORMAT('0**** FOR RUN NO. ',I2,'MILL DEFLECTION = ',G14.6/
0093         $     '                         FREE LENGTH =     ',G14.6)
0094      295 FORMAT('0DO YOU WANT TO STOP ?  ')
0095      300 FORMAT('0DO YOU WANT NEW VALUES FOR L1,L2,R,D,E ?  ')
0096      310 FORMAT('0DO YOU WANT NEW VALUES FOR H0,W ?  ')
0097      320 FORMAT('0DO YOU WANT NEW VALUES FOR K,N,CW ?  ')
0098      330 FORMAT('0DO YOU WANT NEW VALUES FOR HF,U ?  ')
0099      340 FORMAT(' ECHO: ',5G12.4)
0100          END
0101    C
0102    C
0103          SUBROUTINE EPSSIG(K,N,H0,HF,CW,E1,E2,S0,S1,S2)
0104    C.... FLOW STRESS/STRAIN CALCULATIONS ....
0105          REAL K,N
0106          E1 = ALOG(1.0/(1.0-CW))
0107          E2 = E1 + ALOG(H0/HF)
0108          S1 = K*(E1**N)
0109          S2 = K*(E2**N)
0110          S0 = (S1+S2)/2.0
0111          RETURN
0112          END
0113    C
0114          FUNCTION FSW(H,S0,U,CL)
0115    C.... ROLL-SEPARATING FORCE/UNIT LENGHT CALCULATION ....
0116          FSW = (H*S0*(EXP(U*CL/H)-1.0))/U
0117          RETURN
0118          END
0119
EOF
```

B.4 Program Listings 349

```
0001  C.... PROGRAM LATHE
0002  C
0003  C     DATA ACQUISITION FOR DETERMINING METAL REMOVAL RATE,
0004  C     POWER, AND UNIT HORSEPOWER - ME 488 LAB # 2
0005  C
0006  C     A. GALIP ULSOY - 1/82
0007  C
0008  C     THIS PROGRAM COMPUTES THE METAL REMOVAL RATE (Q),
0009  C     HORSEPOWER AT THE CUTTING EDGE (HP), AND THE UNIT
0010  C     HORSEPOWER (HPU) FOR A LATHE INSTRUMENTED WITH A
0011  C     TOOL POST DYNAMOMETER AND A MAGNETIC PROXIMITY PROBE.
0012  C
0013  C     THE FOLLOWING ASSEMBLY LANGUAGE ROUTINES ARE USED:
0014  C           TRAIL-DETECTS PULSE TRAILING EDGE
0015  C           DWNCNT(NP)-COUNTS NP PULSES
0016  C           ADCO(ADVAL)-A/D CONVERT ON CHANNEL 0
0017  C           TIMEIN,TIMOUT(CLKTIC)-TIMING ROUTINES ON MELIB
0018  C
0019  C.... VARIABLE DECLARATIONS
0020        INTEGER ANS
0021        LOGICAL*1 NP
0022        COMMON/INOUT/NI,NO,IOSW
0023        COMMON/PARS/DIA,GAIN,M,DEPTH,FEED
0024        COMMON/CONST/NT,NP,PI
0025  C.... INITIALIZE CONSTANTS AND PARAMETERS THEN PROMPT FOR
0026  C.... ANY NEW VALUES
0027        CALL INIT
0028    100 CALL PROMPT
0029  C.... CALCULATE RESULTS AND DISPLAY
0030        CALL COMPUT
0031  C.... ANOTHER RUN ?
0032        WRITE(NO,300)
0033        READ(NI,200)ANS
0034        IF(ANS.NE.'Y')STOP
0035  C.... CONTINUE WITH ANOTHER RUN
0036        GO TO 100
0037  C.... FORMAT STATEMENTS
0038    200 FORMAT(A1)
0039    300 FORMAT('0ANOTHER RUN (Y/N) ?')
0040        END
0041  C
0042        SUBROUTINE INIT
0043  C.... INITIALIZE ALL PARAMETERS TO DEFAULT VALUES
0044        LOGICAL*1 NP
0045        COMMON/INOUT/NI,NO,IOSW
0046        COMMON/PARS/DIA,GAIN,M,DEPTH,FEED
0047        COMMON/CONSTS/NT,NP,PI
0048  C
0049        NI = 2
0050        NO = 3
0051        IOSW = 0
0052  C
0053        DIA = 2.0
0054        GAIN = 1.0
0055        M = 20
0056        DEPTH = 0.01
0057        FEED = 0.1
0058  C
0059        PI = 3.141592654
0060        NT = 10
0061        NP = 20
0062  C
0063        RETURN
0064        END
0065  C
0066        SUBROUTINE PROMPT
0067  C.... DISPLAYS CURRENT PARAMETER VALUES AND PROMPTS THE
0068  C.... USER FOR ANY NEW VALUES
0069        INTEGER ANS
0070        COMMON/INOUT/NI,NO,IOSW
0071        COMMON/PARS/DIA,GAIN,M,DEPTH FEED
0072  C
0073     10 WRITE(NO,300)DIA,GAIN,M,DEPTH,FEED
```

```
0074            READ(NI,200)ANS
0075            IF(ANS.NE.'Y') GO TO 20
0076            WRITE(NO,310)
0077            READ(NI,210)NUM
0078            WRITE(NO,320)NUM
0079                IF(NUM.EQ.1)READ(NI,220)DIA
0080                IF(NUM.EQ.2)READ(NI,220)GAIN
0081    C.... USE I3 FORMAT FOR NUMBER OF SAMPLES (M)
0082                IF(NUM.EQ.3)READ(NI,230)M
0083                IF(NUM.EQ.4)READ(NI,220)DEPTH
0084                IF(NUM.EQ.5)READ(NI,220)FEED
0085            GO TO 10
0086         20 CONTINUE
0087            RETURN
0088    C.... FORMAT STATEMENTS
0089        200 FORMAT(A1)
0090        210 FORMAT(I1)
0091        220 FORMAT(F10.0)
0092        230 FORMAT(I3)
0093        300 FORMAT('OCURRENT PARAMETER VALUES:'/
0094       $            ' 1)SHAFT DIAMETER (IN)............ =',G12.4/
0095       $            ' 2)A/D GAIN (LBF/VOLT) ........... =',G12.4/
0096       $            ' 3)NUMBER OF SAMPLES ............. =',I3/
0097       $            ' 4)DEPTH OF CUT (IN) ............. =',G12.4/
0098       $            ' 5)FEED (IN/REV) ................. =',G12.4//
0099       $            'OCHANGE PARAMETER VALUES (Y/N) ?')
0100        310 FORMAT('OTYPE NUMBER OF PARAMETER TO BE CHANGED: ')
0101        320 FORMAT('OTYPE NEW VALUE FOR PARAMETER NO. ',I2,': ')
0102            END
0103    C
0104            SUBROUTINE COMPUT
0105    C.... CALCULATES AND DISPLAYS:
0106    C           THE AVERAGE ROTATION SPEED   (IF IOSW = 0)
0107    C           THE AVERAGE CUTTING FORCE    (IF IOSW = 0)
0108    C           THE METAL REMOVAL RATE
0109    C           THE HORSEPOWER
0110    C           THE UNIT HORSEPOWER
0111    C
0112            INTEGER M,CLKTIC
0113            LOGICAL*1 NP
0114            INTEGER*2 ADDATA
0115            DIMENSION RPM(50),FORCE(50)
0116    C
0117            COMMON/INOUT/NI,NO,IOSW
0118            COMMON/PARS/DIA,GAIN,M,DEPTH,FEED
0119            COMMON/CONSTS/NT,NP,PI
0120    C
0121            AVRPM = 0.0
0122            AVFORC = 0.0
0123    C.... CALCULATE AVERAGE RPM AND FORCE VALUES
0124            DO 100 I=1,M
0125            CALL TRAIL
0126            CALL TIMEIN
0127            CALL DWNCNT(NP)
0128            CALL TIMOUT(CLKTIC)
0129            RPM(I) = FUNRPM(NP,CLKTIC,NT)
0130            AVRPM = AVRPM + RPM(I)
0131            CALL ADCO(ADDATA)
0132            FORCE(I) = GAIN*FLOAT(ADDATA)*(10.0/2047.0)
0133            AVFORC = AVFORC + FORCE(I)
0134        100 CONTINUE
0135            AVRPM = AVRPM/FLOAT(M)
0136            AVFORC = AVFORC/FLOAT(M)
0137    C.... DISPLAY RESULTS
0138            OMEGA = 2.0*PI*AVRPM/60.0
0139            VELCTY = PI*DIA*AVRPM/12.0
0140            IF(IOSW.LT.1)WRITE(NO,300)AVFORC,AVRPM,OMEGA,VELCTY
0141            Q = 12.0*VELCTY*FEED*DEPTH
0142            HP = AVFORC*VELCTY/33000.0
0143            HPU = HP/Q
0144            WRITE(NO,310)Q,HP,HPU
0145            RETURN
```

```
0146      C.... FORMAT STATEMENTS
0147        300 FORMAT('0THE AVERAGE FORCE (LBF)........... =',G12.4/
0148          $         ' THE AVERAGE SPEED (RPM)  ......... =',G12.4/
0149          $         '                        (RAD/SEC) ....... =',G12.4/
0150          $         '                        (FT/MIN) ........ =',G12.4/)
0151        310 FORMAT('0METAL REMOVAL RATE (IN**3/MIN) ... =',G12.4/
0152          $         ' POWER (HP) ....................... =',G12.4/
0153          $         ' UNIT HORSEPOWER (HP*MIN/IN**3) ... =',G12.4//)
0154            END
0155      C
0156            FUNCTION FUNRPM(NP,CLKTIC,NT)
0157      C.... CALCULATES THE SHAFT ROTATION SPEED IN RPM
0158            INTEGER CLKTIC
0159            LOGICAL*1 NP
0160            X = FLOAT(CLKTIC)*FLOAT(NT)/1000.0
0161            IPULSE = NP
0162            Y = 60.0*FLOAT(IPULSE)
0163            FUNRPM = Y/X
0164            RETURN
0165            END
0166
EOF
0001              TITLE    ASSEMBLY ROUTINES FOR LAB 2
0002              SUBTTL   DOCUMENTATION
0003      ;
0004      ;-------------------------------------------------------------
0005      ;
0006      ;        This subroutine is a down counting routine where
0007      ;        the occurance of the trailing edge of an external
0008      ;        pulse causes a decrement in the count. The initial
0009      ;        count is specified by the single argument NCOUNT.
0010      ;        When the count is zero there is a return with
0011      ;        NCOUNT unchanged. This routine can be called from
0012      ;        a FORTRAN program as
0013      ;              CALL DWNCNT(NPULSE)
0014      ;        where NPULSE is of type LOGICAL*1 and indicates
0015      ;        the number of pulses to be counted. To call this
0016      ;        routine from another assembly language routine the
0017      ;        linking procedure is
0018      ;              LD      HL,NPULSE
0019      ;              CALL    DWNCNT
0020      ;        where NPULSE is a one byte memory location that
0021      ;        contains the desired pulse count (<128)
0022      ;
0023      ;        EXTERNAL REFERENCES: TRAIL
0024      ;-------------------------------------------------------------
0025      ;        SUBTTL   CODE
0026      ;
0027              CSEG
0028              PUBLIC   DWNCNT
0029      ;       EXTRN    TRAIL
0030      ;
0031      ;        SAVE THE REGISTERS
0032      ;
0033      DWNCNT: PUSH     BC         ;SAVE BC REGISTER
0034      ;
0035      ;        SET UP B REGISTER FOR DOWNCOUNTING
0036      ;
0037              LD       B,(HL)     ;ASSUMES HL PAIR POINTS TO THE
0038                                  ;1ST ARGUMENT,VIZ.,NPULSE, AND
0039                                  ; 0 < NPULSE < 128
0040      ;
0041      ;        DECREMENT THE COUNTER AFTER EACH TRAILING EDGE
0042      ;
0043      LOOP:   CALL     TRAIL      ;WAIT FOR TRAILING EDGE
0044              DJNZ     LOOP       ;NOTE:THIS USES B REGISTER
0045      ;
0046      ;        RESTORE AND RETURN
0047      ;
0048              POP      BC
```

```
0049                RET
0050                TITLE   SUBROUTINE TRAIL
0051                SUBTTL  DOCUMENTATION
0052        ;
0053        ;------------------------------------------------------------
0054        ;
0055        ;       This routine is used to find the trailing edge of
0056        ;       a pulse sent to the 1824 digital I/O interface.
0057        ;       It has no arguments and can be called from FORTRAN
0058        ;       as:             CALL TRAIL
0059        ;       and returns after finding a trailing edge.
0060        ;
0061        ;       Application specific details are the symbols MASK
0062        ;       and IOPORT that indicate the single bit mask that
0063        ;       the pulse comes in on and the 1824 slot position
0064        ;       on the bus.
0065        ;
0066        ;       EXTERNAL REFERENCE: None
0067        ;
0068        ;------------------------------------------------------------
0069                SUBTTL  CODE
0070                CSEG
0071                PUBLIC  TRAIL
0072        ;
0073        ;       SAVE THE REGISTERS
0074        ;
0075        TRAIL:
0076        SAVE:   PUSH    AF
0077                PUSH    BC
0078                PUSH    HL
0079        ;
0080        ;       LOAD REGISTERS FOR ADDRESSING THE I/O PORT AND MASKING
0081        ;
0082                LD      B,MASK          ;LOAD REGISTER B W/ MASK
0083                LD      HL,IOPORT       ;LOAD HL W/ IOPORT ADDRESS
0084        ;
0085        ;       WAIT FOR A LOW TO HIGH TRANSITION AND THEN THE HIGH
0086        ;       TO LOW TRANSITION OON THE TRAILING EDGE
0087        ;
0088        HI:     LD      A,(HL)          ;INPUT BYTE FROM THE IO PORT TO A
0089                AND     B               ;LOGICAL AND W/ B
0090                JR      Z,HI            ;AND LOOP IF ZERO
0091        ;
0092        LO:     LD      A,(HL)          ;AGAIN IOPORT => A
0093                AND     B               ;MASK RESULT
0094                JR      NZ,LO           ;AND LOOP IF NOT ZERO
0095        ;
0096        ;       RESTORE REGISTERS AND RETURN
0097        ;
0098        RESTOR: POP     HL
0099                POP     BC
0100                POP     AF
0101                RET
0102                SUBTTL  USER DEFINED SYMBOLS
0103        ;
0104        ;------------------------------------------------------------
0105        ;
0106        ;       DEFINE SYMBOLIC LABELS FOR CONSTANTS
0107        ;
0108        IOPORT  EQU     0FF32H          ;1824 DIGITAL IO PORT SETUP TO ADDRESS
0109                                        ;SLOT #4, INPUT PORT #2
0110        ;
0111        MASK    EQU     01H             ;MASK FOR SINGLE BIT SETUP
0112                                        ;TO CHECK BIT #0
0113        ;
0114        ;
0115                SUBTTL  DOCUMENTATION
0116        ;
0117        ;------------------------------------------------------------
0118        ;
0119        ;       This routine  performs an A/D conversion on channel 0,
0120        ;       and returns the digital value in register pair  HL.
0121        ;       It can be called from a FORTRAN routine as,
```

```
0122    ;                   CALL ADCO(DATA)
0123    ;              where DATA is of type INTEGER*2
0124    ;
0125    ;         EXTERNAL REFERENCES: None
0126    ;
0127    ;                A. Galip Ulsoy - 1/82
0128    ;
0129    ;------------------------------------------------------------
0130              SUBTTL CODE
0131    ;
0132              CSEG
0133              PUBLIC ADCO
0134    ;
0135    ;         DEFINE ADDRESS LABELS FOR STATUS, ETC.
0136    ;
0137    STATUS    EQU       0FF70H
0138    MULTP     EQU       STATUS+1
0139    DATAREG   EQU       STATUS+2
0140    ADONE     EQU       7H              ;A/D COMPLETE TEST BIT
0141    ;
0142    ;         SAVE THE REGISTERS
0143    ;
0144    ADCO:     PUSH      IX
0145              PUSH      AF
0146              PUSH      BC
0147    ;
0148    ;         CONVERSION IS DONE ON CHANNEL 0
0149    ;
0150              LD        IX,STATUS
0151              LD        (IX+1),0
0152    ;
0153    ;         TEST BIT 7, A 1 => CONVERSION COMPLETE
0154    ;
0155    WAITER:   BIT       ADONE,(IX)
0156              JP        Z,WAITER
0157    ;
0158    ;         SWAP THE HIGH AND LOW ORDER BYTES OF THE RESULT TO
0159    ;         BE COMPATIBLE WITH MEMORY. IX+2 CONTAINS THE HIGH
0160    ;         ORDER BYTE AND IX+3 CONTAINS THE LOW ORDER BYTE.
0161    ;
0162              LD        A,(IX+2)        ;HIGH BYTE INTO A
0163              LD        B,(IX+3)        ;LOW BYTE INTO B
0164              LD        (HL),B          ;LOW BYTE INTO "L"
0165              INC       HL
0166              LD        (HL),A          ;HIGH BYTE INTO "H"
0167    ;
0168    ;         RESTORE THE REGISTERS AND RETURN
0169    ;
0170              POP       BC
0171              POP       AF
0172              POP       IX
0173              RET
0174    ;
0175              END
0176
EOF
0001    C         PROGRAM STEPXY         ME488 LAB#3
0002    C
0003    C         SOFTWARE DDA'S ARE USED TO CONTROL THE MOTION OF A
0004    C         STEPPING MOTOR DRIVEN X-Y TABLE. THE USER CAN SELECT
0005    C         A SINGLE AXIS (X-AXIS) MOVE WITH CONSTANT (BANG-
0006    C         BANG) ACCELERATION, OR A TWO AXIS (X-Y) MOVE WITH
0007    C         CONSTANT VELOCITY.     A.G. ULSOY  2/82
0008    C
0009              INTEGER*2 ARX,VRX,XRX,XCOUNT,ARMAX,VRMAX,NLEFT
0010              INTEGER*2 ARY,VRY,XRY,YCOUNT,ZIP
0011              LOGICAL*1 SEG1,SEG2,SEG3,DONE,TRUE,FALSE,DIRX,DIRY,OUTWRD
0012    C
0013              COMMON/INOUT/NI,NO,IOSW
0014              COMMON/BLKO/PITCH,SPR,CLKFRQ
0015              COMMON/FLAG/SEG1,SEG2,SEG3,DONE,TRUE,FALSE,DIRX,DIRY,ZIP,
0016         $              OUTWRD
```

354 APPENDIX B DESCRIPTION OF A Z80-BASED SYSTEM

```
0017            COMMON/XBLK/ARX,VRX,XRX,XCOUNT,ARMAX,VRMAX,NLEFT
0018            COMMON/YBLK/ARY,VRY,XRY,YCOUNT
0019      C
0020      C.... INITIALIZE VARIABLES AND SET CONSTANTS
0021            CALL INIT
0022      C.... DESCRIBE THE PROGRAM TO THE USER
0023            CALL DSPLAY
0024      C.... PROMPT THE USER FOR THE DESIRED MOVE
0025        1   CALL PROMPT(MENNUM)
0026      C.... MENNUM IS USED TO SELECT SINGLE AXIS MOVE, TWO AXIS MOVE,
0027      C.... OR PROGRAM TERMINATION
0028            GO TO (100,200,300), MENNUM
0029      C.... SINGLE-AXIS MOVE COMES HERE
0030      100   CALL SETX
0031            CALL MOVEX
0032            GO TO 1
0033      C.... TWO-AXIS MOVE COMES HERE
0034      200   CALL SETXY
0035            CALL MOVEXY
0036            GO TO 1
0037      C.... TERMINATE EXECUTION
0038      300   WRITE(NO,310)
0039      310   FORMAT('O*** PROGRAM STEPXY TERMINATED BY USER ***')
0040            STOP
0041            END
0042      C
0043            SUBROUTINE INIT
0044      C
0045            INTEGER*2 ARX,VRX,XRX,XCOUNT,ARMAX,VRMAX,NLEFT
0046            INTEGER*2 ARY,VRY,XRY,YCOUNT,ZIP
0047            LOGICAL*1 SEG1,SEG2,SEG3,DONE,TRUE,FALSE,DIRX,DIRY,OUTWRD
0048      C
0049            COMMON/INOUT/NI,NO,IOSW
0050            COMMON/BLKO/PITCH,SPR,CLKFRQ
0051            COMMON/FLAG/SEG1,SEG2,SEG3,DONE,TRUE,FALSE,DIRX,DIRY,ZIP,
0052         $              OUTWRD
0053            COMMON/XBLK/ARX,VRX,XRX,XCOUNT,ARMAX,VRMAX,NLEFT
0054            COMMON/YBLK/ARY,VRY,XRY,YCOUNT
0055      C
0056            NI=2
0057            NO=3
0058            IOSW=0
0059      C
0060            XRX=0
0061            VRX=0
0062            ARX=0
0063            XCOUNT=0
0064            VRMAX=0
0065            ARMAX=0
0066            NLEFT=0
0067      C
0068            XRY=0
0069            VRY=0
0070            ARY=0
0071            YCOUNT=0
0072      C
0073            SEG1=.TRUE.
0074            SEG2=.FALSE.
0075            SEG3=.FALSE.
0076            DONE=.FALSE.
0077            DIRX=.FALSE.
0078            DIRY=.FALSE.
0079            TRUE=.TRUE.
0080            FALSE=.FALSE.
0081            ZIP=0
0082            OUTWRD=.FALSE.
0083      C
0084            PITCH=40.0
0085            SPR=200.0
0086            CLKFRQ=1000.0
0087      C.... INITIALIZE THE INTERRUPT VECTOR
0088            CALL SETDDA
```

```
0089      C
0090            RETURN
0091            END
0092      C
0093            SUBROUTINE DSPLAY
0094      C.... MESSAGE TO THE USER
0095            COMMON/INOUT/NI,NO,IOSW
0096            WRITE(NO,100)
0097      100   FORMAT('0THIS PROGRAM ALLOWS THE USER TO SPECIFY MOVES '/
0098          $       ' FOR THE STEPPER MOTOR DRIVEN X-Y TABLE. A SINGLE'/
0099          $       ' AXIS (X-AXIS) MOVE WITH CONSTANT ACCELERATION,'/
0100          $       ' OR A TWO-AXIS MOVE WITH CONSTANT VELOCITY CAN'/
0101          $       ' BE SELECTED. YOU WILL BE ASKED TO SELECT THE TYPE'/
0102          $       ' OF MOVE DESIRED, THEN PROMPTED FOR THE REQUIRED'/
0103          $       ' DATA, AND THEN THE MOVE WILL BE EXECUTED.'////
0104          $       ' TYPE <RETURN> TO CONTINUE...')
0105            READ(NI,110)
0106      110   FORMAT(A1)
0107            RETURN
0108            END
0109      C
0110            SUBROUTINE PROMPT(MENNUM)
0111            COMMON/INOUT/NI,NO,IOSW
0112      1     WRITE(NO,100)
0113      100   FORMAT('0SELECT AN OPTION FROM THE MENU BELOW :'//
0114          $       '  1)X-AXIS ONLY MOVE'/
0115          $       '  2)X-Y MOVE'/
0116          $       '  3)QUIT'/)
0117            READ(NI,110,ERR=115)MENNUM
0118      110   FORMAT(I1)
0119            IF(MENNUM.GE.1.AND.MENNUM.LE.3) GO TO 2
0120      115   WRITE(NO,120)
0121      120   FORMAT('0*** INVALID ENTRY - TRY AGAIN ***')
0122            GO TO 1
0123      2     RETURN
0124            END
0125      C
0126            SUBROUTINE SETX
0127      C.... GETS DATA FROM USER AND SETS UP SINGLE AXIS MOVE
0128      C
0129            INTEGER*2 ARX,VRX,XRX,XCOUNT,ARMAX,VRMAX,NLEFT
0130            INTEGER*2 ARY,VRY,XRY,YCOUNT,ZIP
0131            LOGICAL*1 SEG1,SEG2,SEG3,DONE,TRUE,FALSE,DIRX,DIRY,OUTWRD
0132      C
0133            COMMON/INOUT/NI,NO,IOSW
0134            COMMON/BLK0/PITCH,SPR,CLKFRQ
0135            COMMON/FLAG/SEG1,SEG2,SEG3,DONE,TRUE,FALSE,DIRX,DIRY,ZIP,
0136          $       OUTWRD
0137            COMMON/XBLK/ARX,VRX,XRX,XCOUNT,ARMAX,VRMAX,NLEFT
0138            COMMON/YBLK/ARY,VRY,XRY,YCOUNT
0139      C
0140      10    WRITE(NO,100)
0141      100   FORMAT('0X-AXIS ONLY MOVE WITH CONSTANT ACCELERATION'/)
0142            ASSIGN 20 TO I
0143      20    WRITE(NO,110)
0144      110   FORMAT('0DISTANCE TO BE MOVED (INCHES,-2.0<=X<=2.0) :')
0145            READ(NI,120,ERR=990)X
0146      120   FORMAT(F10.5)
0147            DIRX=.FALSE.
0148            IF(X.LT.0.0) DIRX=.TRUE.
0149            X=ABS(X)
0150            IF(X.LE.0.0.OR.X.GT.2.0) GO TO 990
0151            ASSIGN 30 TO I
0152      30    WRITE(NO,130)
0153      130   FORMAT(' MAXIMUM VELOCITY (IN/SEC, 0.0<V<=0.125) :')
0154            READ(NI,120,ERR=990)V
0155            IF(V.LE.0.0.OR.V.GT.0.125) GO TO 990
0156            ASSIGN 40 TO I
0157      40    WRITE(NO,140)
0158      140   FORMAT(' CONSTANT ACCELERATION (IN/SEC/SEC, 0.0<A<=0.1) :')
0159            READ(NI,120,ERR=990)A
0160            IF(A.LE.0.0.OR.A.GT.0.1) GO TO 990
```

356 APPENDIX B DESCRIPTION OF A Z80-BASED SYSTEM

```
0161                 GO TO 999
0162        990      WRITE(NO,995)
0163        995      FORMAT('0*** INVALID ENTRY - TRY AGAIN ***')
0164                 GO TO I
0165        999      CONTINUE
0166     C.... CALCULATIONS FOR SINGLE AXIS MOVE
0167                 ARY=0
0168                 VRY=0
0169                 XRY=0
0170                 YCOUNT=0
0171                 XRX=32767
0172                 VRX=0
0173                 SEG1=.TRUE.
0174                 VS=V*PITCH*SPR
0175                 AS=A*PITCH*SPR
0176                 XCOUNT=IFIX(X*PITCH*SPR + 0.5)
0177                 NLEFT=XCOUNT
0178                 VRMAX=IFIX((XRX*VS/CLKFRQ) + (0.5))
0179                 ARMAX=IFIX((VRMAX*AS/CLKFRQ) + (0.5))
0180                 ARX=ARMAX
0181     C
0182                 RETURN
0183                 END
0184     C
0185                 SUBROUTINE MOVEX
0186     C.... EXECUTES THE SINGLE AXIS MOVE
0187                 CALL ENABLE
0188                 CALL CHKDD1
0189                 CALL DISABL
0190     C
0191                 RETURN
0192                 END
0193     C
0194                 SUBROUTINE SETXY
0195     C.... GETS DATA FROM USER AND SETS UP THE TWO AXIS MOVE
0196     C
0197                 INTEGER*2 ARX,VRX,XRX,XCOUNT,ARMAX,VRMAX,NLEFT
0198                 INTEGER*2 ARY,VRY,XRY,YCOUNT,ZIP
0199                 LOGICAL*1 SEG1,SEG2,SEG3,DONE,TRUE,FALSE,DIRX,DIRY,OUTWRD
0200     C
0201                 COMMON/INOUT/NI,NO,IOSW
0202                 COMMON/BLKO/PITCH,SPR,CLKFRQ
0203                 COMMON/FLAG/SEG1,SEG2,SEG3,DONE,TRUE,FALSE,DIRX,DIRY,ZIP,
0204           $              OUTWRD
0205                 COMMON/XBLK/ARX,VRX,XRX,XCOUNT,ARMAX,VRMAX,NLEFT
0206                 COMMON/YBLK/ARY,VRY,XRY,YCOUNT
0207     C
0208        10       WRITE(NO,100)
0209        100      FORMAT('0X-Y MOVE WITH CONSTANT VELOCITY '/)
0210                 ASSIGN 20 TO I
0211        20       WRITE(NO,110)
0212        110      FORMAT(' X DISTANCE TO BE MOVED (INCHES, -2.0<=X<=2.0) :')
0213                 READ(NI,120,ERR=990)X
0214                 DIRX=.FALSE.
0215                 IF(X.LT.0.0) DIRX=.TRUE.
0216                 X=ABS(X)
0217        120      FORMAT(F10.5)
0218                 IF(X.LE.0.0.OR.X.GT.2.0) GO TO 990
0219                 ASSIGN 30 TO I
0220        30       WRITE(NO,130)
0221        130      FORMAT(' Y DISTANCE TO BE MOVED (INCHES, -2.0<=Y<=2.0) :')
0222                 READ(NI,120,ERR=990)Y
0223                 DIRY=.FALSE.
0224                 IF(Y.LT.0.0)DIRY=.TRUE.
0225                 Y=ABS(Y)
0226                 IF(Y.LE.0.0.OR.Y.GT.2.0) GO TO 990
0227                 ASSIGN 40 TO I
0228        40       WRITE(NO,140)
0229        140      FORMAT(' VELOCITY ALONG THE PATH (IN/SEC, 0.0<=V<=0.125) :')
0230                 READ(NI,120,ERR=990)V
0231                 IF(V.LE.0.0.OR.V.GT.0.125) GO TO 990
0232                 GO TO 999
```

```
0233    990     WRITE(NO,995)
0234    995     FORMAT('0*** INVALID ENTRY - TRY AGAIN ***')
0235            GO TO I
0236    999     CONTINUE
0237  C.... CALCULATIONS FOR X-Y MOVE
0238            XRX=32767
0239            XRY=32767
0240            ARX=0
0241            ARY=0
0242            VS=V*PITCH*SPR
0243            XCOUNT=IFIX(X*PITCH*SPR + 0.5)
0244            YCOUNT=IFIX(Y*PITCH*SPR + 0.5)
0245            FACT=CLKFRQ*SQRT(FLOAT(XCOUNT)**2 + FLOAT(YCOUNT)**2)
0246            VRX=IFIX(XRX*XCOUNT*VS/FACT)
0247            VRY=IFIX(XRY*YCOUNT*VS/FACT)
0248  C
0249            RETURN
0250            END
0251  C
0252            SUBROUTINE MOVEXY
0253  C.... EXECUTES THE TWO AXIS MOVE
0254            CALL ENABLE
0255            CALL CHKDD2
0256            CALL DISABL
0257  C
0258            RETURN
0259            END
0260
EOF


0001            SUBTTL  LABELS AND COMMON BLOCKS
0002  ;
0003  ;         Assembly routines for LAB #3
0004  ;
0005  ;
0006  PLUSX   EQU     01H
0007  MINUSX  EQU     02H
0008  PLUSY   EQU     04H
0009  MINUSY  EQU     08H
0010  ;
0011  IOPORT  EQU     0FF22H
0012  TRAIL   EQU     0H
0013  ;
0014  ;         Common blocks
0015  ;
0016            COMMON/XBLK/
0017  ARX:    DEFS    2
0018  VRX:    DEFS    2
0019  XRX:    DEFS    2
0020  XCOUNT: DEFS    2
0021  ARMAX:  DEFS    2
0022  VRMAX:  DEFS    2
0023  NLEFT:  DEFS    2
0024  ;
0025            COMMON/YBLK/
0026  ARY:    DEFS    2
0027  VRY:    DEFS    2
0028  XRY:    DEFS    2
0029  YCOUNT: DEFS    2
0030  ;
0031            COMMON/FLAG/
0032  SEG1:   DEFS    1
0033  SEG2:   DEFS    1
0034  SEG3:   DEFS    1
0035  DONE:   DEFS    1
0036  TRUE:   DEFS    1
0037  FALSE:  DEFS    1
0038  DIRX:   DEFS    1
0039  DIRY:   DEFS    1
0040  ZIP:    DEFS    2
0041  OUTWRD: DEFS    1
```

358 APPENDIX B DESCRIPTION OF A Z80-BASED SYSTEM

```
0042   ;
0043   ;
0044   ;              INTERRUPT CONTROLLER COMMAND WORDS
0045   ;
0046   ;
0047   ICW1    EQU     010H            ;INTERRUPT INITIALIZATION
0048   ICW2    EQU     011H            ;COMMAND WORDS
0049   ;
0050   OCW1    EQU     011H            ;INTERRUPT OPERATION
0051   OCW2    EQU     010H            ;CONTROL WORDS
0052   ;
0053   MASK    EQU     0FEH            ;INTERRUPT MASK
0054                                   ;(EVERTHING BUT CLOCK)
0055   EOI     EQU     020H            ;END OF INTERRUPT
0056   ;
0057           SUBTTL  CODE SEGMENT
0058   ;
0059           CSEG
0060   ;   Macros for LAB #3: CLKEOI,DDA16,SUB16
0061   ;
0062   CLKEOI  MACRO
0063   ;
0064   ;   This macro issues the interrupt acknowledge
0065   ;   command to the real time clock, and is designed to
0066   ;   be physically last in an interrupt service routine
0067   ;
0068   ;
0069           PUSH    AF
0070           LD      A,EOI           ;ISSUE AN END OF INTERRUPT
0071           OUT     (OCW2),A        ;TO THE CONTROLLER
0072   ;
0073           POP     AF
0074           EI                      ;ENABLE INTERRUPTS
0075           RETI                    ;AND RETURN
0076           ENDM
0077   ;
0078   ;   Macro for 16 bit DDA
0079   ;
0080   DDA16   MACRO   AR,VR,XR,COUNT,PLUS,MINUS,DIR
0081           LOCAL   NEGVAL,NEXT,OUT
0082   ;   Add acceleration to velocity
0083           LD      HL,(VR)
0084           LD      DE,(AR)
0085           ADD     HL,DE           ; no check for VR overflow
0086   ;   Add velocity to position
0087           LD      (VR),HL
0088           LD      DE,(XR)
0089           EX      DE,HL
0090           AND     A
0091           ADC     HL,DE
0092           LD      (XR),HL
0093   ;   Don't send pulse if no overflow
0094           JP      PO,OUT
0095   ;   Pulse is to be sent determine direction
0096           LD      A,(DIR)
0097           CP      0
0098           LD      A,(OUTWRD)
0099           JP      M,NEGVAL
0100           OR      PLUS            ; plus move
0101           JR      NEXT
0102   NEGVAL: OR      MINUS
0103   NEXT:   LD      (OUTWRD),A      ; A contains (OUTWRD)
0104           LD      HL,(COUNT)      ; decrement count
0105           DEC     HL
0106           LD      (COUNT),HL
0107   OUT:
0108           ENDM
0109   ;
0110   ;
0111   SUB16   MACRO   THIS,THAT
0112   ;   Macro for 16 bit subtraction. Result is in HL
0113           LD      HL,(THIS)
```

```
0114            LD      DE,(THAT)
0115            AND     A
0116            SBC     HL,DE
0117            ENDM
0118    ;
0119    ;
0120    ;       THE REAL TIME CLOCK
0121    ;
0122    ;
0123            ASEG
0124            PUBLIC  INTVEC
0125            ORG     04000H          ;MUST START AT A 64 BYTE
0126                                    ;BOUNARY. HERE IT IS
0127                                    ;LOADED IN HIGH MEMORY
0128    ;
0129    ;       INTERRUPT VECTOR
0130    ;
0131    ;
0132    INTVEC:                         ;ONLY FUNCTION IS TO
0133            JP      SERVICE         ;JUMP TO SERVICE ROUTINE
0134    ;
0135    ;
0136    ;
0137            CSEG
0138            PUBLIC  SERVICE,INTINL
0139    ;
0140    ;
0141    ;
0142    INTINL: PUSH    HL
0143            PUSH    AF
0144    ;
0145    ;
0146    ;       INITIALIZE THE INTERRUPT FACILITY AND
0147    ;       SET THE MASK REGISTER TO ENABLE THE REAL TIME
0148    ;       CLOCK (BIT 0 = CLOCK)
0149    ;
0150    ;
0151            LD      HL,INTVEC       ;INTERRUPT VECTOR -> HL
0152            LD      A,L             ;LOWER ROUTINE ADDR -> A
0153            AND     11100000B       ;MASK OFF BITS 0 - 5
0154            OR      00010010B       ;MERGE SO VECTOR INTERVAL = 8 BYTES
0155            OUT     (ICW1),A        ;INITIALIZE
0156            LD      A,H             ;INTERRUPT
0157            OUT     (ICW2),A        ;ADDRESS
0158    ;
0159    ;
0160            LD      A,MASK          ;INITIALIZE MASK
0161            OUT     (OCW1),A        ;FOR REAL TIME CLOCK
0162    ;
0163    ;       SETS INTERRUPT MODE 0 FOR THE CPU
0164    ;       THE USER MUST ISSUE A ENABLE INTERRUPTS
0165    ;       COMMAND TO ACTUALLY START THE CLOCK
0166    ;
0167            IM      0
0168    ;       .
0169    ;       .
0170    ;       EI
0171    ;       .
0172    ;       Clock is ticking from this point on
0173    ;       .
0174    ;       .
0175    ;
0176    ;
0177    ;       RETURN FROM SUBROUTINE
0178    ;
0179            POP     AF
0180            POP     HL
0181            RET
0182    ;
0183    ;
0184    ;       Service routine for the DDA used to control the
0185    ;       X-Y table with stepping motors. This service routine
```

```
0186   ;          executes every millisecond when a clock interrupt
0187   ;          occurs.
0188   ;
0189   ;
0190   SERVICE:
0191   DDASER:
0192              DI
0193              PUSH     AF
0194              PUSH     HL
0195              PUSH     DE
0196   ;
0197              LD       A,TRAIL
0198              LD       (OUTWRD),A      ; (OUTWRD) <- 0
0199   ;          Check XCOUNT, if zero byass macro call for x DDA
0200   DDAX:      SUB16    ZIP,XCOUNT
0201              JR       Z,DDAY
0202   ;          Otherwise call DDA macro for x
0203              DDA16    ARX,VRX,XRX,XCOUNT,PLUSX,MINUSX,DIRX
0204   ;          Check YCOUNT, if zero bypass macro call for y DDA
0205   DDAY:      SUB16    ZIP,YCOUNT
0206              JR       Z,CONTIN
0207   ;          Otherwise call DDA macro for y
0208              DDA16    ARY,VRY,XRY,YCOUNT,PLUSY,MINUSY,DIRY
0209   CONTIN:    LD       A,(OUTWRD)
0210              CP       TRAIL    ; (OUTWRD) in A. Don't send if
0211              JR       Z,DDARET        ; it contains zero
0212   ;          Otherwise send pulse
0213   ;          translator logic requires that we complement
0214              CPL
0215              LD       (IOPORT),A
0216              LD       A,TRAIL
0217              CPL
0218              NOP
0219              NOP
0220              LD       (IOPORT),A      ; send trailing edge
0221   DDARET:    POP      DE
0222              POP      HL
0223              POP      AF
0224              CLKEOI                   ;ACKNOWLEDGE INTERRUPT
0225                                       ;AND RETURN
0226   ;
0227   ;
0228              CSEG
0229   ;
0230              PUBLIC ENABLE,DISABL,SETDDA,CHKDD1,CHKDD2
0231   ;
0232   ;
0233   ;          Routine to enable interrupts
0234   ENABLE:    EI
0235              RET
0236   ;          Routine to disable interrupts
0237   DISABL:    DI
0238              RET
0239   ;          Routine to initialize interrupts and
0240   ;          zero (OUTWRD)
0241   SETDDA:    PUSH     AF
0242              CALL     INTINL
0243              LD       A,TRAIL
0244              LD       (OUTWRD),A
0245              POP      AF
0246              RET
0247   ;          Routine for checking the single axis move
0248   CHKDD1:    PUSH     AF
0249              PUSH     HL
0250              PUSH     DE
0251   ;
0252   TOP:       LD       A,(SEG1)
0253              CP       0
0254              JP       M,PART1
0255              LD       A,(SEG2)
0256              CP       0
0257              JP       M,PART2
```

```
0258    ;
0259    PART3:  SUB16   ZIP,XCOUNT          ; Deceleration part
0260            JR      NZ,PART3            ; Loop until XCOUNT=0
0261            LD      A,(TRUE)
0262            LD      (DONE),A
0263            LD      A,(FALSE)
0264            LD      (SEG3),A
0265            JP      BOTTOM
0266    ;
0267    PART1:  SUB16   VRMAX,VRX           ; Acceleration part
0268            JP      P,PART1
0269            LD      HL,(ZIP)
0270            LD      (ARX),HL            ; (ARX) <- 0
0271            SUB16   NLEFT,XCOUNT        ; Determine N1 = N2
0272            LD      (NLEFT),HL          ; (NLEFT) <- N2
0273            LD      A,(TRUE)
0274            LD      (SEG2),A
0275            LD      A,(FALSE)
0276            LD      (SEG1),A
0277            JP      TOP
0278    ;
0279    PART2:  SUB16   XCOUNT,NLEFT        ; Constant velocity part
0280            JP      P,PART2
0281            SUB16   ZIP,ARMAX
0282            LD      (ARX),HL            ; (ARX) <- -(ARMAX)
0283            LD      A,(TRUE)
0284            LD      (SEG3),A
0285            LD      A,(FALSE)
0286            LD      (SEG2),A
0287            JP      TOP
0288    ;
0289    BOTTOM: POP     DE
0290            POP     HL
0291            POP     AF
0292            RET
0293    ;
0294    ;       Checking routine for two-axis move
0295    ;
0296    CHKDD2: PUSH    AF
0297            PUSH    HL
0298            PUSH    DE
0299    ;
0300    ONE:    SUB16   ZIP,XCOUNT
0301            JR      NZ,ONE
0302    ;
0303    TWO:    SUB16   ZIP,YCOUNT
0304            JR      NZ,TWO
0305    ;
0306            POP     DE
0307            POP     HL
0308            POP     AF
0309            RET
0310    ;
0311    ;
0312            END
0313
EOF


0001    C
0002    C****************************************************************
0003    C
0004    C       PID Controller Routine to control
0005    C       a DC servo motor.
0006    C
0007    C                       J.Z.Raski       ME 488
0008    C                               LAB 4
0009    C
0010    C****************************************************************
0011    C
0012            INTEGER*2 DACVAL(2),ADCVAL(2)
0013            INTEGER*2 R,C0,C1,C2,EK,EK1,EK2
0014            LOGICAL*1 NADC,NDAC,ADCHN(8),DACHN(4),ANSW,RESPON
```

APPENDIX B DESCRIPTION OF A Z80-BASED SYSTEM

```
0015              REAL KP,KI,KD,STOP
0016       C
0017              COMMON/ADCBLK/NADC,ADCHN
0018              COMMON/DACBLK/NDAC,DACHN
0019              COMMON/VALUES/DACVAL,ADCVAL
0020              COMMON/ERROR/EK,EK1,EK2,MK1
0021              COMMON/GAIN/KP,KI,KD
0022              COMMON/REFER/R,IFACT
0023              COMMON/INOUT/IN,IOUT
0024              DATA RESPON,STOP/'Y','STOP'/
0025       C
0026              IN=2
0027              IOUT=3
0028       C
0029              R=0
0030              IFACT=4
0031       C
0032   987        NADC=1
0033              NDAC=1
0034              ADCHN(1)=1
0035              DACHN(1)=1
0036              EK1=0
0037              EK2=0
0038              MK1=0
0039       C
0040              DACVAL(1)=0
0041              CALL DAC(DACVAL)
0042       C
0043              CALL INPUT
0044       C
0045              CALL COVERT
0046       C
0047              CALL INTINL
0048       C
0049              WRITE(IOUT,60)
0050    60        FORMAT(' ',12X,'STOP TO RETURN ')
0051       C
0052              CALL ENABLE
0053       C
0054    61        CONTINUE
0055       C
0056       C
0057       C
0058       C
0059              GO TO 61
0060       C
0061              CALL DISABL
0062       C
0063              DACVAL(1)=0
0064              CALL DAC(DACVAL)
0065       C
0066              WRITE(IOUT,70)
0067    70        FORMAT(//,13X,'ANOTHER REFERENCE OR NEW GAINS? (Y/N) ')
0068              READ(IN,80) ANSW
0069    80        FORMAT(A1)
0070       C
0071              IF(ANSW.EQ.RESPON) GO TO 987
0072       C
0073              STOP
0074              END
0075       C
0076       C
0077       C****************************************************
0078       C
0079              SUBROUTINE PID
0080       C
0081       C****************************************************
0082       C
0083       C
0084              INTEGER*2 ADCVAL(2),DACVAL(2)
0085              INTEGER*2 C0,C1,C2,EK,EK1,EK2,D1,R
0086              LOGICAL*1 NADC,NDAC,ADCHN(8),DACHN(4)
```

```
0087        C
0088                  COMMON/ADCBLK/NADC,ADCHN
0089                  COMMON/DACBLK/NDAC,DACHN
0090                  COMMON/VALUES/DACVAL,ADCVAL
0091                  COMMON/ZTGAIN/C0,C1,C2,D1
0092                  COMMON/ERROR/EK,EK1,EK2,MK1
0093                  COMMON/REFER/R,IFACT
0094                  COMMON/INOUT/IN,IOUT
0095        C
0096                  CALL ADC(ADCVAL)
0097        C
0098                  EK=R-ADCVAL(1)
0099        C
0100                  MK=(C0*EK+C1*EK1+C2*EK2+D1*MK1)/IFACT
0101        C
0102                  IF(MK.GT.2047) MK=2047
0103                  IF(MK.LT.-2048) MK=-2048
0104        C
0105                  DACVAL(1)=MK
0106                  CALL DAC(DACVAL)
0107        C
0108                  EK2=EK1
0109                  EK1=EK
0110                  MK1=MK
0111        C
0112                  WRITE(IOUT,100) ADCVAL(1),EK,MK
0113        100       FORMAT(13X,'Speed= ',I5,' Error= ',I5,' Manipulation= ',I5)
0114        C
0115                  RETURN
0116                  END
0117        C
0118        C*********************************************************
0119        C
0120                  SUBROUTINE INPUT
0121        C
0122        C*********************************************************
0123        C
0124        C
0125                  REAL KP,KI,KD,K
0126        C
0127                  INTEGER*2 R,C0,C1,C2,D1,ANSW
0128                  COMMON/ZTGAIN/C0,C1,C2,D1
0129                  COMMON/GAIN/KP,KI,KD
0130                  COMMON/REFER/R,IFACT
0131                  COMMON/INOUT/IN,IOUT
0132                  DATA PI/3.14159/
0133        C
0134        C
0135                  WRITE(IOUT,10)
0136        10        FORMAT(//,13X,'INPUT REFERENCE ROTATIONAL VELOCITY (rpm) ')
0137                  READ(IN,20) REF
0138        20        FORMAT(F8.3)
0139        C
0140        C         ...Maximum velocity = 2000 rpm
0141        C
0142                  R=IFIX((2047./2000.)*REF)
0143                  IF (R.LT.1) R=1
0144                  WRITE(IOUT,15) R
0145        15        FORMAT(13X,'Digital word= ',I4)
0146        C
0147        25        CONTINUE
0148                  WRITE(IOUT,30)
0149        30        FORMAT(' ',12X,'INPUT THE PROPORTIONAL CONTROLLER GAIN ')
0150                  READ(IN,20) KP
0151        C
0152                  WRITE(IOUT,40)
0153        40        FORMAT(' ',12X,'INPUT THE INTEGRAL CONTROLLER GAIN ')
0154                  READ(IN,20) KI
0155        C
0156                  WRITE(IOUT,50)
0157        50        FORMAT(' ',12X,'INPUT THE DERIVATIVE CONTROLLER GAIN ')
0158                  READ(IN,20) KD
```

```
0159    C
0160    C           ...Assuming a time constant of TAU seconds
0161    C
0162                TAU=0.5
0163                K=1300./2047.
0164                WN=SQRT(KI*K/(TAU+KD*K))
0165                ZETA=(1+KP*K)/(2.0*WN*(TAU+KD*K))
0166                FREQ=WN/(2.0*PI)
0167    C
0168                WRITE(IOUT,60) ZETA,WN,FREQ
0169       60       FORMAT(' Zeta= ',F10.6,' WN = ',F10.6,'rad  Freq.= ',F10.6,'Hz.'
0170    C
0171                READ(IN,61) ANSW
0172       61       FORMAT(A2)
0173                IF(ANSW.NE.'OK') GO TO 25
0174                RETURN
0175                END
0176    C
0177    C
0178    C*********************************************************
0179    C
0180                SUBROUTINE COVERT
0181    C
0182    C*********************************************************
0183    C
0184    C
0185                REAL KP,KI,KD,DELTA
0186                INTEGER*2 D1,C0,C1,C2,R,EK,EK1,EK2
0187                COMMON/ZTGAIN/C0,C1,C2,D1
0188                COMMON/GAIN/KP,KI,KD
0189                COMMON/REFER/R,IFACT
0190                COMMON/INOUT/IN,IOUT
0191    C
0192                DELTA=0.10
0193    C
0194                C0=IFIX((KP+(KI*DELTA)+(KD/DELTA))*FLOAT(IFACT))
0195    C
0196                C1=IFIX((-(2.0*(KD/DELTA)+KP))*FLOAT(IFACT))
0197    C
0198                C2=IFIX(FLOAT(IFACT)*(KD/DELTA))
0199    C
0200                D1=4
0201    C
0202                XC0=FLOAT(C0)/FLOAT(IFACT)
0203                XC1=FLOAT(C1)/FLOAT(IFACT)
0204                XC2=FLOAT(C2)/FLOAT(IFACT)
0205                XD1=FLOAT(D1)/FLOAT(IFACT)
0206                WRITE(IOUT,100) XC0,XC1,XC2,XD1
0207      100       FORMAT(' C0= ',F10.4,' C1= ',F10.4,' C2= ',F10.4,' D1= ',F10.4)
0208    C
0209                WRITE(IOUT,101)
0210      101       FORMAT(///)
0211                READ(IN,102) IDUMMY
0212      102       FORMAT(I1)
0213    C
0214                RETURN
0215                END
0216
EOF
0001    SAVE    MACRO
0002            PUSH    AF
0003            PUSH    HL
0004            PUSH    BC
0005            PUSH    DE
0006            ENDM
0007    ;
0008    RESTO   MACRO
0009            POP     DE
0010            POP     BC
0011            POP     HL
0012            POP     AF
0013            ENDM
```

```
0014  ;
0015  ;
0016  ;           INTERRUPT CONTROLLER COMMAND WORDS
0017  ;
0018  ;
0019  ICW1    EQU     010H            ;INTERRUPT INITIALIZATION
0020  ICW2    EQU     011H            ;COMMAND WORDS
0021  ;
0022  OCW1    EQU     011H            ;INTERRUPT OPERATION
0023  OCW2    EQU     010H            ;CONTROL WORDS
0024  ;
0025  MASK    EQU     0FEH            ;INTERRUPT MASK
0026                                  ;(EVERTHING BUT CLOCK)
0027  EOI     EQU     020H            ;END OF INTERRUPT
0028  ;
0029  CLKEOI  MACRO
0030  ;
0031  ;       This macro issues the interrupt acknowledge
0032  ;       command to the real time clock, and is designed to
0033  ;       be physically last in an interrupt service routine
0034  ;
0035  ;
0036          PUSH    AF
0037          LD      A,EOI           ;ISSUE AN END OF INTERRUPT
0038          OUT     (OCW2),A        ;TO THE CONTROLLER
0039  ;
0040          POP     AF
0041          EI                      ;ENABLE INTERRUPTS
0042          RETI                    ;AND RETURN
0043          ENDM
0044  ;
0045          PUBLIC  PIDSERV
0046          EXT     PID
0047  PIDSERV: DI
0048  ;
0049          CALL    PID
0050  ;
0051          CLKEOI                  ;ACKNOWLEDGE INTERRUPT
0052  ;
0053  ;
0054  ;
0055          CSEG
0056          EXT     INTVEC
0057          PUBLIC  ENABLE
0058  ;
0059  ENABLE:
0060          PUSH    HL
0061          PUSH    IX
0062          LD      IX,INTVEC
0063          LD      HL,PIDSERV
0064          LD      (IX+1),L
0065          LD      (IX+2),H
0066          POP     IX
0067          POP     HL
0068          EI
0069          RET
0070
0071          CSEG
0072          PUBLIC  DISABL
0073
0074  DISABL: DI
0075          RET
0076          END
0077
EOF
```

Index

Actuators, 67–70, 110
Addressing methods, 53–56
Aliasing, 81–82, 112
  anti-aliasing filter, 82, 112
American Standard Code for Information Interchange (ASCII), 51–52
Analog to digital converter (ADC), 15, 80, 84–85, 111–112, 229, 262–264, 331, 353
Arithmetic and logical unit (ALU), 16, 46
Assembly language, 18, 233, 236, 238, 331, 337, 339–340, 346
  instructions and programming, 52–60
Asynchronous, 15, 71
Autocovariance, 118

Binary numbers and arithmetic, 33–37
Bit, 34, 51
Boolean algebra, 25–29
Byte, 51

Causality, 132
Central processing unit (CPU), 16, 46
Characteristic equation, 138
Clock, 15, 30–33, 47, 90–95, 110, 187, 197, 200, 202, 213, 229
Closed loop system, 155, 193, 215
Computer aided design (CAD), 4
Computer aided manufacturing (CAM), 4
Computer integrated manufacturing (CIM), 4–6, 8
Computer numerically controlled (CNC) machine tools, 7, 9
Correlation, 113
Counters, 29–33, 48
Covariance, 113
Cycle time, 17

Decoupling, 132, 139, 166–167
DeMorgan's Theorem, 28

Derivative (D) control, 157–158
Difference equations, 140
Digital to analog converter (DAC), 80, 83–84, 229, 328–329, 331
Digital differential analyzer (DDA), 193–214
Digital Input/Output (I/O), 15–16, 70–80, 229
  parallel I/O, 15–16, 74–80
  serial I/O, 15–16, 71–74
Direct current (dc) motors, 70
  speed control, 161–163, 215–221, 304–329, 361–365
Dynamic system, 131

Eigenvalue assignment, see Pole assignment
Eigenvalue problem, 138–139
Expected value, 113

Flank wear, see Machining
Flip flops, 29–33
Fourier transform, 117
  Discrete Fourier transform (DFT), 119
  Fast Fourier transform (FFT), 121

Grouped bit number systems, 17, 37

Hexadecimal numbers and arithmetic, 17, 37–39

Integral (I) control, 157–158
Intel 8086/8088 microprocessor, 15, 50–60, 229
Interpolation, 195, 204
  circular, 207–208
  linear, 205–207
Interrupts, 9, 85, 88–95, 110

Jerk, 196, 203–204

Kurtosis, 115

Laplace transform, 134, 141
Leadscrew, 68–69
Least squares method, 146
Linear superposition, 132, 137
Logic gates, functions and devices, 25–33

Machining:
    data acquisition in, 115–117, 246–264, 349–353
    economics of, 6–7
    forces, 115–117, 121, 134–135, 181–190
    tool wear in, 6, 115–117, 149–150
Maximum percent overshoot, 161
Mean, 113
Memory, 9, 16, 47
Microcomputers, 9, 13–14
Microprocessors, 15, 45
Minicomputers, 9, 13–14
Mnemonics, 17
Multi input multi output (MIMO), 155, 164–168

Numerically controlled (NC) machine tools, 7
Nyquist frequency, 82

Octal numbers and arithmetic, 17, 37–39
One's complement, *see* Two's complement
Opcodes and pseudo opcodes, 52–60
Operating system, 18
Optical encoders, 65–66

Paper making process, 164–167
Parallel input and output, *see* Digital I/O
Parameter estimation, 144
Pole assignment, 160, 216
Polling, 85–88
Probability density function, 113
Proportional (P) control, 157–158

Proportional plus integral (PI) control, 157–158, 161–163, 215
Pulse, 15, 76–79

Recursive least squares, 148
Registers, 29–33, 47–48
Robots, 7, 9
Rolling mill, 173, 239–246, 347–348
Root mean square (ms), 115

Sampled data system, 155
Sampling, 81–82
    period, 140
Sensors, 65–67, 109
Serial input and output, *see* Digital I/O
Settling time, 161
Single input single output (SISO), 132–133, 143
Spectral analysis, *see* Fourier transform
Stacks, 48
Stability, 138, 156, 159–160
Standard deviation, 113
State:
    equations, 133
    transition matrix, 138–139
Stepping motors, 67–69, 193–214, 264–303; 353–361
Synchronous, 15, 71

Tachometers, 66–67, 216
Tool wear, *see* Machining
Transfer function, 134, 138, 141
Truth tables, 25–26
Turning on a lathe, *see* Machining
Two's complement, 36–37

Variance, 113

Z Transform, 141
Zilog Z80 microprocessor, 15, 48–60, 331

Printed in Singapore by Chong Moh Offset Printing Pte Ltd